Green Engineering for Optimizing Firm Performance

This book offers a detailed examination of how sustainable technologies are reshaping firm performance. Through an integration of empirical research, expert opinions, and case studies, it explores how green management practices are enhancing business outcomes and contributing to sustainable development. It offers an in-depth understanding of how green technologies and practices, such as green engineering, AI/ML applications, green HRM, and green innovation, impact firm performance.

- Explores topics such as green engineering, AI/ML applications, green finance, green HRM, and green innovation, showing their collective impact on business performance
- Presents real-world case studies and empirical findings to demonstrate how organizations across different industries have successfully implemented sustainable technologies
- Examines regional variations in green management practices, offering insights into the impact of economic, regulatory, and cultural contexts on sustainability initiatives
- Critically analyzes contemporary challenges with practical strategies for addressing issues effectively
- Recommends actionable policy and future research directions for sustainable business practices, providing a roadmap for advancing green management

This reference book is for academicians, scholars, and practitioners who are interested in emerging technologies that are reshaping firm performance and impacting sustainability.

Artificial Intelligence, Machine Learning, Data Analytics and Automation for Business Management

Rajesh Dey, Senior Commissioning Editor, CRC Press, Taylor & Francis Group

The series presents a comprehensive exploration of how organisations strategically leverage technology to achieve their goals and maintain competitiveness in rapidly evolving industries. With contributions from experts across multiple disciplines, including computer science, business management, and data science, the series adopts an interdisciplinary approach, offering diverse perspectives and fostering cross-functional collaboration. In an era defined by rapid technological innovation and digital transformation, the book series on Computational Intelligence, Data Analytics, and Emerging Technology for Business Management serves as an indispensable resource for professionals, researchers, and students seeking to harness the power of technology to drive organisational growth and success.

- Includes works on the latest advancements in computational intelligence, data analytics, and emerging technologies to offer invaluable insights into leveraging cutting-edge tools and methodologies to drive organisational success.
- Emphasizes practical applications, featuring real-world case studies, examples, and best practices that demonstrate how these technologies can be effectively implemented to address business challenges and drive innovation.
- A multidisciplinary series draws on principles from management, engineering, economics, and other disciplines to effectively leverage technology for organisational success.
- Comprising professional insights, reference materials, and case studies, this series delves into key concepts essential for effective technology for management success.
- Serves as an indispensable resource for scholars, practitioners, and decision-makers committed to driving innovation and sustainable growth in the digital age.

Digital HR

Technologies for HR Transformation and Performance Improvement

Edited by Deepa Gupta, Mukul Gupta, Balamurugan Balusamy, Rajesh Kumar Dhanaraj, and Parth M. Gupta

Green Engineering for Optimizing Firm Performance

AI and Automation for Sustainable Technologies

Edited by Sonal Trivedi, Balamurugan Balusamy, Liza Macasukit Gernal, and Mahmoud Ahmad Al-Khasawneh

Organizational Excellence

Data, Technology and Leadership

Shyama Prasad Mukherjee

Artificial Intelligence, Machine Learning and IoT for Smart Business Management

Edited by Garima Jain, Ankush Jain, Veena Grover, Balamurugan Balusamy, and Praveen Tomar

Cyber Security in Business Analytics

Edited by Gururaj H L, B Ramesh, Chandrika J, and Hong Lin

For more information about this series, please visit: www.routledge.com/Artificial-Intelligence-Machine-Learning-Data-Analytics-and-Automation/book-series/AIBM

Green Engineering for Optimizing Firm Performance

AI and Automation for Sustainable Technologies

Edited by
Sonal Trivedi, Balamurugan Balusamy, Liza Macasukit Gernal and Mahmoud Ahmad Al-Khasawneh

CRC Press
Taylor & Francis Group
Boca Raton London New York

CRC Press is an imprint of the
Taylor & Francis Group, an **informa** business
A CHAPMAN & HALL BOOK

Designed cover image: ShutterStock

First edition published 2026
by CRC Press
2385 NW Executive Center Drive, Suite 320, Boca Raton FL 33431

and by CRC Press
4 Park Square, Milton Park, Abingdon, Oxon, OX14 4RN

CRC Press is an imprint of Taylor & Francis Group, LLC

ISBN: 9781041040347 (hbk)
ISBN: 9781041040354 (pbk)
ISBN: 9781003626527 (ebk)

DOI: 10.1201/9781003626527

Typeset in Times
by Deanta Global Publishing Services, Chennai, India

Contents

Preface

The rapidly evolving intersection of green management and technological advancements is creating transformative opportunities and challenges for businesses across the globe. This edited volume, *Green Engineering for Optimizing Firm Performance: AI and Automation for Sustainable Technologies*, brings together a diverse array of voices from across the world, each offering unique perspectives on the convergence of sustainability and innovation in business practices. As we confront a future marked by climate change, resource depletion, and environmental degradation, the role of advanced technologies like artificial intelligence (AI), machine learning (ML), data analytics, and automation has never been more critical in driving green practices and organizational performance.

This book is the result of collaborative research by scholars hailing from various countries, showcasing the diverse global landscape of thought leadership in sustainable business management. The contributors come from varied disciplines, including business, law, economics, management, and technology, which enriches the interdisciplinary nature of the discussions within. These authors represent institutions from diverse geographical regions, ranging from India and the United Kingdom to Brunei and Serbia. Such diversity not only highlights the universal relevance of green management but also reflects the different socioeconomic and environmental contexts within which businesses operate.

In Chapter 1, the authors provide a comprehensive literature review on the role of green management and technological advancements in improving firm performance. This lays the foundation for understanding how environmental sustainability and business innovation are inextricably linked. Chapters 2 and 3 explore practical applications of technologies, with data analysis techniques like K-means and AI-driven solutions for greening SMEs, showing how resource efficiency and human resource management are becoming more sustainable through digital tools.

The contributors from diverse cultural and academic backgrounds bring varied insights on how technology can support green transitions in sectors like energy, microfinance, and human resources (HR) management. In Chapters 4, 5, and 6, readers will find deep dives into specific industry sectors and countries, from solar energy projects in India to green HR practices in corporate settings. Authors from both academia and industry share their empirical evidence, offering case studies and frameworks that provide readers practical lessons in integrating sustainability into business strategies.

The inclusion of international perspectives and cross-disciplinary approaches enriches the discussion on how AI, ML, and automation are driving sustainable outcomes in business. In Chapters 7 and 8, the authors examine the intersection of these technologies with green management practices, particularly focusing on the issue of greenwashing and the role of data analytics in promoting transparency and accountability. Chapter 9 focuses on green purchase behavior, and Chapter 10 discusses greenwashing.

Chapter 11, takes a forward-looking approach, providing a bibliometric overview and examining the evolving role of AI in business management and sustainable finance. This comprehensive exploration serves as a vital resource for academics, researchers, and practitioners looking to understand the current trends, challenges, and future directions in green management and technological integration.

The book also addresses critical issues of consumer behavior and green finance, with contributions on consumer skepticism toward sustainability claims (Chapter 12) and the importance of higher education in shaping future leaders capable of driving green transformation (Chapter 13). Additionally, the case studies provided in Chapter 13 highlight how organizations are leveraging technological innovations in data analytics and green finance to foster innovation and promote environmental stewardship.

The final chapter, chapter 14 examines the relationship between green innovation and firm performance.

As the contributors come from different regions and fields of expertise, this book embodies the value of collaborative global thinking. The diversity of the authors and their collective contributions serves as a testament to the shared global responsibility for environmental sustainability and the transformative potential of technology. We hope this volume will inspire further research, practice, and policy development in the field of green management, leading to tangible outcomes in promoting sustainable business practices worldwide.

It is our sincere hope that this book sparks further inquiry and becomes a valuable resource for anyone interested in the intersection of sustainability, technology, and firm performance. The journey toward a greener, more sustainable future requires collective action, and through this book, we hope to contribute meaningfully to that larger conversation.

Editors

Sonal Trivedi is an associate professor with the School of Management & Commerce at Manav Rachna University, Faridabad, Haryana, India.. She has more than twelve years of teaching experience. She has published two books in national publications and three edited books that are Scopus-indexed. She has been granted seven patents. She has twenty Scopus book chapters and six research papers in Scopus-indexed journals. She has twenty-two research papers in journals approved by the University Grants Commission and fifteen papers presented at international and national conferences. She has acted as conference chair in five international conferences. She is a reviewer of four reputed journals. She completed her PhD in management (finance) from the University of Kota, Rajasthan, India, in 2017. She has also worked as a teaching assistant at S.V. National Institute of Technology (NIT), Surat (Institute of National Importance under Ministry of Human Resource Development [MHRD], Government of India). Her areas of interest are blockchain, finance, banking, and insurance.

Balamurugan Balusamy is an associate dean student at Shiv Nadar University, Delhi-NCR, India. Prior to this assignment, he was a professor in the School of Computing Sciences & Engineering and director of international relations at Galgotias University, Greater Noida, India. His contributions focus on engineering education, Blockchain, and data sciences. He has 200-plus high-impact factor papers in Springer, Elsevier, and IEEE. He has published more than 80 edited and authored books and collaborated with eminent professors across the world from top QS-ranked universities. Balusamy has served as ab associate professor in his stint of twelve years of experience with VIT University, Vellore. He has several conferences on his resumé and has published more than 200 journal, conference, and book chapters. He serves on the advisory committee for several start-ups and forums, and does consultancy work for industries on industrial IoT. He has given over 195 talks at various events and symposiums.

Liza Macasukit Gernal is a distinguished academic with over two decades of experience in the field, having contributed her expertise in various countries, including the Philippines, Oman, and the United Arab Emirates. Gernal holds a PhD in management from the University of Liverpool, UK, as well as a doctorate in management from Liceo de Cagayan University, Philippines. To further her academic journey, she is a candidate for the AQ Bridge to Business program at the University of Toledo in Ohio, USA. Gernal is renowned for her groundbreaking research, which delves into the profound impact of human activities on ecosystems. She has also served as a dedicated mentor and supervisor to MBA and DBA students from Universidad de Catholica in Spain. Gernal's diverse research interests encompass environmental sustainability, human resource management, educational sustainability, AI in business, and various other business- and technology-related fields. She is currently connected as an Assistant Professor at University of Fujairah, United Arab Emirates.

Mahmoud Ahmad Al-Khasawneh holds a BSc in computer science from Yarmouk University, Jordan, conferred in 2003. He obtained both his MSc and PhD in computer science from Universiti Teknologi Malaysia (UTM), Johor, Malaysia, in 2013 and 2018, respectively. Currently, he serves as a faculty member in the School of Computing, Skyline University College, Sharjah, UAE. Al-Khasawneh's scholarly pursuits span a diverse array of fields within computer science. He has authored numerous papers in esteemed, peer-reviewed journals across leading publishers. His research interests encompass security, image encryption, wireless networks, blockchain, the Internet of Things, and big data.

Contributors

Vinita Chaudhary
School of Management
Manav Rachna Educational Institutions
Faridabad, Haryana, India

Isha Chhabra
School of Leadership and Management
Manav Rachna International Institute of Research & Studies
Faridabad, Haryana, India

K. Vinodha Devi
Department of Economics
Kristu Jayanti College (Autonomous)
Bengaluru, Karnataka, India

Nand Dhameja
School of Leadership and Management
Manav Rachna International Institute of Research and Studies
Faridabad, Haryana, India

Rajiv Divekar
Symbiosis Institute of Management Studies
Symbiosis International (Deemed University)
Pune, Maharashtra, India

Vandana Gaur
Department of Management
Delhi Institute Higher Education
Delhi, India

Priyanka Ghosh
Department of Commerce
Kristu Jayanti College (Autonomous)
Bengaluru, Karnataka, India

Divya Gogia
School of Management
K.R. Mangalam University
Gurugram, Haryana, India

Khushboo Gulati
School of Management & Commerce
Manav Rachna University
Faridabad, Haryana, India

Viktoriia Hamaiunova
Law School Newcastle
University Law School
United Kingdom

Shivani Inder
Chitkara Business School
Chitkara University
Rajpura, Punjab, India

Ivana Janjić
Innovation Centre
University of Nis
Republic of Serbia

Sonal Jauhari
Department of Management
Delhi Institute of Higher Education
Delhi, India

Rameesha Zahir Khan
School of Business and Economics
Universiti Brunei Darussalam
Muara, Gadong, Brunei

Pravin Kumar
Symbiosis Institute of Management Studies
Symbiosis International (Deemed University)
Pune, Maharashtra, India

Shehnaz Lockhat
Symbiosis Institute of Management Studies
Symbiosis International (Deemed University)
Pune, Maharashtra, India

J. Malarvizhi
Department of Commerce with Bank Management
Nadar Saraswathi College of Arts and Science
Theni, Tamil Nadu, India

Sanjay Kumar Mangla
Symbiosis Institute of Management Studies
Symbiosis International (Deemed University)
Pune, Maharashtra, India

Geetha Manoharan
School of Business
SR University
Warangal, Telangana, India

Nandita Mishra
Department of Management and Engineering
Linkoping University
Sweden

Navtika Singh Nautiyal
School of Law, Forensic Justice and Policy Studies
National Forensic Sciences University
Gandhinagar Gujarat, India

L.R. Niranjan
School of Business and Management
CHRIST (Deemed to Be University)
Bangalore, Karnataka, India

M. Nivedha
School of Creative & Cultural Business
Robert Gordon University
Aberdeen, Scotland, United Kingdom

Bhartrihari Pandiya
School of Management Studies
National Forensic Sciences University
Gandhinagar, Gujarat, India

Anand Patil
School of Business and Management
CHRIST (Deemed to Be University)
Bangalore, Karnataka, India

Gamini Premaratne
School of Business and Economics
Universiti Brunei Darussalam
Muara, Gadong, Brunei

Lutfi Abdul Razak
School of Business and Economics
Universiti Brunei Darussalam
Muara, Gadong, Brunei

Paul Ritika
Symbiosis Institute of Management Studies
Symbiosis International (Deemed University)
Pune, Maharashtra, India

Žarko Rađjenović
Innovation Centre
University of Nis
Republic of Serbia

Manisha Raj
Symbiosis Centre for Management Studies
Symbiosis International (Deemed University)
Pune, Maharashtra, India

Lakshmikanth Raja
School of Business
SR University
Warangal, Telangana, India

Shivakami Rajan
Faculty of Management and Commerce
Ramaiah University of Applied Sciences
Bangalore, Karnataka, India

Ashish Saraswat
Vivekananda School of Law
Vivekananda Institute of Professional Studies, Technical Campus
New Delhi, India
and
Bennett School of Law
Bennett University
Greater Noida, India

Anjali Sharma
Mittal School of Business
Lovely Professional University
Phagwara, Punjab, India

Swathi Shekar
School of Business and Management
CHRIST (Deemed to Be University)
Bangalore, Karnataka, India
and
Jain (Deemed to Be University)
Bangalore, Karnataka, India

Ashwitha Shetty
Department of Commerce
Kristu Jayanti College (Autonomous)
Bengaluru, Karnataka, India

Jaswinder Pal Singh
Chitkara Business School
Chitkara University
Rajpura, Punjab, India

Soumya Singhal
Trinity Institute of Professional Studies
Guru Gobind Singh Indraprastha University (GGSIPU)
New Delhi, India

Ankita Srivastava
School of Management Studies
National Forensic Sciences University
Gandhinagar, Gujarat, India

Miljana Talić
Innovation Centre
University of Nis
Republic of Serbia

Vijayta Taneja
Department of Journalism
Delhi Institute of Higher Education
Delhi, India

Garima Tiwari
National Law University Delhi
India

Sonal Trivedi
School of Management & Commerce
Manav Rachna University
Faridabad, Haryana, India

Jyoti Verma
Chitkara Business School
Chitkara University
Rajpura, Punjab, India

Nirmaljeet Virk
School of Management
K.R. Mangalam University
Gurugram, Haryana, India

1 Literature Review of Green Management and Advanced Technologies for Firm Performance

Rameesha Zahir Khan, Lutfi Abdul Razak, and Gamini Premaratne

1.1 INTRODUCTION

An organizational management strategy known as "green management" aims to increase profitability and efficiency while minimizing the adverse effects of commercial activities on the environment (Rauf, Wang, and Voinea 2024). Sustainability is the main focus of green management, which entails making choices and acting in socially, ecologically, and economically sound ways. This article will review green management, including its significance, advantages, and difficulties. It will also cover methods for implementing green business management techniques (Huang and Huang 2024).

Any business, including manufacturing, retail, hospitality, and transportation, can implement green management techniques. Businesses that implement green management techniques can reap many advantages, such as lower costs, higher customer loyalty, improved brand perception, and enhanced profitability (Abbas et al. 2024).

Due to its potential to build a sustainable future for our world, green management is becoming increasingly important. With a growing number of people worldwide consuming more resources, the need for sustainable business practices has become more critical than ever in recent years. Failure to implement sustainable practices could result in severe consequences for an organization, such as lost revenue, harm to its reputation, and legal repercussions (Abbas et al. 2024).

Green management has the potential to generate significant business opportunities in addition to reducing risk. Companies that can meet this need are expected to see a rise in sales and revenue as the market for sustainable goods and services grows. Reducing waste, energy use, and other resource usage are different ways that using green management techniques can save costs (Kaur and Dharni 2024).

Moreover, several academics have argued in favor of empirical studies that look at how pro-environmental workplace behaviors of employees fit into turning green methods (Nureen, Liu, Irfan, and Sroufe 2023). However, little research has

DOI: 10.1201/9781003626527-1

examined the significance of workers' pro-environmental behaviors in developing nations, where environmental issues and difficulties are critical to small and medium-sized enterprises (SMEs) in emerging markets (Siedschlag and Yan 2023). Thus, the hotel sector and travel agencies understand the need for more research examining the outcomes that ecological management can achieve regarding the triple bottom line (TBL), as SMEs may face additional constraints due to a lack of knowledge and resources to invest in green management strategies: social, environmental, and economic performance (Hartono, Siagian, and Tarigan 2023).

Reducing the environmental impact of a company's activities through proactive management is known as green management. Reducing waste, conserving energy and natural resources, and minimizing pollution are techniques and practices that must be implemented. Playing a leading role in environmental sustainability and stewardship is part of green management, which extends beyond merely following ecological standards (Rauf, Wang, and Voinea 2024; Huang and Huang 2024).

SMEs contribute to the gross domestic product in developing economies by as much as 40%. Adding informal SMEs to the mix makes these numbers even higher. These SMEs also contribute 60–70% of global pollution (Li et al. 2024; Abbas et al. 2024; Kaur and Dharni 2024). Green management practices critically impact sustainable corporate performance (SCP), defined as "environmental performance, financial performance, and social performance." However, this relationship has not been thoroughly studied, particularly in emerging markets, developing countries, and small and medium-sized hospitality businesses (e.g., hotels and travel agencies) (Glowka et al. 2024).

1.1.1 Conceptual Framework of Green Management

To take advantage of different features and develop their competitive edge, today's businesses have built international networks of suppliers. Supply chain managers are faced with balancing low costs and developing new ideas to support economic and environmental sustainability. Green supply chain management, or GSCM, has evolved to assist managers in striking a balance between these competing demands. Due to its international reach, numerous stakeholders, and unclear consequences, environmental practices in supply chain management (SCM) provide intrinsic complexity and have historically proven difficult for researchers to study. Supporting the creation of management strategies to transform the entire supply chain into a green supply chain requires much research. Ad hoc, fragmented, and incomplete research has been done on GSCM (Hartono, Siagian, and Tarigan 2023; Siedschlag and Yan 2023). Thus, more integrative efforts are required, particularly in best practice distribution. Furthermore, the conceptual framework of our work is detailed later.

1.1.2 Industry 4.0 and Technological Innovation

SMEs must create collaborative platforms in order to develop smart goods that exhibit a high degree of technical and financial viability, according to Guo et al. (2023).

Since it will help them get over the financial barriers of purchasing smart technology, the skill and expertise to create such goods are uncommon in a single firm (Cherian et al. 2024).

According to Lin et al. (2023), the outcomes of this kind of integration are enormous. The many stakeholders must, however, work closely together (Gauthier 2013; Söderholm 2020).

There are many obstacles to SMEs implementing smart manufacturing technologies, which might help them control production costs, increase productivity, and reduce their environmental impact (Yang 2014).

Big data is an intriguing field of smart technology. Although it is believed that SMEs may not require big data because they have fewer clients and products, this field has grown quickly in recent years as a result of the internet's ability to provide businesses with access to previously unheard-of volumes of data that cannot be processed by conventional methods and tools (Allam et al. 2022). Culot et al. (2020) contend that SMEs can gain a great deal by implementing big data techniques and technologies. They clarify that there are numerous methods that SMEs can use big data to attain sustainability. Big data, for instance, encourages information to be accessible and transparent, which can assist managers in reaching decisions more quickly and efficiently. They have the ability to store data electronically, which speeds up access. Reports and decision-making tools for more precise forecasts of consumer behavior could be produced using customer-related data. Big data may help SMEs run their supply chains, logistics, and customer support more effectively.

Cloud computing is a well-established smart technology that SMEs can also use in larger enterprises. Whether SMEs embrace cloud computing early on or switch to it later in their development, Jin and Chen (2015) found that it offers numerous advantages. Saving money is the main advantage since SMEs can spend less on hardware—like servers, which take up a lot of empty capacity—and avoid wasting money while their business is still expanding. In addition, SMEs can attain flexibility with regard to their IT resources and capabilities by making precise strategic choices about their IT investments. The carbon footprint of their IT activities is lessened, nevertheless.

According to Liang et al. (2020), cloud computing benefits SMEs in a number of ways. Teams working on various projects can access the data more easily, regardless of whether they have network or hardware access. Because they may continue to work remotely, this greatly boosts their productivity (Rost et al. 2014). Such adaptability is crucial for SMEs' growth goals, particularly when the company is targeting new product development, competitor analysis, or client acquisition. The fact that small firms only pay for the resources they use and may readily acquire more resources as needed is one of the main advantages of cloud computing, according to Ling et al. (2013).

Although there have been indications of improvements in certain SMEs' operational domains, the results of adoption of the Internet of Things (IoT) in this setting have been conflicting. To maximize processing speed, for example, it is suggested to incorporate automation into supply chains. Furthermore, by promoting productive interactions with customers and suppliers, which lead to waste reduction, effective supply chains, and environmental sustainability, IoT indirectly enhances the sustainability of SMEs generally. According to Ayaz et al. (2019), SMEs have distinct resources and market characteristics compared to large organizations, making it challenging to transfer sustainability practices from large enterprises to the SME sector. This may not be possible for all SMEs, especially those in developing countries, even if IoT necessitates an initial financial commitment to develop and integrate the technology into the business. Yet, government support can help allay this concern to a great extent and enable SMEs to profit from sustainability and business growth by implementing smart technology (Hsu et al. 2023).

By incorporating AI into their operations and customer-facing activities, large corporations have been able to attain a fair degree of sustainability (Alhalaseh and Alasasfeh 2020; Kanawaday and Sane 2018). According to Fragopanagos and Taylor (2005) and Mnih et al. (2015), the implementation of chatbots reduced the requirement for specialized human personnel to handle customer inquiries and feedback while maintaining customer satisfaction.

Accordingly, Jin and Chen (2015) contend that SMEs in the manufacturing industry in particular have enormous potential to leverage AI to achieve their sustainable objectives. For these businesses, like SMEs, to adopt such systems, they must determine the factors that facilitate and hinder change inside their organizations (Ahsan and Siddique 2022; Ortt et al. 2020).

An intelligent decision system that runs on an AI model is fed manufacturing data collected by the Internet of Things and AI cooperatively. This produces a comprehensive solution that SMEs can use to streamline and organize their data collection and analysis procedures. For their own decision-making requirements, each team member—from thc operator to the manager and team leader—can access data at any time within this system (Posevkin and Bessmertny 2015).

AI can assist SMEs in creating business models that allow them to attain outcomes in terms of productivity and size, per an Organisation for Economic Co-Operation and Development research study from 2021. This invariably results in less of an impact on the environment and more effective use of resources.

Using sensors, machine learning, and other AI components, (Posevkin and Bessmertny 2015) show that AI is the most important trend for SMEs since computers were first used in the workplace. Big data, automation, natural language processing, and facial recognition systems are some of the most advantageous uses of AI for both large and small businesses (Abbas et al. 2024).

1.1.3 Technology and Environmental Sustainability Relationship

In the field of supply chain management, AI systems—such as agent-based systems, genetic algorithms, and expert systems—are becoming more popular (Guo et al. 2023).

Systems can be systematically planned, controlled, and managed with the help of AI systems. Due to the complexity of GSCM, the system as a whole may not produce the intended outcomes if sufficient planning and control are not implemented. Gathering data from each stage of the process is essential to obtaining important information. According to Kunrath et al. (2023), supply chain systems can be planned and controlled for sustainability using intelligent technology for data collection. There is a gap in the literature, as there isn't a thorough integration of AI systems in the GSCM procedure. The scope of AI's use in information processing in GSCM-practicing automotive and related manufacturing companies requires more research. Technology is defined as an artifact that aims to give users a great deal of control and interaction capabilities. In other words, the technology goes beyond simple consumer information searching or transmitting and receiving. Smart technologies are defined as possessing the capabilities of self-monitoring, analysis, and reporting. It follows that intelligent technologies engage with users and influence their actions (Matsukawa et al. 2020).

According to Qayyum et al. (2015) and Shafik (2023) there are three categories of smart technologies: IoT devices, smart connected devices, and smart devices themselves. Smart devices can also be programmed with an easy-to-use user interface and feature basic automation. Smart-connected gadgets use Bluetooth, Wi-Fi, or other technologies to connect or monitor them remotely. However, IoT devices are software-defined goods that are more valuable than linked or smart equipment. Intelligent technologies are, therefore, future-ready, automated, scalable, and upgradeable.

By analyzing its characteristics, numerous researchers have attempted to define the idea of smart technologies. According to Qayyum et al. (2015) and Soare Senior et al. (n.d.), smart technologies may respond to messages individually and accept new reconfigurations, making them programmable and addressable. There are ramifications as smart technologies change to serve new purposes or engage with people (Aslam et al. 2024).

The traits of digital technologies were determined by the researchers. They therefore used them to improve people's quality of life and set goals for all parties involved in an organization. According to Söderholm (2020), smart technology can help businesses increase income, improve customer satisfaction, and increase employee contentment.

In order to increase output and productivity, it is reorganizing its organizational environments (Gupta and Gupta 2021). The next step in the technological era is technological innovation.

As a result, it will help address a number of issues that human society and the economy face (Siyal et al. 2019; Sokienah 2023).

1.1.2 Theoretical Perspectives on Green Management and Firm Performance

The last 15 years have seen a global rise in the importance of strategic environmental management, or SEM, as a business discipline. Its basic idea is that businesses can find ways to make money while lessening their environmental impact by increasing revenues or decreasing expenses. By incorporating impact-reducing product and process design into businesses' fundamental strategic aspirations, SEM aims to balance ecological objectives with financial objectives. However, pre-SEM practice may have missed out on potential profits if corporate greening or individual SEM projects don't always have to be expensive and can boost profitability in regular, non-exceptional situations. The neoclassical theoretical framework, which advocates for enterprises to utilize all available information in order to maximize profits rationally, would not support such behavior (for an earlier discussion of this issue, see Yue, Huo, and Ye 2023; Nureen, Liu, Irfan, Malik, et al. 2023). However, evolutionary economics—which holds that organizations' capacities change in peculiar ways over time and rely only on a small portion of supposedly accessible information to function—offers a systematic framework for comprehending SEM as opposed to rejecting it. For corporate strategy and policy purposes, theoretical viewpoints on SEM are crucial. As shown in the following, organizations that take action to lessen their environmental impacts should generally anticipate that doing so will increase costs and decrease profits, according to a neoclassical understanding of the company. This premise implies treating the claims made by SEM practitioners with a considerable deal of skepticism. It aligns with the traditional ways top business managers see the economic and ecological relationship. It will be claimed, however, that SEM might make perfect sense when viewed through the prism of firm theories that center on organizational skills.

1.2 PRINCIPLES OF GREEN MANAGEMENT

To oversee the sustainability initiatives of their businesses, sustainability managers follow ten principles (activities). Altogether, these ten principles imply that modern sustainability managers need to perform standard management tasks like organizing, planning, directing, and controlling, but also go above and beyond to use effective leadership to influence people and deeply ingrained systems and practices. The principles are arranged in the approximate chronological sequence that a manager entrusted with implementing sustainable business practices inside an organization might encounter them (Arbabzadeh, Lewis, and Keoleian 2019).

1.2.1 Resource-Based Views and Green Management

Growing worries about how to present production models fail to account for environmental issues and long-term sustainability, which come from concerns that ecosystems are disappearing. Companies are under more pressure than ever to follow environmental regulations and disclose their pollutant emissions due to these worries. The governments of the United States, Canada, and Mexico, for instance, mandate the creation of pollutant release and transfer registers (PRTRs), for example, the Toxic Release Inventory (TRI), National Pollutant Release Inventory (NPRI), and the Registro de Emisiones y Transferencia de Contaminantes (RETC), respectively. These environmental management standards require many manufacturing facilities to lessen their environmental impact and demonstrate their achievements. Some opponents of this requirement call for a "race to the bottom" push that would lead some industries to outsource their polluting operations (Khanra et al. 2022; Gavronski et al. 2011). The management of a factory or corporation can enhance its operational environment when this occurs. Still, there is no overall benefit to the

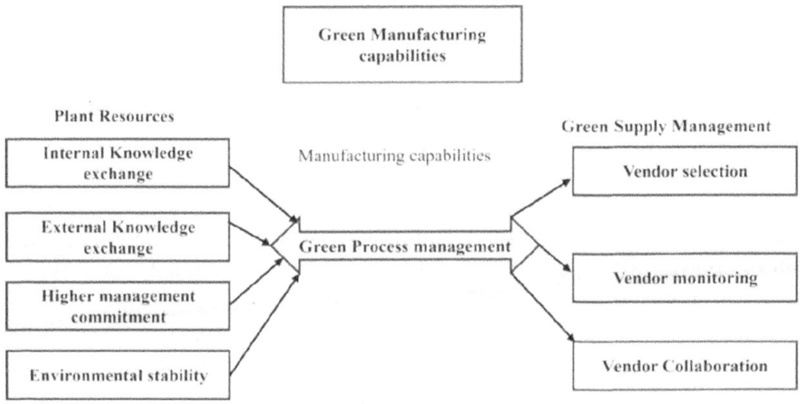

FIGURE 1.1 Green source management theoretical model.

manufacturing process along the supply chain (or even a negative one if transportation emissions are taken into account).

One way a corporation can cut down on pollution emissions is to outsource all of its metal surface treatment activities. This would lead to the plating process being closed. Still, emitting the same amount of pollutants is another facility further up the supply chain, possibly located in a nation with laxer environmental rules.

The environmental performance of the entire supply chain will eventually be impacted by the enhanced monitoring that governments, large customers, and environmental activists will likely start doing soon. This increased monitoring will focus on the output of specific factories or enterprises.

Monitoring supply chain performance is feasible despite not being done often. Businesses may be forced to submit updated environmental reports using life cycle analysis (LCA) or other systematic techniques in response to ecological monitoring constraints. Figure 1.1 explains the theoretical model of green manufacturing capabilities and required resources, which are useful for the progress of complex capabilities needed for green manufacturing implementation.

1.2.2 Stakeholder Theory and Green Management

The idea behind stakeholder theory (Mahajan et al. 2023) is that corporate duty should now be to all stakeholders rather than just shareholders, who have a financial stake in the company's success. By doing this, the scope of corporate responsibility is intended to be expanded from individuals who directly gain from corporate actions to include everyone who can influence or be impacted by such acts (Siems, Seuring, and Schilling 2023; Laplume, Sonpar, and Litz 2008; DesJardine, Zhang, and Shi 2023). Stakeholder theory is viewed as a strategic and normative imperative to involve stakeholders in business choices, either to boost long-term results strategically for the organization or because it is normatively recognized as the right thing to do in socially responsible ways.

Owing to corporate governance, the ties between people and groups connected to the corporation (business entity) are the primary emphasis for identifying pertinent stakeholders. The government and society must abide by its laws and regulations and be able to pay taxes and other obligations. This leads to a classification based on the categories "customers, suppliers, employees, shareholders and communities" or "shareholders and investors, employees, customers and suppliers" (Nureen, Xin, et al. 2023). Additionally, we classify public interest groups as those that provide market infrastructure. This viewpoint, frequently used to discuss economic exchange or regulatory control, links decision outcomes to quantifiable stakeholder relationships (Nureen, Liu, Irfan, and Sroufe 2023).

According to Mahajan et al. (2023), stakeholders are abstract ideas defined by their roles rather than real people with moral interests. As a result, the organizational categories that include stakeholders in the business environment form the foundation of social systems. A stakeholder-centered theoretical framework may cause the term "stakeholder" to lose its meaning (DesJardine, Zhang, and Shi 2023; Govindan and Bouzon 2018). Consistent with their studies, Bhattacharya and Fayezi (2021) and de Gooyert et al. (2017) contend that stakeholders should be conceptually categorized according to their social group rather than the function they perform inside the organization.

1.3 METHODOLOGY OF THE LITERATURE REVIEW

We worked on a comprehensive review of green management and firm performance. We reviewed articles in green management up to the day we are writing this work. We preferred quality and reputed journals, conferences, and book chapters for this. We used MS Word for writing, Mendeley for citations, and Google Scholar for easy access to current and quality work.

We carried out an extensive study across several top published databases with a solid reputation in the research community in order to examine and identify the reviewed research on green management and advanced technology performance. We started by using the Scopus database platform, which systematically incorporates databases from Elsevier, IEEE, Web of Science, ACM, IET, and many more. Priority was given to research published between 2019 and 2024; however, some earlier work was also taken into account for background and database purposes. Elsevier, Science Direct, IEEE Explore, IET Digital Library, Wiley, Taylor & Francis, and Springer were among the databases whose research was being reviewed in the field of computer science and engineering. Additionally, high-quality preprints from TechRiv and arXiv were taken into account. Publications written in English were preferred. Furthermore, the refining process is shown in Figure 1.2. Exclusion was based on duplication, irrelevance, and publication in languages other than English. Inclusion was based on publication year, similarity to the proposed work, and quality. The quality criteria were that the works were close to our proposed literature and have a minimum two contributions. Works with fewer than two contributions were part of the exclusion criteria.

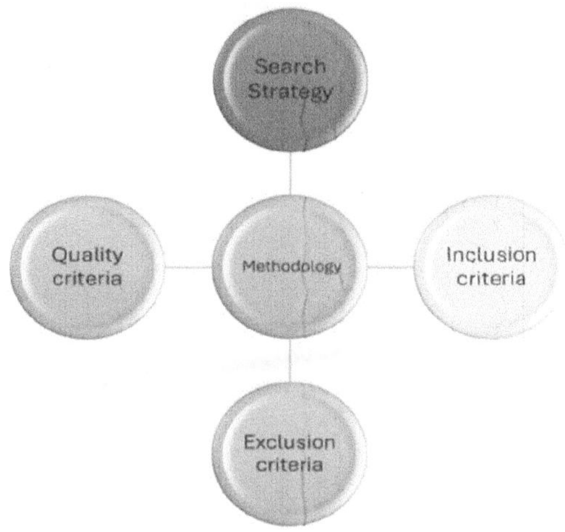

FIGURE 1.2 Review methodology.

1.3.1 Green Management Practices and Firm Performance

Geographic indicators are advancements in hardware or software that are connected to environmentally friendly products or practices (Al Halbusi, Klobas, and Ramayah 2023; Nureen, Xin, et al. 2023). These advancements include waste recycling, energy-saving technology, pollution avoidance technology, sustainable product design technology, and business environment regulations. Cities differ in what they invest in R&D, how technology spreads, and how green innovation entities like businesses, academic institutions, and research centers come together to innovate (Novitasari et al. 2023; Riaz et al. 2024). They also contend that geographical indications give companies a competitive edge by granting them advantages like brand recognition and product distinction, which open doors to international market expansion and eventually foster company expansion. Companies can enhance their products' longevity and recyclable nature, reduce raw material consumption, select eco-friendly raw materials, and steer clear of the use and manufacture of hazardous materials by taking part in geographical indications. According to Mahsina and Soewarno (2024), management of businesses dealing with escalating difficulties and environmental pressures must recognize the significance of geographic indicators in satisfying stakeholder expectations, satisfying market demand, and enhancing business performance through sustainable development. Generally speaking, geographic indicators contribute significantly to GSCM and fights for sustainable advantage, which brings SCP to the company. Figure 1.3 is the representation of GSCM planning.

1.3.2 Sustainable Corporate Performance

For businesses to succeed, they must minimize their environmental impact while concentrating on social, environmental, financial, and economic aspects. This is how SCP is summarized. Consequently, stockholders can also accomplish this objective

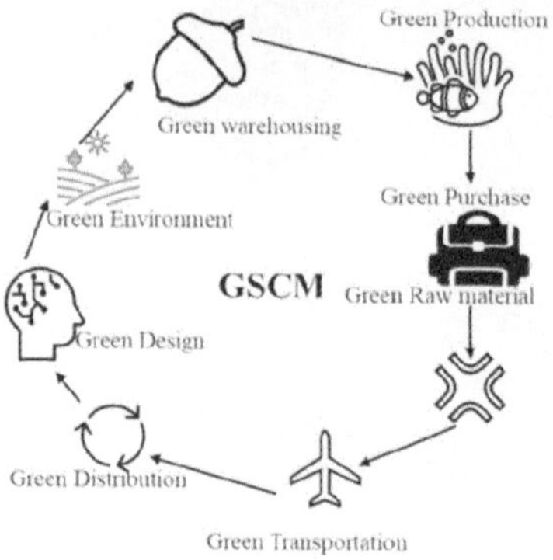

FIGURE 1.3 GSCM planning.

(Novitasari et al. 2023; Riaz et al. 2024). SCP is thought to be a very good predictor of many organizational outcomes (Mahsina and Soewarno 2024; Riaz et al. 2024). SCP measures a business's financial, social, and environmental performance (Huang and Huang 2024; Chotiyanon and Ueasangkomsate 2024). According to Chotiyanon and Ueasangkomsate (2024), a firm's SCP comprises multidimensional performance indicators considering nonfinancial and economic factors. Growth metrics for SCP design include income, profitability, rivalry, return on equity, and raised environmental awareness. Businesses can choose their performance with greater knowledge because of SCP design.

1.4 ENVIRONMENTAL AND FINANCIAL PERFORMANCE

1.4.1 Corporate Financial and Environmental Performance, and Green Purchasing

Green purchasing entails supply chain recycling and purchase reduction (Huang and Huang 2024; Gupta and Gupta 2021). Similarly, Gauthier (2013) and Kauppi (2013) define green purchasing as encouraging product recycling and reducing waste sources. Implementing green purchasing practices reduces pollution control costs, enhances an organization's reputation (financial performance) in the marketplace, and improves the organization's environmental performance (Chotiyanon and Ueasangkomsate 2024) research. According to Kauppi (2013), green purchasing is more common in large organizations than in small ones, and with the help of ecological protection awareness, this practice spreads across the organization and advances it.

According to Gavronski et al. (2011), green purchasing enhances corporate performance, safeguards transitory resources, and creates a competitive edge. Environmentally friendly (green) purchases have both direct and indirect positive effects on business performance (Khanra et al. 2022; Gavronski et al. 2011). It has a good effect on environmental performance by shielding the environment from dangerous pollutants. Green procurement significantly impacts an organizat ion's financial and environmental performance since it is a dependable instrument for pollution reduction (Kauppi 2013). Additionally, it enhances the company's favorable perception and reputation in the marketplace (Arbabzadeh, Lewis, and Keoleian 2019; Li et al. 2024). Following is our hypothesis:

- Green purchasing has a beneficial impact on how environmentally conscious a firm is.
- The financial performance of enterprises is positively impacted by green purchasing.

1.4.2 Organizational Environment, Financial Performance, and Eco-Design

Certain companies use eco-friendly components in their product designs to gain a competitive edge and set themselves apart from rivals (Glowka et al. 2024; Gavronski et al. 2011). Internal, supplier, and customer management all play a significant role in eco-design (Rauf, Wang, and Voinea 2024). Organizations can cut the adverse environmental effects of their processes and products by 80% by implementing green design approaches effectively (Nureen, Liu, Irfan, and Sroufe 2023). Environmentally friendly design techniques have significantly decreased their unfavorable impact on the environment and enhanced these organizations' contributions to sustainable development, according to Hartono, Siagian, and Tarigan (2023).

Organizations can avoid the use of hazardous materials in products and cut down on the amount of resources used in product production by placing a strong emphasis on eco-design. Furthermore, green design makes reusing, reproducing, recycling, and disassembling products easier. Customers are eager to purchase environmentally friendly products to receive cost savings and safety benefits for the environment, and eco-design is significant and favors regional performance (Siedschlag and Yan 2023). The study's findings demonstrate that eco-design goods give a business a competitive edge and favorably correlate with its financial and environmental performance. Adopting eco-design methods can also enhance an organization's standing in the community (Silva, Nuzum, and Schaltegger 2019; Laplume, Sonpar, and Litz 2008). Additionally, eco-design products considerably raise the worth of the product and save money. Following is our hypothesis for this:

- The environmental performance of an organization is favorably correlated with eco-design.
- Eco-design has a good correlation with the financial performance of enterprises.

1.5 GREEN INNOVATION AND COMPETITIVE ADVANTAGE

Eco-design techniques include systematic steps in designing products and processes to lessen their lifetime impact on the environment. Additionally, eco-design seeks to minimize the energy and resources used in the production process (Gavronski et al. 2011); minimize or entirely eliminate the use of hazardous materials (both during the process and in the final product); and incorporate secondhand goods, recycle them, or remanufacture them into new products. In the Finnish pulp and paper industry, for instance, several factors that encourage and hinder the adoption of eco-design techniques have been noted and examined to improve the sustainability score and obtain a competitive edge. Anajemba et al. (2022) examined obstacles to eco-design and their sluggish implementation in nine industrial enterprises across five nations.

Green process methods cover a range of process characteristics that reduce environmental effects over time. These characteristics include lean manufacturing (Nureen, Xin, et al. 2023), agile manufacturing (Huang and Huang 2024), quality control processes that reduce or eliminate damaged parts, usually using statistical detection methods and automation (Nureen, Xin, et al. 2023; Novitasari et al. 2023), innovation in the development of raw materials, and a decrease in waste and by-product production. Businesses must employ green marketing techniques, such as informing customers about the environmental elements of their products, in order to comprehend client demands and create long-lasting partnerships. Official policies focusing on or consuming post-recycled materials are examples of other green marketing methods (Huang and Huang 2024; Chotiyanon and Ueasangkomsate 2024). Eco-labeling is another technique.

The implementation of eco-design was also found to be challenging in five categories (Alsharif et al. 2024; Qamar et al. 2020): strategies, tools, cooperation, management, and knowledge. Furthermore, the Ecodesign Maturity Model in Practice was introduced by Glowka et al. (2024). This framework was developed to support the eco-design implementation process by means of process improvement using a continuous improvement method. Abbas et al. and (2024) and Li et al. (2024) established a business case framework for the application of eco-design using a simulation-based approach based on the ecological design maturity model (EcoM2) proposed by Huang and Huang (2024). Features of products that reduce their environmental impact over the course of their life cycle are referred to as "green product practices." Examples of these features include lowering product weight and size (Glowka et al. 2024), increasing energy efficiency (Eltayeb et al., 2014), employing nonpolluting materials (Eltayeb et al., 2011), and creating specifications for easily disassembled, recyclable, and reusable modular components (Guo et al. 2023).

Ecological innovation is associated with the number of patents on products and techniques meant to address environmental challenges (DesJardine, Zhang, and Shi 2023; Siems, Seuring, and Schilling 2023). In order to reduce their ecological liability, businesses innovate by changing their products and procedures (Siems, Seuring, and Schilling 2023). This is due to the fact that these changes need the creation of new methods to deal with the sources of damage. Our research takes a more comprehensive approach to green innovation in order to enhance process efficiency while reducing the environmental impact of the process. The following contexts have been recognized by previous study (Yue, Huo, and Ye 2023) as conducive to green innovation: green market, ecological design, green products (Zhu et al. 2012), green process (Hartono, Siagian, and Tarigan 2023); and green marketing (Arbabzadeh, Lewis, and Keoleian 2019).

Green market practices pertain to consumer behaviors that involve switching to a more sustainable lifestyle, which may include paying a premium for environmentally friendly items (Huang and Huang 2024). These measures include assisting distributors (Chotiyanon and Ueasangkomsate 2024; Huang and Huang 2024) and retailers on ecologically friendly issues, safeguarding the consumers' and employees' health and safety (Gauthier 2013), and strictly adhering to regulations, which call for a thorough understanding of current local legislation (Khanra et al. 2022). Lastly, environmental technology includes the use of newer equipment that requires less energy to operate and emits less CO_2 into the environment (Arbabzadeh, Lewis, and Keoleian 2019); it also provides for the use of less trash, rework, burr, and scrap (Glowka et al. 2024). Technical actions such as using software, heuristics, and models to enhance the environmental performance of transportation routes and production processes are also included in the category of ecological technology (Kauppi 2013; Gupta and Gupta 2021; Chotiyanon and Ueasangkomsate 2024).

1.5.1 Drivers and Barriers of Green Management Adoption

Researching the factors that encourage and hinder the application of GSCM in various contexts and sectors has gained momentum in recent years. Several scholars (Al Halbusi, Klobas, and Ramayah 2023; Nureen, Xin, et al. 2023; Novitasari et al. 2023; Riaz et al. 2024; Mahsina and Soewarno 2024) looked for reasons why companies should or shouldn't implement green supply chain methods. Concurrently, businesses typically design environmental policies due to the increasing awareness of contemporary ecological issues (Laosuwan, Gomasathit, and Rotjanakusol 2017). As a result, various variables drive their employment of green supply chain techniques. Following this pattern, several empirical research studies (Chotiyanon and Ueasangkomsate 2024; Huang and Huang 2024; Gupta and Gupta 2021; Gauthier 2013; Kauppi 2013) have identified and classified driving forces into two groups: internal and external. The former group includes driving factors from within the organization itself (Gauthier 2013). These drivers include the company's environmental strategy (corporate social responsibility), organizational values, employees, internal policies and corporate awareness, middle and senior managers' goals and commitments, the drive to improve performance and cut costs, and investor pressure. Alternatively, the latter can bc described as pressure from outside sources, such as laws, public opinion, rival businesses, suppliers, users, and partners of global corporations (Gauthier 2013; Kauppi 2013; Khanra et al. 2022). As a result, the drivers of GSCM deployment are ranked, categorized, and summarized. On the other hand, numerous barriers impede the adoption of green supply chain management strategies. Unlike drivers, they are forces that obstruct the successful use of green measures (Gauthier 2013). It is also separated into internal and external categories (Gavronski et al. 2011; Arbabzadeh, Lewis, and Keoleian 2019; Li et al. 2024; Glowka et al. 2024). Internal forces include manager–employee relations, employee resistance, organizational structure, lack of funding (costs associated with implementation), ignorance, and inadequate training. External barriers arise from the company's external environment, including a lack of government oversight and a reluctance to conduct information transactions, lack of commitment, lack of supplier awareness, lack of competitiveness, and other factors between the business and its suppliers.

Many authors (Rauf, Wang, and Voinea 2024; Huang and Huang 2024) highlight the advantages of internal finances.

Practices like recycling waste, using less paper, and turning off electronics when not in use can all help save money. For instance, it was discovered that while implementing green business practices typically entails additional expenditures, they can also result in observable advantages like decreased material costs and higher profitability (Abbas et al. 2024; Huang and Huang 2024). Notwithstanding these possible benefits, SMEs frequently lack understanding regarding financing environmentally related benefits initiatives (Rauf, Wang, and Voinea 2024; Glowka et al. 2024). The level of sustainability of SMEs can also be influenced by the personal values of business leaders, as well as by their intimate relationships and internal duties toward employees and communities (Kaur and Dharni 2024; Nureen, Liu, Irfan, and Sroufe 2023).

1.5.2 Barriers and Challenges in Implementing Green Management

Environmental issues are typically seen as the most crucial in the building sector. The authors of the sustainable building framework recommended a set of requirements for contract chapters and specifications about the implementation of environmental instruments, such as environmental impact and environmental management systems, that need careful consideration of environmental issues (Al Halbusi, Klobas, and Ramayah (2023). The authors believe that "green" design and standards are essential for building projects, particularly those subject to environmental evaluation (Al Halbusi, Klobas, and Ramayah (2023). The authors highlighted three primary challenges with the standards utilized in sustainable engineering (Al Halbusi, Klobas, and Ramayah (2023). They think that the most significant barriers to the growth of green buildings in China are economic rather than technological or policy related (Al Halbusi, Klobas, and Ramayah (2023). The rapid economic growth and urban development in China, according to Al Halbusi, Klobas, and Ramayah (2023) has given rise to new concerns with energy saving in buildings.

Building technologies suited for various settings and sustainable building development are crucial in China because of the country's high building density, high population density, and low amount of renewable energy per square meter of building area (Nureen, Xin, et al. 2023; Novitasari et al. 2023). Riaz et al. (2024) and Mahsina and Soewarno (2024) noted that practitioners in the Chinese construction industry are more driven to implement green techniques to receive incentives and assistance from the Chinese government. Many developers are hesitant to incorporate sustainability into their projects due to a lack of awareness and cost-cutting measures taken in developing nations like China. The rate of action for sustainable applications is determined by one's knowledge, understanding, and comprehension of the repercussions of their actions (Chotiyanon and Ueasangkomsate 2024; Huang and Huang 2024). And when it comes to adopting green practices, contractors think that management aspects are most significant (Gupta and Gupta 2021; Gauthier 2013). The aforementioned research, in brief, distills the primary barriers to green construction into four fundamental domains: management, technology, economy, and awareness.

1.5.1.2 Studies on Environmental and Social Responsibility

According to Glowka et al.'s (2024) study on small and medium-sized businesses in Latin America, profit is crucial to involvement. Green practices should be encouraged in addition to following the legislation (Glowka et al. 2024). Others point to improved ties with the public sector and society and a desire to strengthen client relationships.

Their study further reveals the reasons for implementing green company practices that are most similar to ours; survey methodologies are also employed (Glowka et al. 2024). While studies have long held that intrinsic motivation is the primary force behind adopting green business practices, the findings remain somewhat inconsistent.

1.5.1.1 Studies on German SMEs' Use of Renewable Energy

Research shows entrepreneurs' inner sense of responsibility for their surroundings reflects their personality. Adoption is a fundamental component (Khanra et al. 2022). A case study examining how green practices have evolved (Gavronski et al. 2011) supports this idea in microbusinesses. The study demonstrates that green strategies can be created on a small scale.

Ethical ideals are intimately associated with businesses. Corporate executives frequently engage in moral behavior. Landmark research by Rauf, Wang, and Voinea (2024) and Li et al. (2024) examined why businesses should go green.

The factors that motivate corporate environmental responsibility are split into four primary sections: legislation, pressure from stakeholders, economic opportunities, and ethical motivations. Numerous additional ones (Abbas et al. 2024; Huang and Huang 2024; Kaur and Dharni 2024) likewise include economic interests.

An additional reason why companies should use green practices is the notion that more businesses embracing sustainability to achieve sustainable growth lends credence to this. Other reasons include identifying new market benefits, encouraging individuals to reduce pollution, and communicating innovatively to build image and reputation (Glowka et al. 2024; Rauf, Wang, and Voinea 2024).

1.6 INDUSTRY-SPECIFIC PERSPECTIVES ON GREEN MANAGEMENT

The phrase "green management" refers to managing strategies and laws that attempt to address environmental issues while making a profit, satisfying customer demands, and advancing business objectives (Gauthier 2013; Kauppi 2013). A business seeks to accomplish its goals with a green marketing strategy or plan while reducing adverse environmental effects. This is applied across the board to the marketing plan in an ecologically responsible and sustainable manner, based on the marketing mix elements of place (distribution), price, product, and promotion (Gavronski et al. 2011; Arbabzadeh, Lewis, and Keoleian 2019).

Since eco-design and green purchasing are intimately tied to one another, green marketing is crucial to the overall sustainable development of commercial enterprises (Khanra et al. 2022). Pro-environmental behavior and psychological research on consumer behavior focus on purchasing, utilizing, and removing items that affect the environment.

Citation of academic frameworks frequently used to explain sustainable consumer behavior adhere to moral principles while discussing pertinent sustainable consumer behavior and environmental concerns (Huang and Huang 2024). Consumers today don't always act pro-environmentally, even if they claim to support ecological values (Nureen, Xin, et al. 2023). Several theoretical models have also been empirically tested, but the models are not very reliable. Remember that consumers have varying experiences while making purchases and that these experiences will shape their values and how they will make purchases in the future.

Al Halbusi, Klobas, and Ramayah (2023) emphasize the importance of choice for businesses. While discussing non-green standards in green products and their distinctive marketing qualities, it is important to create a positive experience for potential users. Non-green requirements include features, customs, and hobbies of the product.

Customers tend to undervalue the importance of green standards. Consumers value design, color, or fit more than green standards (Riaz et al. 2024; Mahsina and Soewarno 2024). Brand, concrete product features, price, initial product experience, dependability, and appearance are some factors that consumers tend to discount.

The following are the pillars of green management:

- The corporate perspective
- Government and trade association role
- Consumer access to green management practices

1.6.1 The Corporate Perspective

Studies show that implementing green business strategies has a beneficial impact and that there is a difference in managers' and customers' opinions about green efforts (Chotiyanon and Ueasangkomsate 2024; Mahsina and Soewarno 2024). The views of hotel owners or managers about green integration and facts from the company are presented in this part. Water conservation measures include reusing towels and sheets, using energy-efficient light bulbs, wastewater treatment, and planning.

Green practices are most often used in the lodging sector (Huang and Huang 2024; Gupta and Gupta 2021). Some findings from the research on hotel managers and owners assert that the best way to encourage the adoption of green practices is to reduce costs (Guo et al. 2023). However, the majority of studies note that the cost-benefit analysis of green practices is a crucial component of green integration (Novitasari et al. 2023).

Hotel owners also emphasized the possible drawbacks. Research on consumers' willingness to pay more for green hotels is inconsistent, and support for each green measure raises room rates by 5.15% (Wang et al. 2023). Green practices are not reflected on hotel websites despite the added impact that green initiatives have on business image (Hsu et al. 2023). Regarding hotel management's role, Al Halbusi, Klobas, and Ramayah (2023) emphasize the importance of top management's dedication to achieving green integration.

The study findings unequivocally demonstrate that hotels in wealthier nations are more committed to implementing green practices (Nureen, Xin, et al. 2023). Thus, it is possible to conclude that while hotel managers support being green in the business, their primary incentive is the financial gain that comes with it rather than the environmental one.

Depending on the kind of hotel, the results indicate that bigger establishments typically invest more effort in creating a thorough green model (Novitasari et al. 2023) and disseminating it via the hotel website Green Practices (Novitasari et al. 2023; Nureen, Xin, et al. 2023). In contrast to large hotels, the use of green practices in small hotels is contingent upon the environmental convictions of the proprietors (Nureen, Xin, et al. 2023). However, Riaz et al. (2024) discovered that there was no scale effect in the adoption of green practices.

However, chain hotels and independent hotels are distinguished from one another. Novitasari et al. (2023) highlighted that more luxurious hotels use greener measures. Additionally, there is a significant discrepancy between intended and implemented green measures in contrast to smaller lodging establishments. We can speculate that this result might serve as a recommendation to refocus the research into a more thorough analysis of different kinds of hotels.

1.6.2 Government and Trade Association Roles

Scholarly investigations underscore the function of governmental and political authority in mandating or motivating hotel operations to adopt more ecologically sustainable practices (Mahsina and Soewarno 2024; Chotiyanon and Ueasangkomsate 2024). Instead of restricting the type of environmental action (Appiah et al. 2023; Liu, Chen, and Zhu 2023). Managers are encouraged to be more motivated by volunteering. Remarkably, just one study was carried out.

The strongest motivator is green integration per se. Despite the lack of representation of trade associations and nongovernmental organizations in the literature, their impact on public opinion and belief systems may be significant.

Combining the aforementioned corporate perspectives, we may conclude that while the primary motivation for implementing green initiatives is their favorable effect on business performance, trade associations, the government, and other institutions may also play a role in fostering green integration.

Wang et al. (2023) and Hsu et al. (2023) highlight the importance of environmental reporting in determining corporate performance. Still, they also stress the need to reevaluate the metrics used to determine the degree of an integrated hotel environment to use a more trustworthy green model (Al Halbusi, Klobas, and Ramayah 2023; Nureen, Xin, et al. 2023).

1.6.3 Consumer Access to Green Management Practices

The opinions of customers on the integration of green lodging are grouped into the following three categories in this section:

- Research that details how particular green behaviors affect consumers' attitudes and behavior or divides consumers into groups according to how they behave sustainably
- Research on the efficiency of green initiative corporate communication
- Research that broadens the paradigms of customer behavior in green lodging

1.6.3.1 Analysis of Stakehloders

Stakeholder theory, which emphasizes public participation in decision-making and is based on the notion that including multiple voices in the process can lead to better outcomes for people and the environment, has been modified from management and commercially applied to environment and natural resource management (Velyvis et al. 2019; You and Ke 2023) with increasing pressure for public participation. In the developing discipline of participatory ENRM decision-making, stakeholder theory concentrates corporate management on quantifiable stakeholder interactions. In the broad framework of public engagement in NRM, stakeholder theory is divided into two primary parts: analysis, which involves comprehending stakeholders to inform decision-making, and participation, which entails including stakeholders in the decision-making process (Velyvis et al. 2019; You and Ke 2023).

Understanding the social aspects of governance, environmental, and risk concerns may be the only objective of a stakeholder analysis. Stakeholder involvement, on the other hand, typically comes after stakeholder analysis. Particularly in communities, interest groups, or both, this is true. Stakeholder engagement is an ongoing process of information sharing about a particular project or issue between stakeholders and decision-makers (de Gooyert et al. 2017; Mahajan et al. 2023). Stakeholders participate in several activities throughout the process that are intended to get their feedback and promote communication between them and decision-makers about the project or problem. Examples of activities include serving on committees, answering questions in surveys and polls, taking part in forums, composing briefs, and obtaining information. Depending on how the process is set up, stakeholders may be "empowered" to the point where they can make decisions (Aaltonen and Kujala 2016; Laplume, Sonpar, and Litz 2008).

Stakeholder participation can occasionally be tokenistic, which means that their opinions are either disregarded or gathered but not thoroughly considered when making decisions (; Johnson et al. 2016). This demonstrates how the nature of the issue, the time and resources available, the power relationships between stakeholders and essential decision-makers, and other variables may all affect how involved specific stakeholders are.

TABLE 1.1
Current Review on Staceholder

Reference	Stakehloder Connection	Scope	Citations
(Aaltonen and Kujala 2016)	In project management	Only for *International Journal of Managing Projects in Business*, from 1995 to 2015	214
(Bhattacharya and Fayezi 2021)	In supply chain management	Unclear	62
(de Gooyert et al. 2017)	In operation management	Only the years 2013 to 2014 were covered by decision support systems	236
(Govindan and Bouzon 2018)	In reverse logistics	2004–2015	258
(Laplume, Sonpar, and Litz 2008)	In mainstream management	1984–2007	1831
(Silva, Nuzum, and Schaltegger 2019)	In evaluating and measuring sustainability performance	Up to 2017	206
(DesJardine, Zhang, and Shi 2023)	In theoretical lenses and perspectives	Recognizing the diversity of shareholders and how they affect other stakeholders	26
(Siems, Seuring, and Schilling 2023)	How stakeholders and their roles are related to sustainable supply chain management practices	2000–2020	19
(Mahajan et al. 2023)	Introduction	1969–2021	50

Despite the frequent confusion between these words, some authors clearly distinguish between stakeholder involvement, analysis, and public participation (Siems, Seuring, and Schilling 2023; Silva, Nuzum, and Schaltegger 2019). Stakeholder analysis and engagement seek to evaluate and include individuals and organizations based on their direct link with global environmental risks management issues, unlike public engagement, which involves all members of society regardless of their relationship to GERM concerns. The term "stakeholder" is defined by DesJardine, Zhang, and Shi (2023) as any individual who can impact or influence a decision.

Table 1.1 provides a comparison of current reviews on stakeholders. According to Google Scholar, the citations are from the day we wrote this work (May 2, 2024).

1.6.3.2 Institutional Theory and Green Management

In contrast to earlier studies (Gupta and Gupta 2021; Gauthier 2013) that concentrated on environmental management, corporate social responsibility (CSR), etc., we employ institutional theory as a theoretical framework. We are just beginning to apply this organizational theory to supply chain management. Primarily because eco-friendly supply chains and sustainability are the main topics, while this study looks at supply networks made up of numerous organizations, previous institutional theory research has examined organizations. The management of organizational environments has been the subject of extensive research using institutional theory. One of institutional theory's main advantages is its capacity to clarify why certain behaviors are selected without obvious financial benefits (Kauppi 2013; Khanra et al. 2022).

Through the theoretical lens of institutional theory, researchers can discern and examine the factors—such as culture, social environment, regulations (including the legal environment), tradition and history, and economic incentives—that support the survival and legitimacy of organizational practices. Recognizing resources is also essential. In this context, legitimacy refers to the application of sustainable techniques that stakeholders deem appropriate and suitable (Gauthier 2013). Traditionally, institutional theory has concentrated on how groups and organizations can better maintain their standing and legitimacy by adhering to laws, regulations, court orders, professional codes, regulatory structures, and other social and cultural practices that compel compliance.

According to institutional theory, external social, political, and economic pressures influence the strategic and organizational decisions of firms when firms seek to adopt legitimate practices or legitimize their practices in the eyes of other stakeholders (Gupta and Gupta 2021).

The way that decisions on environmentally friendly activities and sustainable green activities are influenced by societal values, technological advancements, and regulations can be explained by institutional theory (Huang and Huang 2024; Gupta and Gupta 2021). Various organizations have been studied using institutional theory regarding how to affect choices on green and sustainable activities. The plan culminates with the application of environmental management techniques. Government rules and large supply chain corporations are the primary forces behind changes in the green regulatory landscape.

Institutional theory describes three types of drivers as producing isomorphism in organizational processes, structures, and strategies. These are mimetic, normative, and coercive impulses (Al Halbusi, Klobas, and Ramayah 2023). In this situation in the dairy supply chain, the influence of those in positions of authority leads to coercion. To advance sustainable development and environmental stewardship, coercive pressure is necessary. For enterprises to be perceived as acting lawfully, normative drivers ensure they conform. It has been discovered that normative constraints push businesses to become more environmentally sensitive. Institutional research is necessary to comprehend new social norms (such as moral principles and ecological thinking) and organizational responses to environmental problems. Normative drives, then, have an impact because people follow social duties that stem from societal demands or what is right for organizations or individuals to do. When businesses mimic the actions of their successful competitors in the market, such as specializing in sustainable milk supply for supermarkets, to emulate their route to success and acquire legitimacy, this is known as the mimetic isomorphic drive.

In addition to forming a logic in which laws, rules, and accepted norms of behavior appear natural and permanent, institutions also set expectations that define what constitutes appropriate behavior for organizations (Nureen, Xin, et al. 2023;

Al Halbusi, Klobas, and Ramayah 2023). Thornton et al. (2012) said: "Institutional logic, when it becomes dominant, influences organizational decision-making … by focusing executive attention on problems and solutions consistent with the dominant logic and away from those inconsistent with the dominant logic."

Consequently, organizations have the authority to establish what is proper or lawful, or, in other words, what conduct is permissible, disregarding or rejecting other behaviors. This will impact the way the company makes decisions. Here is an outline of the many players' roles in creating a sustainable supply chain and their contributions to establishing coherence.

The institutional viewpoint makes it possible to focus on how social pressure, regulation, and integration influence organizational behavior (Novitasari et al. 2023; Nureen, Xin, et al. 2023; Al Halbusi, Klobas, and Ramayah 2023). To determine whether sustainability is essential from a strategic standpoint, this study examines the roles played by various supply chain participants and their sustainability strategies (Novitasari et al. 2023; Riaz et al. 2024). We perform this action to learn about the strategies and actions of stakeholders to enhance energy efficiency. The primary obstacles to creating sustainable strategy—in this example, energy-saving strategies—are also covered in this chapter.

1.7 STRATEGIES FOR IMPLEMENTING EFFECTIVE GREEN MANAGEMENT PRACTICES

It should be understood that integrating sustainability, the environment, and human resource management (HRM) is crucial in putting the "green" idea into practice. According to many human resources experts, systems must be in sync to boost the organization's chances of implementing a green plan. Numerous studies have demonstrated that hiring, training, employee motivation, and rewards are vital human elements supporting employees in implementing green management ideas (Al Halbusi, Klobas, and Ramayah 2023). Human resource management must make green adjustments or modifications to guarantee that the organization receives the correct green input from employees and green work performance. Green HRM aims to develop a green workforce that recognizes and values the company's green culture. Ensuring that the entire human resource management process is environmentally friendly is the primary objective of this "green culture" in HRM (Riaz et al. 2024). An organization's human resources division can contribute significantly to creating a green culture by managing procedures, including hiring, screening, training and development, and remuneration (Harmon et al. 2010). According to Chotiyanon and Ueasangkomsate (2024), human resources are the most crucial element in implementing green policies. According to Liu, Chen, and Zhu (2023), key components of an effective organization are employee relations, training, centralized remuneration, performance appraisal, and selection procedures. Thus, it is proposed that the human resource function is a crucial strategic issue and that it is necessary to accomplish organizational change that adjusts to the new needs of the organization. Coordination between the domains of sustainability, the natural environment, and human resource management is necessary to implement a green strategy (Nureen,

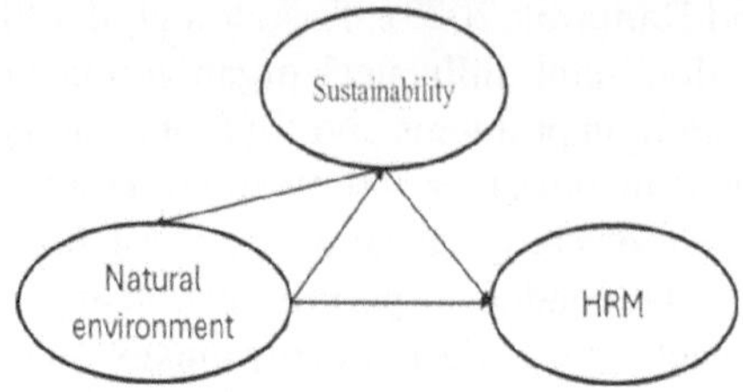

FIGURE 1.4 Sustainability and HRM relation.

Xin, et al. 2023; Nureen, Liu, Irfan, and Sroufe 2023). The execution of a green strategy depends heavily on factors such as hiring, training, employee motivation, and rewards—all of which are significant indicators of human resource management (Gavronski et al. 2011; Arbabzadeh, Lewis, and Keoleian 2019). It is necessary to change the human resource management function to be green to guarantee that the company has a green output (work performance) and green input (appropriate people) (Gavronski et al. 2011; Arbabzadeh, Lewis, and Keoleian 2019). Figure 1.4 shows the relationship between sustainability and HRM.

1.7.1 Requirements for Green Management

Attracting skilled personnel to fill roles is a significant challenge for firms in this "war for talent." Creating a reputation for being a green employer can help draw in new hires. Graduates consider environmental performance and company reputation when deciding which jobs to apply for (Gavronski et al. 2011; Hsu et al. 2023). According to a UK Carbon Trust survey of 1018 workers, 75% of employees said companies needed to have proactive measures to decrease carbon emissions (Gough and Shackley 2005). Adopting a green workplace policy will enhance employees' perceptions of the company and be a powerful recruiting tool for suitably qualified workers. This section now focuses on how businesses may incorporate their green hiring practices into the individual and employee hiring processes. "The practice of ensuring that newcomers are familiar with the culture of the organization's environment and can uphold its environmental values" is how Nureen, Sun, et al. (2023) and Wang et al. (2023) define green recruiting. To guarantee that new hires possess the knowledge and dispositions necessary to adhere to the company's green strategy, it is recommended that enterprises integrate green job descriptions with green job specifications (Appiah et al. 2023; Liu, Chen, and Zhu 2023). New members must be capable of recycling and possess a resource-saving mindset, saving electricity, water, paper, and other materials. However, from an organizational standpoint, we also need to consider green recruitment. Organizations should try to make hiring a green procedure at the managerial and administrative levels. Here, it is imperative that the entire recruitment process, or at least all of the activities undertaken throughout it, align with the organization's green initiatives. Instead of traditional paper-based work, firms can employ human resource information systems (HRIS), paperless interviews, and "online recruitment" in their recruitment mix. The use of web-based tools or technologies to help with the recruitment process is known as online recruitment. The company's corporate website or a job site, portal, or websites such as www.bdjobs.com, www.prothom-alojobs.com, or monster.com, among others, might serve this function. The risk of keeping paper records is decreased by the HRIS, a computer-based information system that contains all HR-related data methodically for decision-making purposes. As a result, hiring procedures themselves may serve as green functions.

1.7.2 Green Deployment and Training

The terms "development and training" are sometimes used synonymously, although they have a tiny distinction in human resource management. Giving new or existing employees the skills they need to do their jobs is known as training. On the other side, development is a learning process that aims to enhance employees' general development and prepare them for new challenges. Green training and development is the process of giving staff members the knowledge, abilities, and attitudes they need to follow green methods. Developing abilities in resource conservation, recycling, and environmentally conscious thinking is the principal objective of green training. Green training is necessary whenever any green technology is implemented, ensuring that staff members are knowledgeable about it and that the company reaps its most significant benefits. Green training focuses on work practices that minimize waste, save energy, and involve staff members in resolving environmental issues. The dissemination and transmission of greening power also raise staff members' awareness of the values and diverse facets of environmental management.

1.8 GREEN PERFORMANCE MANAGEMENT

Performance management, or PM, is the process of routinely giving employees performance feedback, responsibility, and documentation to inspire them to enhance their work and better utilize their abilities to meet corporate objectives. Organizational performance management ultimately arises from an understanding of corporate strategy. Global business strategy is influenced by environmental management (EM), and the green wave also has a favorable impact on performance management. Several businesses are now using company-wide environmental performance criteria as part of their green performance management program to motivate staff members to follow the company's green initiatives (Hsu et al. 2023; Al Halbusi, Klobas, and Ramayah 2023). Employee compliance with green policies and procedures is positively motivated when green standards are established and assessed using these standards. Many businesses use green information systems, green audits, and company-wide environmental performance standards to address PM issues and gather relevant environmental performance data (Novitasari et al. 2023; Riaz et al. 2024). Performance evaluation is the most crucial component of performance management. The most significant contribution we can make to executing a green strategy is through performance appraisal, a potent tool for human resource management (Mahsina and Soewarno 2024; Chotiyanon and Ueasangkomsate 2024). The performance appraisal method identifies the issues or facets of an employee's job that are given the most attention. Performance evaluation criteria shape employee performance and attitudes. As a result, adding different environmental performance requirements to the performance evaluation system will encourage staff members to follow the company's green policies. Therefore, good performance reviews can assist the ongoing improvement of organizational environmental outcomes and give employees insightful feedback in addition to satisfying the requirements of validity, reliability, and fairness (Huang and Huang 2024). HR managers should review it and include characteristics like creativity, diversity, teamwork, collaboration, recycling, and achieving environmental obligations that evaluate people's abilities. This capability will undoubtedly reinforce the organization's basic principles (Gauthier 2013; Gupta and Gupta 2021). For staff members to perform better in various environmental aspects, managers should also keep them updated on performance standards and continuous performance status throughout the year.

1.9 CONCLUSION

In conclusion, the integration of green management and advanced technologies presents a promising pathway for firms aiming to achieve sustainable growth and enhanced performance, ultimately contributing to a more sustainable future. Furthermore, the review points to a growing need for further research to explore the specific mechanisms through which green management and advanced technologies interact to drive performance. Longitudinal studies could provide valuable insights into the causal relationships and long-term impacts of these practices, offering a roadmap for companies seeking to navigate the complexities of sustainability in today's business landscape. The findings underscore the transformative potential of advanced technologies, such as IoT, AI, and big data analytics, in facilitating effective green management practices. These technologies enable firms to optimize resource usage, reduce waste, and enhance operational efficiency, thereby strengthening their overall performance. As stakeholder expectations for corporate responsibility rise, the adoption of green management techniques becomes increasingly vital for maintaining a positive brand image and fostering customer loyalty. This chapter has presented a comprehensive literature review on the interplay between green management, advanced technologies, and firm performance. The analysis highlights how the integration of sustainable practices and cutting-edge technologies can significantly enhance business outcomes. Through various studies, it is evident that green management not only contributes to environmental sustainability but also plays a critical role in improving financial performance, innovation, and competitive positioning in the market.

REFERENCES

Aaltonen, Kirsi, and Jaakko Kujala. 2016. "Towards an Improved Understanding of Project Stakeholder Landscapes." *International Journal of Project Management* 34 (8): 1537–52. https://doi.org/10.1016/j.ijproman.2016.08.009.

Abbas, J., D. Balsalobre-Lorente, M. A. Amjid, K. Al-Sulaiti, I. Al-Sulaiti, and O. Alderea. 2024. "Financial Innovation and Digitalization Promote Business Growth: The Interplay of Green Technology Innovation, Product Market Competition and Firm Performance." *Innovation and Green Development* 3 (1): 100111. https://doi.org/10.1016/j.igd.2023.100111.

Abbas, Jaffar, Daniel Balsalobre-Lorente, Muhammad Asif Amjid, Khalid Al-Sulaiti, Ibrahim Al-Sulaiti, and Osama Aldereai. 2024. "Financial Innovation and Digitalization Promote Business Growth: The Interplay of Green Technology Innovation, Product Market Competition and Firm Performance." *Innovation and Green Development* 3 (1): 100111. https://doi.org/10.1016/j.igd.2023.100111.

Ahsan, M. M., and Z. Siddique. 2022. "Industry 4.0 in Healthcare: A Systematic Review." *International Journal of Information Management Data Insights* 2 (1). https://doi.org/10.1016/j.jjimei.2022.100079.

Alhalaseh, R., and Alasasfeh, S. 2020. "Machine-Learning-Based Emotion Recognition System Using EEG Signals." *Computers* 9 (4): 1–15. https://doi.org/10.3390/computers9040095.

Allam, Z., A. Sharifi, S. E. Bibri, D. S. Jones, and J. Krogstie. 2022. "The Metaverse as a Virtual Form of Smart Cities: Opportunities and Challenges for Environmental, Economic, and Social Sustainability in Urban Futures." *Smart Cities* 5 (3): 771–801. https://doi.org/10.3390/smartcities5030040.

Alsharif, Mohammed H., Abu Jahid, Raju Kannadasan, and Mun Kyeom Kim. 2024. "Unleashing the Potential of Sixth Generation (6G) Wireless Networks in Smart Energy Grid Management: A Comprehensive Review." *Energy Reports* 11 (June): 1376–98. https://doi.org/10.1016/J.EGYR.2024.01.011.

Anajemba, Joseph Henry, Celestine Iwendi, Imran Razzak, James AduAnsere, and Izuchukwu Michael Okpalaoguchi. 2022. "A Counter-Eavesdropping Technique for Optimized Privacy of Wireless Industrial IoT Communications." *IEEE Transactions on Industrial Informatics* 18 (9): 6445–54. https://doi.org/10.1109/TII.2021.3140109.

Appiah, Michael Karikari, Aloysius Sam, Emmanuel Twum, and Enchill Godslove. 2023. "Modelling the Influencing of Green Entrepreneurship Orientation on Sustainable Firm Performance: A Moderated Mediation Model." *Economic Research-EkonomskaIstraživanja* 36 (2). https://doi.org/10.1080/1331677X.2023.2179094.

Arbabzadeh, Maryam, Geoffrey M. Lewis, and Gregory A. Keoleian. 2019. "Green Principles for Responsible Battery Management in Mobile Applications." *Journal of Energy Storage* 24 (August): 100779. https://doi.org/10.1016/j.est.2019.100779.

Aslam, Muhammad Muzamil, Rufai Yusuf Zakari, Ali Tufail, Sajid Ali, Kassim Kalinaki, and Wasswa Shafik. 2024. "Introduction to Industry's Fourth Revolution and Its Impacts on Healthcare." *Digital Transformation in Healthcare 5.0: Volume 1: IoT, AI and Digital Twin* 33.

Ayaz, M., M. Ammad-Uddin, Z. Sharif, A. Mansour, and E. H. M. Aggoune. 2019. "Internet-of-Things (IoT)-Based Smart Agriculture: Toward Making the Fields Talk." *IEEE Access* 7: 129551–83. https://doi.org/10.1109/ACCESS.2019.2932609.

Bhattacharya, Ananya, and Sajad Fayezi. 2021. "Ameliorating Food Loss and Waste in the Supply Chain through Multi-Stakeholder Collaboration." *Industrial Marketing Management* 93 (February): 328–43. https://doi.org/10.1016/j.indmarman.2021.01.009.

Cherian, A. K., A. Singh, and A. Singh. 2024. "Sustainability-Mobility Solutions: A Pathway to Sustainable Urban Development." *2024 10th International Conference on Communication and Signal Processing (ICCSP)*, 1377–82. https://doi.org/10.1109/ICCSP60870.2024.10543902

Chotiyanon, Phasawit, and Pittawat Ueasangkomsate. 2024. "The Effect of Innovative Corporate Social Responsibility in Supply Chain Management on Firm Performance." In *2024 Joint International Conference on Digital Arts, Media and Technology with ECTI Northern Section Conference on Electrical, Electronics, Computer and Telecommunications Engineering (ECTI DAMT & NCON)*, 198–203. IEEE. https://doi.org/10.1109/ECTIDAMTNCON60518.2024.10480098.

Culot, G., G. Nassimbeni, G. Orzes, and M. Sartor. 2020. "Behind the Definition of Industry 4.0: Analysis and Open Questions." *International Journal of Production Economics* 226. https://doi.org/10.1016/j.ijpe.2020.107617.

DesJardine, Mark R., Muhan Zhang, and Wei Shi. 2023. "How Shareholders Impact Stakeholder Interests: A Review and Map for Future Research." *Journal of Management* 49 (1): 400–29. https://doi.org/10.1177/01492063221126707.

Eltayeb, A. O., M. C. Stern, H. Herzog, and T. A. Hatton. 2014. "Energetics of Electrochemically-Mediated Amine Regeneration." *Energy Procedia* 63: 595–604.

Fragopanagos, N., and J. G. Taylor. 2005. "Emotion Recognition in Human-computer Interaction." *Neural Networks* 18 (4): 389–405. https://doi.org/10.1016/j.neunet.2005.03.006.

Gauthier, J. 2013. "Institutional Theory and Corporate Sustainability: Determinant Versus Interactive Approaches." *Organization Management Journal* 10 (2): 86–96. https://doi.org/10.1080/15416518.2013.801741.

Gavronski, Iuri, Robert D. Klassen, Stephan Vachon, and Luis Felipe Machado do Nascimento. 2011. "A Resource-Based View of Green Supply Management." *Transportation Research Part E: Logistics and Transportation Review* 47 (6): 872–85. https://doi.org/10.1016/j.tre.2011.05.018.

Glowka, Gundula, Robert Eller, Mike Peters, and Anita Zehrer. 2024. "Risk Perception in Small- and Medium-Sized Hospitality Family Enterprises." *Tourism Review*, February. https://doi.org/10.1108/TR-06-2023-0421.

Gooyert, Vincent de, Etiënne Rouwette, Hans van Kranenburg, and Edward Freeman. 2017. "Reviewing the Role of Stakeholders in Operational Research: A Stakeholder Theory Perspective." *European Journal of Operational Research* 262 (2): 402–10. https://doi.org/10.1016/j.ejor.2017.03.079.

Gough, Clair, and Simon Shackley. 2005. "An Integrated Assessment of Carbon Dioxide Capture and Storage in the UK." Technical Report, Tyndall Project T2/2i, October 2005.

Govindan, Kannan, and Marina Bouzon. 2018. "From a Literature Review to a Multi-Perspective Framework for Reverse Logistics Barriers and Drivers." *Journal of Cleaner Production* 187 (June): 318–37. https://doi.org/10.1016/j.jclepro.2018.03.040.

Guo, Y., M. N. Irshad, M. M. Aslam, A. Qurban, and A. Raza. 2023. "IHGWO-Based Optimization of IoT Wireless Sensor Networks." *Proceedings of the 19th ACM International Symposium on QoS and Security for Wireless and Mobile Networks*, 97–104. https://doi.org/10.1145/3616391.3622769.

Gupta, Amit Kumar, and Narain Gupta. 2021. "Environment Practices Mediating the Environmental Compliance and Firm Performance: An Institutional Theory Perspective from Emerging Economies." *Global Journal of Flexible Systems Management* 22 (3): 157–78. https://doi.org/10.1007/s40171-021-00266-w.

Halbusi, Hussam Al, Jane E. Klobas, and T. Ramayah. 2023. "Green Core Competence and Firm Performance in a Post-conflict Country, Iraq." *Business Strategy and the Environment* 32 (6): 2702–14. https://doi.org/10.1002/bse.3265.

Hartono, Bryan Young, Hotlan Siagian, and Zeplin Jiwa Husada Tarigan. 2023. "The Effect of Knowledge Management on Firm Performance. Mediating Role of Production Technology, Supply Chain Integration, and Green Supply Chain." *Uncertain Supply Chain Management* 11 (3): 1133–48. https://doi.org/10.5267/j.uscm.2023.4.009.

Harmon, J., K. D. Fairfield, and J. Wirtenberg. 2010. "Missing an Opportunity: HR Leadership and Sustainability." *People & Strategy*, 33 (1).

Hsu, Chin-Chun, Keah Choon Tan, Brent A. Hathaway, and Suhaiza Zailani. 2023. "Business Networking Orientation, Green Operations Practices and Firm Performance." *Journal of Manufacturing Technology Management* 34 (3): 455–75. https://doi.org/10.1108/JMTM-09-2022-0341.

Huang, Yi-Chun, and Chih-Hsuan Huang. 2024. "Exploring Institutional Pressure, the Top Management Team's Response, Green Innovation Adoption, and Firm Performance: Evidence from Taiwan's Electrical and Electronics Industry." *European Journal of Innovation Management* 27 (3): 800–24. https://doi.org/10.1108/EJIM-03-2022-0126.

Jin, J. X., and X. Y. Chen. 2015. "Cooperative Operation of Superconducting Fault-Current-Limiting Cable and SMES System for Grounding Fault Protection in a LVDC Network." *IEEE Transactions on Industry Applications* 51 (6): 5410–14. https://doi.org/10.1109/TIA.2015.2438252.

Johnson, Alistair E. W., Mohammad M. Ghassemi, Shamim Nemati, Katherine E. Niehaus, David Clifton, and Gari D. Clifford. 2016. "Machine Learning and Decision Support in Critical Care." *Proceedings of the IEEE* 104 (2): 444–66. https://doi.org/10.1109/JPROC.2015.2501978.

Kanawaday, A., and Sane, A. 2018. "Machine Learning for Predictive Maintenance of Industrial Machines Using IoT Sensor Data." *Proceedings of the IEEE International Conference on Software Engineering and Service Sciences, ICSESS, 2017-November*, 87–90. https://doi.org/10.1109/ICSESS.2017.8342870.

Kauppi, Katri. 2013. "Extending the Use of Institutional Theory in Operations and Supply Chain Management Research." *International Journal of Operations & Production Management* 33 (10): 1318–45. https://doi.org/10.1108/IJOPM-10-2011-0364.

Kaur, Charanjeet, and Khushdeep Dharni. 2024. "Influence of Performance Measures and Sample Characteristics on the Relationship Between Corporate Environmental Management and Firm Performance: A Meta-analytic Review." *Business Strategy & Development* 7 (2). https://doi.org/10.1002/bsd2.367.

Khan, N., Ahmed, J., Nawaz, M., and Zaman, K. 2015. "The Socio-Economic Determinants of Crime in Pakistan: New Evidence on an Old Debate." *AEBJ* 10 (2): 73–81. https://doi.org/10.1016/j.aebj.2015.01.001.

Khanra, Sayantan, Puneet Kaur, Rojers P Joseph, Ashish Malik, and Amandeep Dhir. 2022. "A Resource-based View of Green Innovation as a Strategic Firm Resource: Present Status and Future Directions." *Business Strategy and the Environment* 31 (4): 1395–413. https://doi.org/10.1002/bse.2961.

Kumar, P., and U. C. Pati. 2017. "IOT Based Monitoring and Control of Appliances for Smart Home." *2016 IEEE International Conference on Recent Trends in Electronics, Information and Communication Technology, RTEICT 2016 – Proceedings, 769008*, 1145–1150. https://doi.org/10.1109/RTEICT.2016.7808011.

Kunrath, T. L., A. Dresch, and D. R. Veit. 2023. "Supply Chain Management and Industry 4.0: A Theoretical Approach." *Brazilian Journal of Operations and Production Management* 20 (1). https://doi.org/10.14488/BJOPM.1263.2023.

Laosuwan, Teerawong, Torsak Gomasathit, and Tanutdech Rotjanakusol. 2017. "Application of Remote Sensing for Temperature Monitoring: The Technique for Land Surface Temperature Analysis." *Journal of Ecological Engineering* 18 (3): 53–60. https://doi.org/10.12911/22998993/69358.

Laplume, André O., Karan Sonpar, and Reginald A. Litz. 2008. "Stakeholder Theory: Reviewing a Theory That Moves Us." *Journal of Management* 34 (6): 1152–89. https://doi.org/10.1177/0149206308324322.

Li, Mimi, Wenqing Xu, Weiwei Liu, and Huiyi Cao. 2024. "Networking for Innovation Dynamics: A Design-Driven Approach in the Hospitality Industry." *International Journal of Contemporary Hospitality Management* 36 (2): 632–60. https://doi.org/10.1108/IJCHM-10-2022-1259.

Liang, J., M. Zhang, and V. C. M. Leung. 2020. "A Reliable Trust Computing Mechanism Based on Multisource Feedback and Fog Computing in Social Sensor Cloud." *IEEE Internet of Things Journal* 7 (6): 5481–90. https://doi.org/10.1109/JIOT.2020.2981005.

Lin, Y., Z. Gao, Y. Tu, H. Du, D. Niyato, J. Kang, and H. Yang. 2023. A Blockchain-Based Semantic Exchange Framework for Web 3.0 Toward Participatory Economy. *IEEE Communications Magazine* 61 (8): 94–100. https://doi.org/10.1109/MCOM.003.2200817.

Ling, L., Ma, X., and Huang, Y. 2013. CDN Cloud: A Novel Scheme for Combining CDN and Cloud Computing." *Proceedings of 2013 2nd International Conference on Measurement, Information and Control, ICMIC 2013*, 1: 687–690. https://doi.org/10.1109/MIC.2013.6758055.

Liu, Junjun, Yuan Chen, and Qinghua Zhu. 2023. "Green Supplier Governance and Firm Performance: A Comprehensive Understanding of Three Governance Approaches." *International Journal of Physical Distribution & Logistics Management* 53 (9): 1073–100. https://doi.org/10.1108/IJPDLM-07-2022-0232.

Mahajan, Ritika, Weng Marc Lim, Monica Sareen, Satish Kumar, and Rajat Panwar. 2023. "Stakeholder Theory." *Journal of Business Research* 166 (November): 114104. https://doi.org/10.1016/j.jbusres.2023.114104.

Mahsina, Mahsina, and Noorlailie Soewarno. 2024. "The Mediation Effect of Firm Performance on the Association between Two-Tier Independent Boards and Green Innovation Practices: Evidence from Indonesia." *Intangible Capital* 20 (1): 126. https://doi.org/10.3926/ic.2367.

Matsukawa, H., Minner, S., and Nakashima, K. 2020. "Editorial: Industry 4.0 and Production Economics." *International Journal of Production Economics* 226: 107666. https://doi.org/10.1016/j.ijpe.2020.107666.

Mnih, V., K. Kavukcuoglu, D. Silver, A. A. Rusu, J. Veness, M. G. Bellemare, A. Graves, M. Riedmiller, A. K. Fidjeland, G. Ostrovski, S. Petersen, C. Beattie, A. Sadik, I. Antonoglou, H. King, D. Kumaran, D. Wierstra, S. Legg and D. Hassabis. 2015. "Human-Level Control Through Deep Reinforcement Learning." *Nature* 518 (7540): 529–33. https://doi.org/10.1038/nature14236.

Novitasari, Maya, Anggita Langgeng Wijaya, Nindya Mariana Agustin, Ardi Gunardi, and Leo-Paul Dana. 2023. "Corporate Social Responsibility and Firm Performance: Green Supply Chain Management as a Mediating Variable." *Corporate Social Responsibility and Environmental Management* 30 (1): 267–76. https://doi.org/10.1002/csr.2353.

Nureen, Naila, Da Liu, Muhammad Irfan, and Robert Sroufe. 2023. "Greening the Manufacturing Firms: Do Green Supply Chain Management and Organizational Citizenship Behavior Influence Firm Performance?" *Environmental Science and Pollution Research* 30 (31): 77246–61. https://doi.org/10.1007/s11356-023-27817-1.

Nureen, Naila, Da Liu, Muhammad Irfan, Maida Malik, and Usama Awan. 2023. "Nexuses among Green Supply Chain Management, Green Human Capital, Managerial Environmental Knowledge, and Firm Performance: Evidence from a Developing Country." *Sustainability* 15 (6): 5597. https://doi.org/10.3390/su15065597.

Nureen, Naila, Huaping Sun, Muhammad Irfan, Alina Cristina Nuta, and Maida Malik. 2023. "Digital Transformation: Fresh Insights to Implement Green Supply Chain Management, Eco-Technological Innovation, and Collaborative Capability in Manufacturing Sector of an Emerging Economy." *Environmental Science and Pollution Research* 30 (32): 78168–81. https://doi.org/10.1007/s11356-023-27796-3.

Nureen, Naila, Yongrong Xin, Muhammad Irfan, and Shah Fahad. 2023. "Going Green: How Do Green Supply Chain Management and Green Training Influence Firm Performance? Evidence from a Developing Country." *Environmental Science and Pollution Research* 30 (20): 57448–59. https://doi.org/10.1007/s11356-023-26609-x.

Ortt, R., C. Stolwijk, and M. Punter. 2020. Implementing Industry 4.0: Assessing the Current State. *Journal of Manufacturing Technology Management* 31 (5): 825–836. https://doi.org/10.1108/JMTM-07-2020-0284.

Posevkin, R., and I. Bessmertny. 2015. "Translation of Natural Language Queries to Structured Data Sources." *9th International Conference on Application of Information and Communication Technologies, AICT 2015 – Proceedings*, 57–59. https://doi.org/10.1109/ICAICT.2015.7338516.

Qamar, Faizan, Maraj Uddin Ahmed Siddiqui, MHD Nour Hindia, Rosilah Hassan, and Quang Ngoc Nguyen. 2020. "Issues, Challenges, and Research Trends in Spectrum Management: A Comprehensive Overview and New Vision for Designing 6g Networks." *Electronics (Switzerland)* 9 (9): 1–39. https://doi.org/10.3390/electronics9091416.

Qayyum, F. A., M. Naeem, A. S. Khwaja, A. Anpalagan, L. Guan, and B. Venkatesh. 2015. "Appliance Scheduling Optimization in Smart Home Networks." *IEEE Access* 3: 2176–90. https://doi.org/10.1109/ACCESS.2015.2496117.

Rauf, Fawad, Wanqiu Wang, and Cosmina L. Voinea. 2024. "Interaction of Corporate Social Responsibility Reporting at the Crossroads of Green Innovation Performance and Firm Performance: The Moderating Role of the Enterprise Life Stage." *Sustainability* 16 (5): 1821. https://doi.org/10.3390/su16051821.

Riaz, Adil, Martin Cepel, Alberto Ferraris, Khurram Ashfaq, and Shafique Ur Rehman. 2024. "Nexus among Green Intellectual Capital, Green Information Systems, Green Management Initiatives and Sustainable Performance: A Mediated-Moderated Perspective." *Journal of Intellectual Capital*, March. https://doi.org/10.1108/JIC-03-2023-0063.

Rost, P., C. Bernardos, A. de Domenico, M. Girolamo, M. Lalam, A. Maeder, D. Sabella, and D. Wübben, D. 2014. "Cloud Technologies for Flexible 5G Radio Access Networks." *IEEE Communications Magazine* 52 (5): 68–76. https://doi.org/10.1109/MCOM.2014.6898939.

Shafik, W. 2023. "IoT-Based Energy Harvesting and Future Research Trends in Wireless Sensor Networks." In *Handbook of Research on Network-Enabled IoT Applications for Smart City Services*, edited by K. Hemant Reddy, Diptendu Roy, Tapas Mishra, and Mir Hussain, 282–306. Hershey, PA: Information Science Reference.

Siedschlag, Iulia, and Weijie Yan. 2023. "Do Green Investments Improve Firm Performance? Empirical Evidence from Ireland." *Technological Forecasting and Social Change* 186 (January): 122181. https://doi.org/10.1016/j.techfore.2022.122181.

Siems, Erik, Stefan Seuring, and Lara Schilling. 2023. "Stakeholder Roles in Sustainable Supply Chain Management: A Literature Review." *Journal of Business Economics* 93 (4): 747–75. https://doi.org/10.1007/s11573-022-01117-5.

Silva, Samanthi, Anne-Katrin Nuzum, and Stefan Schaltegger. 2019. "Stakeholder Expectations on Sustainability Performance Measurement and Assessment. A Systematic Literature Review." *Journal of Cleaner Production* 217 (April): 204–15. https://doi.org/10.1016/j.jclepro.2019.01.203.

Siyal, A. A., M. Bhatti, M. Munir Babar, K. Ansari, Rubab Saher, and S. Ahmed. 2019. *World Environmental and Water Resources Congress*. 727–36. https://earthexplorer.usgs.gov/.

Soare Senior, S. R., A. Analyst, and J. Burton. n.d. *Smart Cities, Cyber Warfare and Social Disorder*. London: Routledge.

Söderholm, P. 2020. "The Green Economy Transition: The Challenges of Technological Change for Sustainability. *Sustainable Earth* 3 (1): 6.

Sokienah, Y. 2023. "Exploring the Integration of IoT Systems in Interior Design and the Built Environment: A Systematic Review.' *Heliyon* 9 (12): e22869. https://doi.org/10.1016/j.heliyon.2023.e22869.

Thornton, P. H., W. Ocasio, and M. Lounsbury. 2012. *The Institutional Logics Perspective: A New Approach to Culture, Structure, and Process*. Oxford University Press.

Velyvis, K., T. Coen, I. Cheban, N. Dorsey, and A. Mamun. 2019. Interim Report on Grants under the ENRM Project in Malawi (No. 07c35e466e314984944b4b2f43ebd28b). Mathematica Policy Research.

Wang, Jinfeng, Lei Zhu, Lijie Feng, and Jian Feng. 2023. "A Meta-Analysis of Sustainable Supply Chain Management and Firm Performance: Some New Findings on Sustainable Supply Chain Management." *Sustainable Production and Consumption* 38 (June): 312–30. https://doi.org/10.1016/j.spc.2023.04.015.

Yang, R. 2014. "Making Sustainability Sustainable: Challenges in the Design of Eco-Interaction Technologies." *CHI '14: Proceedings of the SIGCHI Conference on Human Factors in Computing Systems*, 823–32.

You, Guoqiao, and Yilong Ke. 2023. "ENRM: An Alternative Tool for Studying Dynamical Systems." *Chaos, Solitons & Fractals* 174 (September): 113889. https://doi.org/10.1016/j.chaos.2023.113889.

Yue, Xiaochen, Baofeng Huo, and Yuxiao Ye. 2023. "The Impact of Coercive Pressure and Ethical Responsibility on Cross-Functional Green Management and Firm Performance." *Journal of Business & Industrial Marketing* 38 (5): 1015–28. https://doi.org/10.1108/JBIM-09-2021-0446.

2 Data Analysis on Greening SMEs through Resource Efficiency *K-Means Algorithm*

Žarko Rađenović, Miljana Talić, and Ivana Janjić

2.1 INTRODUCTION

Environmental pollution and degradation are core global challenges. The green economy (GE) and resource efficiency (RE) practices are widely accepted solutions ensuring the decrease of the negative impact on the environment. With rapid economic development, increased waste pollution, and severe emissions such as carbon, companies struggle to foster green innovation practices (Musaad et al. 2020). Although there are already legal regulations that oblige business entities to implement certain sustainable practices in business, this should not be their main motivation. There are also numerous other reasons and benefits that they can realize if they turn to a more sustainable way of doing business (OECD 2021).

In the 20th and 21st centuries, the concept of the greening economy and green growth can be linked to intensive green transformation processes. Green growth is recognized as the basic way of economic development that can oppose the exploitation and degradation of natural resources; endangerment of ecosystems; reduction of energy, water, and food resources; unsustainable production and consumption; and climate change. In today's competitive market, research and development (R&D) is an essential contributor to gaining a competitive advantage (Krstić et al. 2021).

In globalized and modern conditions, SMEs can improve the competitiveness of an economy through the assimilation and execution of circular economy mechanisms. Thus, SMEs are observed as the major driver of economic and industrial development, providing over two-thirds of the total annual industrial production. SMEs are often characterized as initiators of change and generators of innovation (Talić et al. 2023b; European Commission 2020). Their impact on the environment can be large, and accordingly, in recent years, there have been increasing demands for them to get involved and make their contribution to sustainable development (Talić et al. 2023a). The excessive demand for resources forces SMEs to change their production policies so that they are as efficient as possible and to adapt their products to the new markets that are opening by applying certain innovations. Resource efficiency for SMEs appears to be the new business opportunity for achieving green

DOI: 10.1201/9781003626527-2

production. Additionally, greening production through infrastructure efficiency, material savings, and manufacturing efficiency can decrease costs.

The chapter aims to identify and describe the process of SMEs greening through resource efficiency and implementing K-means clustering. The chapter includes four main sections. Following the Introduction, Section 2.2 highlights the main determinants of the concepts of the green economy and resource efficiency. Section 2.3 presents the new directions of SMEs' development by the green economy and resource efficiency. Section 2.4 describes the application of the K-means algorithm for introducing sustainable business initiatives and optimal utilization of resources in SMEs. Following are the results and discussion. The last section provides an overall conclusion and possible policy implications.

2.2 DETERMINING THE CONCEPTS OF GREEN ECONOMY AND RESOURCE EFFICIENCY

2.2.1 The Importance of Green Economy in Modern Conditions

The disturbance of the global economy and intensive economic, social, and environmental trends resulted in the creation of essentially new models of the economy. In response to the global crisis of climate change and degradation of the human environment, excessive exploitation of resources, pollution, and inadequate treatment of nature, the green economy developed. In the late 1980s, a green economy emerged as an aspect of sustainable development (Trushkina 2022, p. 1). The green economy plays an important role in solving environmental, social, and economic problems and adjusting the economy, the environment, and human society.

The greening economy aims to provide a simultaneous solution to economic and environmental challenges and a pathway to sustainable development. A greener economy is described as an economy that causes the least damage to the natural environment, minimizes ecosystem degradation processes, strengthens social well-being, and promotes economic growth (Gasparatos et al. 2017).

To motivate countries to recognize the significance of strategic policies in the context of a sustainable and green economy, the United Nations Program for Environmental Development (UNEP) in 2008 developed green economic initiatives. According to UNEP (2011, p. 9), a green economy refers to an economy that seeks to optimize natural asset stocks, reduce environmental impacts, overcome ecological challenges, and lead to human well-being and social equity. A green economy leads to the growth of resource productivity, efficiency of resource usage, strengthening of new economic activities, and the development of competitiveness.

Due to the growing public demand for more environmentally conscious businesses, the greening of the economy has received much attention from researchers and policymakers (Purwandani and Michaud 2021; Milanović et al. 2022). The transition to the green economy means the redefinition of business models and strategies, market approaches, and product and service portfolios of SMEs. Mitigation of climate change and the depreciation of environmental degradation are considered as green growth. The green economy, as a catalyst, seeks to raise the productivity of resources used, and increase the stability of the ecosystem as well as social welfare (Brears 2018). The main green economic principles are focused on: (1) recognizing natural capital and values through integrating economic development, growth models, and promoting resource and energy efficiency; (2) protecting biodiversity and ecosystems through investing in sustained natural capital, recognizing and respecting ecological limits and planetary boundaries; and (3) poverty reduction, social protection, and access to essential services (Adamowicz 2022).

2.2.2 The Concept of Resource Efficiency

The constant growth in demand for natural resources at the global level, caused by their increased use, is the reason for numerous problems. From increasingly noticeable climate changes to problems related to the scarcity of resources (ISSA 2014; Panthai and Kungwalsong 2024), price volatility and increased competition also affect increased opportunities to cause instability in certain regions of the world (European Commission 2017). Accordingly, there are increasing pressures on all sectors of the economy that demand improvement of production processes, reduced exploitation of raw materials, and greater engagement in terms of resource recycling, all in accordance with the principles of sustainability (Panthai and Kungwalsong 2024).

Using resources in an efficient way can bring benefits to the whole society and economy, especially if the effects are considered in the long term. While, on the one hand, it affects the protection of the environment, on the other hand, it reduces the economy's dependence on natural resources (Van Ewijk 2018; European Commission 2011). Making efforts to increase the efficiency of resource use mainly includes all three aspects of sustainability: ecological, social, and economic (Faulkner and Badurdeen 2014; Panthai and Kungwalsong 2024). In fact, by maintaining a balance between environmental, economic, and social goals in a sustainable way, aims such as creating new jobs and profits are achieved, but not necessarily by using more natural resources (EEA 2016). Some of the goals of resource efficiency and eco-innovation should be the replacement of limited resources with renewable ones and a limited impact on the environment (EEA 2023). Achieving the stated goals is conditional on the existence of a strategic approach, support for innovation, and the introduction of specific measures to reduce the impact on the environment and improve the social outcome (Van Ewijk 2018).

There are numerous doubts and disagreements regarding the definition of resource efficiency, as well as terms that are used as synonyms for resource efficiency. The European Commission (2011) defines it as "improving economic performance while reducing the pressure on natural resources through their efficient use." Resource efficiency has its roots in the long history of the development of waste management and resource management concepts (Van Ewijk 2018). Resource efficiency and waste management are very closely related concepts, which through the circular economy can significantly contribute to modern living and business conditions (EEA 2016). Accordingly, measuring and assessing circularity is key to monitoring progress and assessing the impact of circular economy initiatives (Rađenović and Rajić 2024). An economy can be said to be resource efficient when it minimizes the absolute amount of resources used to produce a unit of output, that is, when it maximizes production for each unit of resources used (European Commission 2017).

In order to achieve appropriate results in the area of efficient use of resources, it is necessary to engage political actors and adopt adequate public policy measures (UNEP 2017). Concern about environmental degradation, but also about increasing prices and decreasing availability of raw materials, has contributed to greater political interest in the concept of resource efficiency (EEA 2016; Terzić 2024; Van Ewijk 2018).

It is important to apply measures at multiple levels—international, national, and local. In recent decades, intensive work has been done on developing complex mechanisms to achieve the most efficient use of resources, reduce the impact on the environment, and create good foundations for sustainable development and a better quality of life (UNEP 2017).

2.3 NEW DIRECTIONS OF SME DEVELOPMENT IN ACCORDANCE WITH THE CONCEPTS OF GREEN ECONOMY AND RESOURCE EFFICIENCY

Business transformations have been a major consequence of the "greening" industry through the implementation of green activities and innovations. The concept of greening the economy is the process of transition from the industrial-market paradigm to the paradigm of inclusive development, based on sustainability and ecological balance (Adamowicz 2022). Through conducting environmental activities, SMEs are the main drivers of a green economy and green and inclusive growth. By achieving the objectives of green growth, SMEs present the major source of win-win effects and green transformation.

SMEs as the engines of macroeconomic growth and development of societies and economies bring positive but also significant negative impacts on the environment. The enhancement of SMEs' responsibilities toward the environment is one of the major strands that signify the efficiency of greening the economy. According to Miller et al. (2011), out of all the European countries, SMEs account for 60%–70% of industrial pollution. Due to this, the government has chosen to enhance these enterprises to decrease their pollution and sustain economic stability (Rodrigues and Franco 2023).

The key to achieving resource efficiency and at the same time reducing the excessive exploitation of resources and environmental destruction is the reorientation from current business models to more sustainable forms—those that involve the development of production and consumption processes according to the principles of the circular economy. In such an economy, the value of resources is used to the maximum, products are kept in use as long as possible, and waste production is minimal. That is, a less intensive economy is sought, which requires investment in eco-innovation that brings with them numerous advantages (European Commission 2017).

The responsibility of SMEs to minimize the impact of their industrial activities on the environment increases due to awareness and pressure from various stakeholders and governments (Rodrigues and Franco 2023). The contribution of small and medium-sized enterprises to achieving efficient use of resources is increasingly emphasized (Patricio et al. 2018). Efficient use of resources and sustaining the growth of the economy create pressure on SMEs to actively participate in environmentally friendly environmental business activities with high economic returns. Economic value and environmental efficiency increase with the growing social consideration of environmentally (green) innovative practices. Additionally, studies show that enterprises that introduce product innovation or process innovation are among the most productive forms of entrepreneurship, especially in developed countries (Ivanović-Đukić et al. 2021). To gain a competitive advantage over larger companies, SMEs are trying to adopt and implement green innovative practices (Mathiyazhagan et al. 2013).

Small and medium-sized enterprises are expected to have "cleaner" production processes and activities, which they can achieve by investing more in resource efficiency measures (Özbuğday et al. 2020), but which also requires certain changes through industrial processes, as well as organizational structures (Caporale et al. 2023). Their important role in this process is indicated by the growing number of guidelines and policy measures in this area that relate to SMEs (OECD 2021; Talić et al. 2023a). One of them, at the level of the European Union, is the European SME Strategy (COM 2020) announced in 2020 with the aim to contribute to the objectives of the European Green Deal and to achieve a climate-neutral, resource-efficient, and agile digital economy (European Commission 2022). Apart from the legal ones, there are other reasons that stimulate small and medium-sized enterprises to efficiently manage their resources and start implementing circular practices in their business.

Research showed that more than half of SMEs in the EU have already invested or plan to invest in the implementation of environmental practices, while two-thirds of SMEs already practice resource efficiency activities (European Commission 2022). Small and medium-sized enterprises usually implement activities such as reducing waste; saving energy, materials, and water; and reusing materials or waste (European Commission 2022).

2.3.1 Green Innovations in the SME Sector

In the conditions of dynamic evolution and growth of the industrial and corporate sectors, enhancing global sustainability among SMEs is essential (Lufti et al. 2023). Due to the rapidly growing industrialization, SMEs are vital participants in promoting green innovation to achieve sustainable development (Song et al. 2023). SMEs are smaller compared to large companies, but through funds and subsidies, they can reduce their expenditures and be more willing to participate in green and social activities (Rodrigues and Franco 2023). According to some authors (Hasan and Rahman 2023), the incorporation of green innovation would improve the SMEs' image on the market, as well as their economic and financial performance.

Since the beginning of the 1990s, green innovation has been considered one of the emerging fields for many researchers. Among scholars, green innovation has been acknowledged and utilized interchangeably with other terms, namely, eco-innovation, environmental innovation, and sustainable innovation (Arfi et al. 2018; Almeida and Wasim 2023). OECD (2009) defines green innovation as "the creation or implementation of new or significantly improved products (goods and services), processes, marketing methods, organizational structures and institutional arrangements which—with or without intent—lead to environmental improvements compared to relevant alternatives." Green innovation is environmentally friendly innovation through which environmental risks, pollution, and negative impacts on resources can be reduced (Kemp 2010). Eco-innovation becomes an important aspect in the process of shifting from a linear to a circular business perspective (Rađenović et al. 2024, p. 189).

Transforming to a sustainable business model requires redefining business processes and operations. This includes the application of green innovation practices such as improving the energy efficiency of production systems, improving packaging, and numerous other innovations (Peiró-Signes and Segarra-Oña 2018).

The application of green technologies involves the adoption of green practices (green product innovation, green process innovation, and green marketing innovation) as well as the practice of green human resource management (recruitment, selection, education, training, and development) (Wang et al. 2021). Green recruitment is the "process of hiring individuals with knowledge, skills, approaches and behaviours that identify with environmental management systems within an organization" (Ahmad 2015, p. 6). Green selection is based on selecting the best environment-friendly employees and detecting an applicant's knowledge, awareness, and attitude toward the environment (Faisal and Naushad 2020). Developing skills, attitudes, and awareness about environmental protection; learning how to reduce waste; and implementing eco-friendly practices is referred to as employee green training.

Green product innovation is defined as innovation that leads to reduced raw materials, increased recyclability of products using renewable free and biodegradable materials, and removing harmful substances (Durif et al. 2010). Green product innovation aims to decrease the negative impact on the environment during the entire life cycle of the product, using eco-friendly materials, non-toxic compounds, or biodegradable resources; replacing organic solvents with water; and improving energy efficiency. By creating and implementing green product innovation as the highest level of green technology innovation, SMEs can improve their social reputation (Wang et al. 2021) and financial performance. Green process innovation can be considered an effective and crucial strategy for SMEs to improve their financial performance and promote sustainable advantage. Green process innovation includes systematic improvements to the operational and managerial processes by reducing air and water emissions and enhancing efficiency in resource and energy use for producing and promoting green products (Lin et al. 2017). Green marketing innovation has been recognized as a tool for SMEs to promote sustainable consumption of green products by sustainable packaging and effective recycling (Kar and Harichandan 2022).

Green innovative practices decrease environmental harm through business activities and greener behaviors. Some authors (Ashton et al. 2017) have found that energy efficiency in facilities, and recycling materials such as paper and plastics, were the most commonly adopted green innovative practices among small manufacturing firms in the Greater Chicago area. Chang and Slaubaugh (2017) discovered that green innovative practices implemented by U.S. enterprises, which increase firms' financial performance, were related to adopting energy-efficient technologies and conserving water.

2.3.2 Drivers of the Transition of SMEs to GE and RE

As small and medium-sized enterprises differ from large ones in terms of their characteristics, the question of the main motives for their transition to resource-efficient practices, as well as the barriers they face on the way to transition, is often asked (Talić et al. 2023; Green Win 2018).

The strategic capability of SMEs to facilitate and implement green innovation is influenced by internal factors (knowledge, organizational culture, organizational learning, financial resources, technology capability) and external drivers (customer demand, environmental regulation, and peer pressure from the other related organizations) (Singh and Chakraborty 2016; del Rosario et al. 2017; Khan et al. 2021). Wong et al. (2012) stated that effective knowledge sharing encourages green innovation. Some scholars (Abbas and Sagsan 2019; Shahzad et al. 2020) stated that knowledge management practices have noticeable effects on the level of green innovation. Some researchers (El-Kassar and Singh 2019) pointed out that the most significant factors that affect a firm's adoption of green innovations in SMEs are knowledge acquisition, knowledge dissemination, and knowledge responsiveness. Besides knowledge of employees, organizational culture has a significant impact on the adoption of green innovation, primarily a culture of adhocracy in which egalitarianism in the distribution of power is valued, and thus employee participation and decentralization (del Rosario et al. 2017). To overcome global societal challenges and to create and maintain the competitive advantage of enterprises, R&D and innovation ensure the base of technical solutions (Janjić et al. 2022; Janjić et al. 2021) in SMEs. Organizational learning is a significant driver that affects the emergence of green innovations. One of the necessary preconditions for the creation and implementation of green innovation is adequate promotion and support of R&D, education, and training of employees (De Jesus and Mendonca 2018). R&D activities are seen as one of the significant driving forces in implementing innovative activities in enterprises and national economies (Rađenović et al. 2023, p. 35). Innovation involves simultaneous engagement of R&D and all functional areas within the enterprise (Janjić and Rađenović 2019).

SMEs' technology capabilities consist of intangible knowledge and experience and tangible technical infrastructure. Technology allows firms to design and develop eco-friendly products and green manufacturing processes (Hasan and Rahman 2023).

An organization's tendency to implement green technologies depends on financial resources, which are the main predictor of enhancing sustainable performance through green innovation (Khan et al. 2021, p. 37). It will be easier for small and medium-sized enterprises to make the decision to introduce measures that achieve resource efficiency if they are convinced of their positive impact on the company's operations and performance (Özbuğday et al. 2020). One of the reasons for implementing resource efficiency measures that is very often cited, and is the subject of numerous scientific articles, is financial. Namely, numerous previous studies have suggested that the implementation of sustainable practices could have a significant positive financial effect on the companies that implement them (Rosa and Paula 2023; Kurapatskie and Darnall 2013). Financial benefits would be reflected through cost savings due to less need for resources, as well as based on the sale of products in new markets (Wijkman and Skånberg 2015; Rizos et al. 2016; Taranic et al. 2016; OECD 2021). Cost reduction based on more efficient use of resources provides an opportunity for companies to direct their funds to expand sales or to other, more productive activities, such as the development of new products and services (Özbuğday et al. 2020). The connection between the implementation of environmental activities and financial performance can also be recognized through a potential increase in profits, achieving a better position in the market and competitive advantages (Morić et al. 2020; OECD 2011).

The various external drivers that influence the feasibility of SMEs in implementing green innovation can include market force pressure relating to customers, competitive pressure within the marketplace, costs of innovation, supply chain pressures, and government regulations (Rodríguez-Rebés et al. 2024; Arranz et al. 2019). Among those portrayed by scholars, market demand defines the process that compels SMEs to adopt environmentally friendly behavior and engage in green innovation (Arranz et al. 2019). Therefore, another reason that can be a driver and stimulate a company to practice resource efficiency measures is greater consumer satisfaction (UNEP 2017; EC 2011). In addition, compliance with environmental standards can serve as a quality badge for SME products, which can increase consumer interest and sales (Özbuğday et al. 2020). In recent years, consumer awareness of the problems facing the world has been growing, which affects their choice of products and services but also creates certain pressure on producers (Shao and Ünal 2019; Green Win 2018). Changing consumer behavior patterns and their demands for products that last longer and/or can be easily recycled and reused require a different approach to production and product improvement.

Competition is a key motivator of green innovation initiatives for firms dealing with ecologically sensitive customers (Hasan and Rahman 2023). SMEs must innovate eco-friendly processes that can yield lower-cost production; therefore, cost savings in production are essential for green innovation and sustainable development (Hojnik and Ruzzier 2016). SMEs are part of a larger supply chain managed by larger firms. SMEs may be forced to collaborate with other suppliers to fulfill specific environmental standards and supply chain requirements for green innovations (Rodríguez-Rebés et al. 2024). Regulation and government intervention can encourage firms to undertake environment-friendly initiatives (Hasan and Rahman 2023).

In general, the key drivers for eco-innovation in SMEs are legislation and regulation, availability of financial (grants, subsidies, or eco-innovative public procurement) or non-financial support, cooperation with others, and, in certain cases, market demand (European Commission 2020). However, it should be taken into account that reasons, which include market and organizational failures, the economic and environmental benefits resulting from the implementation of a resource-efficient concept are often difficult to achieve in practice (UNEP 2017).

2.3.3 Barriers to the Transition of SMEs to Green and Resource-Efficient Practices

Previous research has tried to examine the factors that can hinder the adoption of green innovative practices in SMEs. According to some authors (Marin et al. 2015), adopting green innovation may be hampered by insufficient information about technology and markets, a lack of qualified staff, high innovation costs, uncertain demand for innovative goods, lack of internal funds, limited access to foreign information and knowledge, weak collaboration with universities, and a lack of suitable business partners. Gupta and Baruas (2018) have established that the key challenges in adopting green innovation among Indian SMEs include inadequate R&D capabilities, inadequate new technological and skills development for innovation, and market risks and uncertainty. Some researchers (Simms et al. 2020) identified that the main obstacles that hinder the adoption of green innovation in the UK and the Netherlands are a lack of environmental and production capabilities, high risk of unproven technologies, and high switching and unit costs.

As changes involving resource efficiency and green practices require financial resources, this can be a significant problem for SMEs. Compared to large companies, SMEs have difficulties in securing financial resources, which can significantly affect their transition to a sustainable business. Resource-efficient innovations certainly require costs and risks (Cai and Li 2018; EC 2022). As it is difficult for SMEs to access external sources of capital, financial resources for innovations that will help them to be more sustainable are mainly provided through bank loans or venture capital funds. As these forms of financing are generally unfavorable, SMEs often give up on the implementation of eco-innovations due to a lack of financial resources (Talić et al. 2020). Small and medium-sized enterprises rely heavily on their own funds to finance investments in cleaner technology (Ghisetti et al. 2016; Caporale et al. 2023), and if this is the only source of financing for these measures, then their inclination toward investments of this type is smaller (Kalantzis and Revoltella 2019).

Although it has already been stated that the economic benefits of resource-efficient practices can be significant, for SMEs it is very difficult to predict how much this benefit would be. Eco-innovation requires additional investments, which may be excessive for SMEs (Dalhammar 2016). They imply the adaptation and introduction of specific procedures for management and measurement, which can significantly burden the company (Konar and Cohen 2001;Pinget et al. 2015). Thus, introducing resource-efficient practices can incur high costs for SMEs, whose resources are generally limited, which can be a great challenge for them.

In addition, the implementation of resource-efficient practices generally does not show results in the short term and it takes time to see the benefits. The transition requires radical and structural changes that imply a long-term process (Chatzistamoulou and Tyllianakis 2022). That is why SMEs generally opt for resource and energy-saving activities that are related to their core business or if it is very likely that these activities will lead to lower costs in the short term (Green Win 2018).

Although policies that support the transition to a green economy have been discussed before, and the number of countries applying them is increasing, in many cases, the policies are not adapted to small and medium-sized enterprises. Existing policies mostly provide general information and advice without taking into account the heterogeneity of the SME sector. It is not possible to define a universal policy that will suit everyone, but the existing approaches do not consider the opportunities and challenges faced by SMEs on the way to greening and, therefore, do not provide them with the necessary support (OECD 2021; Talić et al. 2023a). The impression is that governments avoid increasing external pressure on SMEs through environmental legislation and that it is easier to attribute negative impacts to large companies (Green Win 2018). Many SMEs see inadequate regulations and complicated administrative procedures as the biggest barrier (European Commission 2020).

Some barriers may be related to the individual characteristics of the SME owner. The implementation of resource-efficient practices and the inclusion of the concept of sustainability in the company's business strategy may depend on personal inclinations and preferences. In addition to the lack of capacity and time to consider environmental requirements, awareness of environmental impact may also be underdeveloped. Moreover, there may be an unwillingness to take risks and uncertainty as to whether adequate results and benefits will be achieved from the implementation of ecological practices (Green Win 2018).

The mentioned problem is also related to the lack of information. Entrepreneurs often do not have enough information and knowledge about the requirements and needs of the environment, or about the importance of support in solving them (OECD 2021). Lack of information about financial incentives, training, or support programs for the implementation of eco-innovations, especially about costs and benefits for SMEs, can represent a significant barrier. The lack of expertise, i.e., technical skills, is also recognized among SMEs as a major obstacle to the adoption of resource efficiency measures (Backman 2017). Prior engagement in or familiarity with resource-efficient practices also determines whether SMEs perceive regulations and administrative burdens as barriers to resource-efficient practices (Garcia-Quevedo et al. 2020).

Therefore, some of the key barriers for small and medium-sized enterprises are primarily financial (lack of funds, high costs of implementing eco-innovations, uncertainty of return on investment), but there are also regulatory, legal, and administrative obstacles, and technical barriers (European Commission 2020; Talić et al. 2023a). According to a survey (European Commission 2022), the most common difficulty encountered by SMEs in European countries when establishing actions for resource efficiency is the complexity of administrative or legal procedures, the costs of environmental actions, and the lack of supply of necessary materials, parts, products, or services.

2.3.4 Measuring Resource Efficiency

Measuring, monitoring, and improving resource efficiency can be one of the steps in solving the issue of resource scarcity. In order to improve resource efficiency, it is necessary to include appropriate measures or indicators of resource efficiency. There are very few studies in the literature dealing with the measurement of resource efficiency at the level of companies. Gharfalkar et al. (2018) in their paper analyzed some of the existing measures and indicators of resource efficiency and effectiveness (REMI) and proposed an indicator for measuring "operational resource efficiency" that would be intended for production units. There is a need to develop robust and easy-to-understand indicators for measuring resource efficiency progress. The European Commission (2011) suggests that states motivate companies to "systematically measure, benchmark and improve their resource efficiency."

Indicators for measuring resource efficiency exist and are measured at the level of the national economy. In order to make adequate decisions and create policies related to resource efficiency, measures and indicators of resource efficiency are of great importance to policymakers.

To measure the progress of EU member states in the direction of achieving resource efficiency, the Resource Efficiency Scoreboard was created (European Commission 2016). It represents a set of indicators for monitoring resource use and resource productivity that have been monitored since December 2013. The instrument was designed based on the 2011 Resource Efficiency Roadmap for Europe. The Resource Efficiency Scoreboard has as its main part a dashboard of indicators that include water, land, materials, and carbon, while a section containing thematic indicators assesses priority policy areas (European Commission 2017; EEA 2016).

As stated earlier, eco-innovations are an important instrument for greening the economy, so it is equally important to deal with the issue of their monitoring and measurement (Rađenović et al. 2024). One of the very important parts of measuring eco-innovations relates to resource efficiency and includes indicators from this area that are considered an essential factor for achieving national competitiveness and sustainability (European Commission 2022; European Environment Agency 2023; Horbach 2016). These indicators are part of the Eco-Innovation Index developed for the purpose of measuring and monitoring eco-innovations at the national level, as well as for their comparison (EC 2022). It is a composite index based on five groups of indicators. One group refers to resource efficiency outcomes and includes the following indicators: material productivity, water productivity, energy productivity, and greenhouse gas emission productivity (EIO 2013).

The framework for monitoring the circular economy action plan adopted by the European Commission is also significant. The aim is to assess progress toward a circular economy in the EU countries (European Commission 2018). This framework, in its original version, did not contain indicators directly related to resource efficiency, but it did emphasize self-sufficiency in raw materials. However, the framework was revised in 2023, when two new indicators were added. The first indicator, Material Footprint and Resource Productivity, monitors material efficiency, and the second indicator, Consumption Footprint, tracks consumption in the EU (European Commission 2023).

2.4 METHODOLOGY

2.4.1 K-Means Algorithm as an Integral Part of Machine Learning

Bearing in mind machine learning workflows, the K-means clustering algorithm has been used as a data preprocessing instrument to identify outliers and eliminate complexity. Regarding this machine learning unsupervised framework for gaining meaningful insights from various data-driven applications, this algorithm could contribute to solving big data issues (Mohiuddin et al. 2020). Additionally, dimensionality reduction and anomaly detection as K-means algorithm features could improve the simplicity and scalability of this hierarchical clustering computation. As mentioned earlier, this versatile and widely used algorithm is an integral part of machine learning, providing a specific toolbox for recognizing patterns in unlabeled data. The K-means clustering approach exhibited relatively strong performance in optimizing the data training set, showing that the cluster-then-predict methodology could group even data with non-uniform distribution. The main goal of any cluster analysis is to separate similar datasets into one group whose members are much more similar than the members of some other separated groups. This statistical method, in the form of an unsupervised machine learning-based algorithm, has several different types, but all subtypes of cluster analysis share the scalability to handle extensive databases, high dimensionality, multidata kinds usability, and dealing with unstructured data, as well as comprehensible, usable, and understandable outcomes. Furthermore, the distance between data within an adequate number of iterations initiates the discovery of similarities between observed objects by creating specific interconnections within one of the clusters. Throughout an exhaustive examination of the gathered data, the author employed a cluster methodology that uses the principles of K-means clustering (Oti et al. 2021). By reviewing the available literature, the authors believe that the K-means algorithm is well-suited for addressing the problem, considering that it does not require pre-labeled data or predefined groups, and that it is a method of non-hierarchical clustering of data that seeks to partition the existing data in the form of one or more clusters. The forecasting capabilities of the K-means algorithm foster the improvement of the overall performance of SMEs (Ikotun et al. 2022). Although there is no official best form of the K-means algorithm as a rule of thumb, the fact that the mentioned algorithm is the partitioning-based clustering method that uses centroids for cluster presentation can be accepted, whose quality is measured through the within-cluster squared error criterion.

2.4.2 Data Collection Preparation and Methodology Structure

Given that a large number of SMEs operating in the territory of the European Union implement certain activities that lead to the efficient use of resources, in this research, the methodological framework is based on the grouping of European Union members by similarity in the application of the mentioned activities. Determining homogeneous groups based on the number of SMEs implementing resource-efficient activities was carried out with the help of cluster analysis as a multivariate statistical technique. In this case, the collection of data and the creation of a database on green SMEs is based on the available data from Eurostat for 2022 from the Eurobarometer

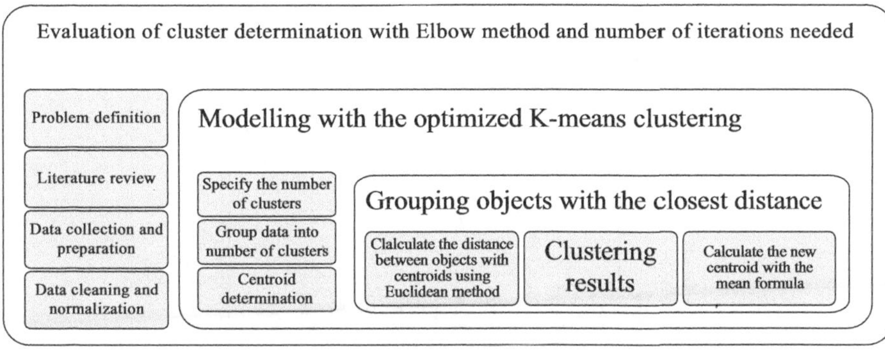

FIGURE 2.1 Methodological framework for the K-means clustering of the current research.

report. The grouping of European Union members by similarity in resource-efficient allocation is made possible by data processing through established patterns of exploratory cluster analysis.

According to Forgy, the initial phase of this algorithm begins with a series of iterations based on the minimization of the average squared Euclidean distance between the data points and the cluster's center known as the centroid, where the centroid is the center of a geometric object and is seen as a generalization of the mean. The obtained outputs of K-means cluster analysis will be determined by the elbow and silhouette methods, within-cluster sum of squares, Dunn index, Adjusted Rand Index, ANOVA, before-cluster model-based density and classification matrices, and other K-means instruments in the R statistical software package. The understanding of the applied methodological framework in the form of the K-means procedure can best be described in Figure 2.1, where the entire flow of the research process and its interpretation are shown.

Descriptive statistics of the number of SMEs by analyzed EU27 countries are shown in Table 2.1. The number of companies that in their operations have activities that contribute to the more efficient allocation of resources within the framework of descriptive statistics is presented with the help of mean, median, standard deviation, and variance values, as well as the minimum and maximum numbers of companies for all EU27 member countries.

After collecting data from the Eurobarometer report (Figure 2.2), the authors determined nine variables by which European Union member states should be grouped based on the number of SMEs that apply green postulates in their activities (European Commission 2018):

- Saving water (SW)
- Saving energy (SE)
- Using predominantly renewable energy (e.g., including own production through solar panels) (RE)
- Saving materials (SM)
- Switching to greener suppliers of materials (GS)
- Minimizing waste (MW)
- Selling your residues and waste to another company (SRW)

TABLE 2.1
Descriptive Statistics for the Analyzed Number of European Members SMEs According to Determined Variables

Statistics		SW	SE	RE	SM	GS	MW	SRW	RECY	MRR
N	Valid	27	27	27	27	27	27	27	27	27
Mean		223.59	304.85	91.04	280.96	164.15	307.11	119.33	220.70	128.56
Median		230.00	304.00	60.00	301.00	152.00	309.00	130.00	230.00	132.00
Std. Deviation		82.263	96.804	71.381	106.870	67.222	122.574	41.627	93.101	57.145
Variance		6767.251	9371.054	5095.268	11421.268	4518.823	15024.487	1732.846	8667.755	3265.564
Minimum		93	95	20	108	43	87	26	78	25
Maximum		417	470	274	504	318	509	206	457	260
Sum		6037	8231	2458	7586	4432	8292	3222	5959	3471
Percentiles	25	152.00	235.00	44.00	203.00	110.00	211.00	93.00	134.00	88.00
	50	230.00	304.00	60.00	301.00	152.00	309.00	130.00	230.00	132.00
	75	273.00	384.00	132.00	356.00	214.00	423.00	149.00	273.00	160.00

SW, saving water; SE, saving energy; RE, using predominantly renewable energy (e.g., including own production through solar panels); SM, saving materials; GS, switching to greener suppliers of materials; MW, minimizing waste; SRW, selling your residues and waste to another company; RECY, recycling, by reusing material or waste within the company; MRR, designing products that are easier to maintain, repair or reuse.

- Recycling, by reusing material or waste within the company (RECY)
- Designing products that are easier to maintain, repair, or reuse (MRR)

A crucial and fundamental step for the successful determination of the required number of clusters or groups of analyzed observations in the K-means cluster procedure requires the application of the elbow method. Determining the results in principle observes the most drastic decrease in calculating the sum of squared errors (SSE) per cluster until the most extreme difference forms the angle of the elbow, which depicts the optimal number of clusters (Umargono et al. 2020). The evaluation of the results obtained by the elbow method is accompanied by a visual representation of the optimal number of clusters. The elbow method is a very important phase of K-means cluster analysis that determines the further course and schedule of other analytical methods, which in the last instance aim to form the cluster membership of the observed objects in order not to randomize the determination of the number of clusters. Based on the obtained results from Figure 2.3, it can be concluded that the optimal number of clusters is four.

On the other hand, to determine the mutual distance of the neighboring resulting clusters as well as their two-dimensional arrangement, the silhouette method is used as part of the K-means cluster analysis. As can be seen from Figure 2.3, the silhouette plot shows the silhouette width si values for the observed clusters and the objects they are composed of, in the form of smaller bar charts. The values of the silhouette width parameter range between –1 and 1 (Scikit-learn 2019). High values of the mentioned parameter, which are closer to 1, indicate that the object has a strong connection with its parent cluster and that it has no connection with neighboring clusters. Based on this, it can be concluded that the value of silhouette width for the first cluster is 0.32, for the second cluster 0.16, for the third cluster 0.46, and for the fourth cluster 0.27. Besides the fact that the average silhouette score values show the strong

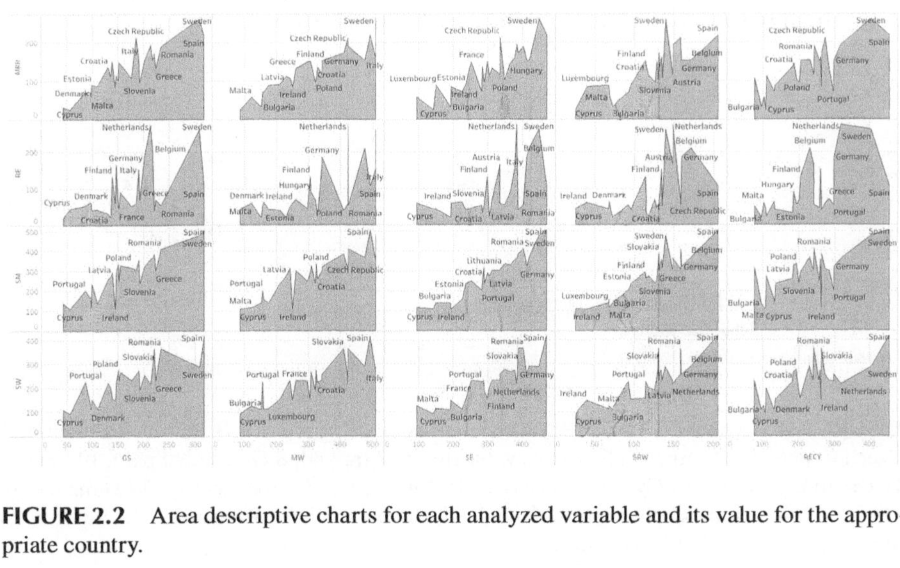

FIGURE 2.2 Area descriptive charts for each analyzed variable and its value for the appropriate country.

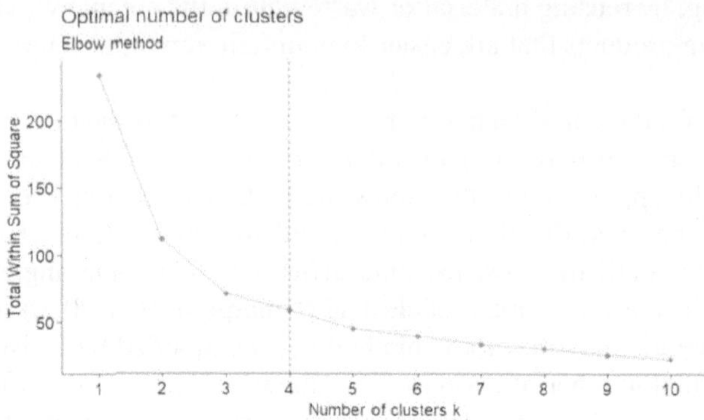

FIGURE 2.3 Optimal number of clusters using elbow method.

quality of the cluster if they are higher than 0.5, which is commonly set, it is worth mentioning that the silhouette score threshold can be changed based on the unique features and performance of the analyzed dataset, such as cluster shape and size and specific dimensionality (Kaoungku et al. 2018). Therefore, it can be concluded that it has a very strong and reasonable quality of cluster and provides cluster number 3 whose silhouette score threshold is nearly 0.5. The average model silhouette width threshold is 0.33 (Figure 2.4) which denotes pretty fair clustering quality.

In order to confirm the correctness and validity of the cluster analysis in this research, a type of pre-research density-based clustering was carried out to determine the distribution of clusters in space in relation to the applied variables. In this way, the arbitrary shape and noise in real-life data of the analyzed SME variable and formed clusters according to their values are observed (Figure 2.5). Density-based spatial clustering has the function of presenting clusters as specific regions in the space determined by the values of the analyzed variables, providing signals about the optimal number of clusters and their spatial dispersion.

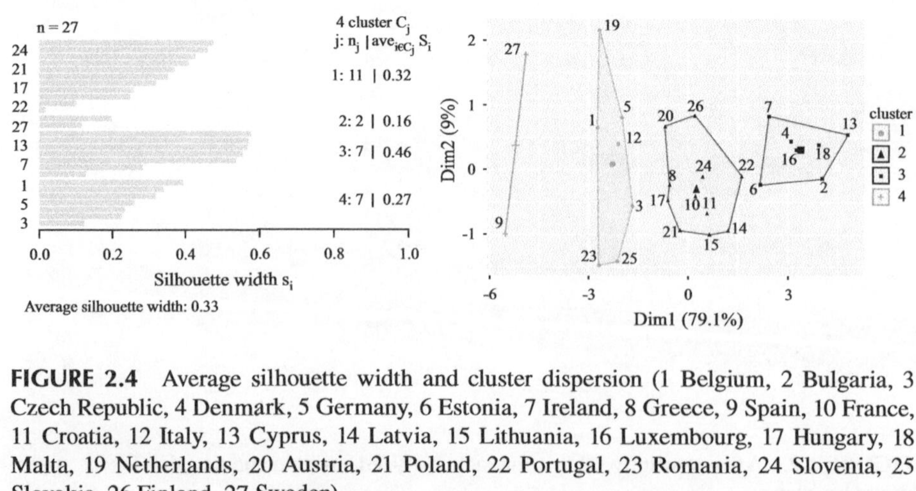

FIGURE 2.4 Average silhouette width and cluster dispersion (1 Belgium, 2 Bulgaria, 3 Czech Republic, 4 Denmark, 5 Germany, 6 Estonia, 7 Ireland, 8 Greece, 9 Spain, 10 France, 11 Croatia, 12 Italy, 13 Cyprus, 14 Latvia, 15 Lithuania, 16 Luxembourg, 17 Hungary, 18 Malta, 19 Netherlands, 20 Austria, 21 Poland, 22 Portugal, 23 Romania, 24 Slovenia, 25 Slovakia, 26 Finland, 27 Sweden).

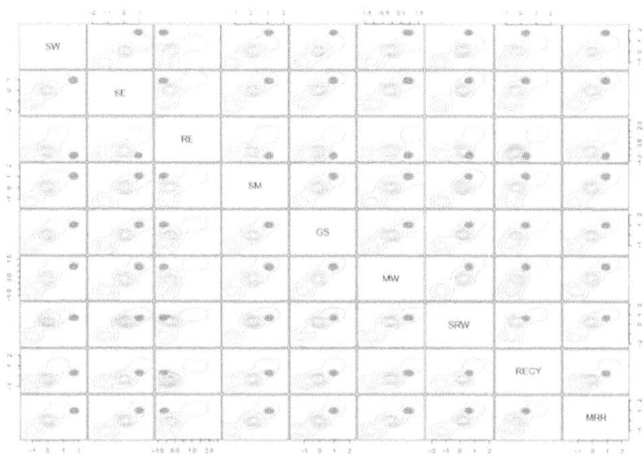

FIGURE 2.5 Cluster density matrix according to the values of the analyzed variables.

By conceiving the presented methodological framework in Figure 2.1, it can be concluded that the conditions have been met for the application of the process to calculate the distance between objects with centroids using the Euclidean method and the realization of the grouping of objects with the closest distance phase in order to achieve real and reliable cluster results.

2.5 RESULTS AND DISCUSSION

The results of the cluster analysis based on the K-means algorithm show the descriptive statistics of the mean value by clusters for each variable individually (Table 2.2). Descriptive intercluster analysis uses Z-scores threshold values for each individual variable. Additionally, the highest values for the analyzed SMEs green variables in 2022 refer to the third cluster consisting of Denmark, Malta, Cyprus, Estonia, Ireland, Luxembourg, Malta, and Bulgaria. In addition to the descriptive cluster statistics, the value of the Dunn Index measures pairwise distance as the inter-luster separation. The growing value of the Dunn Index shows the improved quality of the cluster analysis (Misuraca et al. 2018). Adjusted Rand Index, a symmetric measure from –1 to 1, has good recovery value for this analysis (0.6421) (Warrens and van der Hoef 2022). Another confirmation of a successfully conducted cluster analysis is the within sum of squares value, which is over 76% for the set model.

An interesting example of the green revolution in SMEs on the territory of the European Union is certainly the second cluster in which Spain and Sweden are grouped. Considering that the cluster analysis does not take into account the size of the country and the number of inhabitants, it can be said that given the economic strength of these countries, the second cluster can be a leader in the field of efficient allocation of resources to green activities. Certainly, the fact that the third cluster is currently the most dominant, as the results show, it can be concluded that the K-means cluster analysis also this time shows a degree of inconsistency in the

TABLE 2.2
Descriptive Cross-Cluster Statistics with Validity Parameters

Cluster	SW	SE	RE	SM	GS	MW	SRW	RECY	MRR
1	0.0005	0.0062	–0.2782	0.0012	–0.1550	–0.1908	0.1340	–0.1863	–0.1593
2	0.7881	0.8663	0.6779	0.7200	0.6736	1.0002	0.7882	0.5203	0.8277
3	**1.5609**	**1.5924**	**1.3023**	**1.9607**	**2.2218**	**1.5777**	**1.2892**	**2.2588**	**1.93277**
4	–1.2349	–1.3311	–0.6129	–1.2820	–1.0648	–1.1512	–1.3670	–0.8730	–1.1297
Dunn Index									0.3562
Adjusted Rand Index									0.6421
Within Sum of Squares									0.7642

Cluster 1: Austria, Greece, Finland, France, Croatia, Hungary, Lithuania, Latvia, Poland, Portugal, Slovenia. Cluster 2: Spain, Sweden. Cluster 3: Denmark, Malta, Cyprus, Estonia, Ireland, Luxembourg, Malta, Bulgaria. Cluster 4: Belgium, Czech Republic, Germany, Italy, Netherlands, Romania, Slovakia.

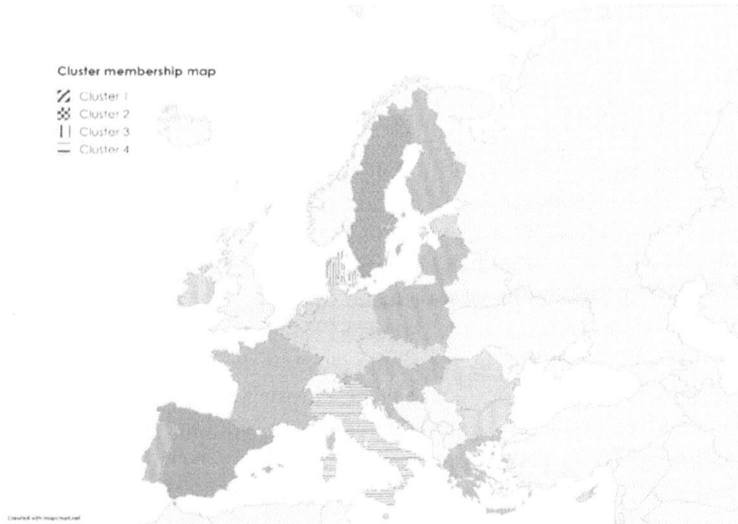

FIGURE 2.6 Cluster membership map.

observation of individual objects. In the continuation of the analysis, a K-means cluster map chart was created that shows the relationship between four clusters based on intercluster descriptive statistics (Figure 2.6).

The validation of the cluster analysis results and the confirmation of the dominance of the third cluster were additionally verified by the ANOVA procedure as a homogeneity of variance test (refer to Table 2.3). Statistically significant differences

TABLE 2.3
ANOVA Procedure and Final Decision Cluster Approval

	Cluster		Error			
Variable	Mean Square	df	Mean Square	df	Significance	Final Decision
SW	42811.366	3	2065.844	23	.000	Cluster verification approved
SE	70444.010	3	1405.016	23	.000	Cluster verification approved
RE	20113.498	3	3136.368	23	.003	Cluster verification approved
SM	88819.115	3	1325.896	23	.000	Cluster verification approved
GS	29972.996	3	1198.714	23	.000	Cluster verification approved
MW	107003.870	3				Cluster verification approved
SRW	12536.919	3				Cluster verification approved
RECY	50576.671	3				Cluster verification approved
MRR	22299.399	3				Cluster verification approved

*Significance level is at 0.05.

in average values can be stated for the variables as seen in the significance column where $P < 0.05$ (Rađenović et al. 2022).

2.6 CONCLUSION

Although, historically, the K-means algorithm is one of the most commonly used methods in cluster analysis, it can be said that its inconsistency and approximation represent key drawbacks. In the previously conducted research, a further increase in the number of clusters and stratification of observations would certainly lead to a deterioration in the quality of the cluster analysis, which would be reflected in the values of the Dunn Index and the Adjusted Rand Index. The validity of the capacitive analysis was confirmed by the high values of the previously mentioned indices, which speaks in favor of the fact that the cluster analysis of SMEs with resource-efficient activities has a very significant contribution in the future assessment of this analytical area. The practical implications of this analysis are reflected in looking at the leadership position of certain clusters and their members, which enables policymakers to redirect their decisions toward practices that produce results. Thus, the dominance of the third cluster must be used when it comes to determining the trajectory of movement of other clusters into this green area. The disadvantage of this type of cluster analysis is that it does not take into account the size of the object it is analyzing, but in its iterations it looks at the objects based on their given values.

Thus, Spain and Sweden are undeservedly in a subordinate position, although they have recorded significant growth in the application of resource-efficient activities in their SMEs, unlike Bulgaria, which has stagnated even though it is in a dominant cluster. A going-green country like Denmark justifies the position of the third cluster considering the very high percentage of applying the resource efficiency postulate and making decisions about removing barriers in the implementation of the green concept in its SMEs. Similar to previous investigations by Eurostat, over one-third of SMEs (36%) believe that grants or subsidies would help their business the most to be more resource-efficient. A large proportion of SMEs (26%) believe that increased cross-sector collaboration would be beneficial for developing novel waste and by-product reuse techniques. This is a higher percentage of SMEs than in 2017 (+6 pp). Over three-quarters of respondents in Slovakia (78%), Sweden (75%), and Spain (78%) said they have decreased their energy use in order to use resources more wisely. A smaller percentage of SMEs in more member countries claim the same. The probability that SMEs in Sweden and the Netherlands would primarily utilize renewable energy is highest in the EU (43% and 45%, respectively). The percentage of SMEs utilizing mostly renewable energy stays below 25% in the majority of other nations. In the next two years, 4% of SMEs in Spain and 45% of SMEs in Denmark do not intend to adopt further resource optimization activities. Moreover, a third of respondents in Germany, Finland, and Estonia (between 34% and 36%) said they would not implement any new resource efficiency initiatives. Last, 6% of SMEs in Bulgaria and 43% of SMEs in Sweden are creating items that are simpler to reuse, repair, or maintain. The countries closest to Sweden include Spain and Czech Republic, where 35%–36% of SMEs report this activity. In the vast majority of the individual member states, there has been a decline in the percentage of SMEs not implementing any resource efficiency initiatives. The countries with the biggest drops in this percentage include Greece (–13 pp, to 13%), Romania (–23 pp, to 13%), Bulgaria (–15 pp, to 19%), Estonia (–30 pp, to 23%), and Lithuania (–24 pp, to 8%). Therefore, future research in this area could be based on the application of two-step cluster analysis and the agglomerative scheme in order for Spain and Sweden to change their cluster position.

ACKNOWLEDGMENTS

This research was financially supported by the Ministry of Science, Technological Development, and Innovation of the Republic of Serbia (Contract No. 451-03-66/2024-03/200371).

REFERENCES

Abbas, Jawad, and Mustafa Sağsan. "Impact of Knowledge Management Practices on Green Innovation and Corporate Sustainable Development: A Structural Analysis." *Journal of Cleaner Production* 229 (2019): 611–20. https://doi.org/10.1016/j.jclepro.2019.05.024

Adamowicz, Mieczysław. "Green Deal, Green Growth and Green Economy as a Means of Support for Attaining the Sustainable Development Goals." *Sustainability* 14, no. 10 (2022): 5901. https://doi.org/10.3390/su14105901

Ahmad, Shoeb. "Green Human Resource Management: Policies and Practices." *Cogent Business & Management* 2, no. 1 (2015): 1–13. https://doi.org/10.1080/23311975.2015.1030817

Almeida, Fernando, and Jahangir Wasim. "Eco-Innovation and Sustainable Business Performance: Perspectives of SMEs in Portugal and the UK." *Society and Business Review* 18, no. 1 (2022): 28–50. https://doi.org/10.1108/sbr-12-2021-0233

Arfi, Wissal Ben, Lubica Hikkerova, and Jean-Michel Sahut. "External Knowledge Sources, Green Innovation and Performance." *Technological Forecasting & Social Change/Technological Forecasting and Social Change* 129 (2018): 210–20. https://doi.org/10.1016/j.techfore.2017.09.017

Arranz, N., M. F. Arroyabe, A. Molina-García, and J. C. Fernandez De Arroyabe. "Incentives and Inhibiting Factors of Eco-Innovation in the Spanish Firms." *Journal of Cleaner Production* 220 (May 1, 2019): 167–76. https://doi.org/10.1016/j.jclepro.2019.02.126

Ashton, Weslynne, Suzana Russell, and Elizabeth Futch. "The Adoption of Green Business Practices among Small US Midwestern Manufacturing Enterprises." *Journal of Environmental Planning and Management* 60, no. 12 (2017): 2133–49. https://doi.org/10.1080/09640568.2017.1281107

Backman, Fredrik. "Barriers to Energy Efficiency in Swedish Non-Energy-Intensive Micro- and Small-Sized Enterprises: A Case Study of a Local Energy Program." *Energies* 10, no. 1 (2017): 100. https://doi.org/10.3390/en10010100

Brears, Robert C. "The Green Economy and the Water-Energy-Food Nexus." In *The Green Economy and the Water-Energy-Food Nexus*, 23–50, London: Palgrave Macmillan, 2017. https://doi.org/10.1057/978-1-137-58365-9_2

Cai, Wugan, and Guangpei Li. "The Drivers of Eco-Innovation and Its Impact on Performance: Evidence from China." *Journal of Cleaner Production* 176 (2018): 110–18. https://doi.org/10.1016/j.jclepro.2017.12.109

Chang, Otto H., and Michael D. Slaubaugh. "Sustainable Business Practices in the United States: A Survey on Implementation." *Journal of Management and Sustainability* 7, no. 3 (2017): 1. https://doi.org/10.5539/jms.v7n3p1

Chatzistamoulou, Nikos, and Emmanouil Tyllianakis. "Commitment of European SMEs to Resource Efficiency Actions to Achieve Sustainability Transition. A Feasible Reality or an Elusive Goal?" *Journal of Environmental Management* 321 (2022): 115937. https://doi.org/10.1016/j.jenvman.2022.115937

COM. An SME Strategy for a Sustainable and Digital Europe, COM/2020/103 Final. 2020.

Dalhammar, Carl. "Industry Attitudes Towards Ecodesign Standards for Improved Resource Efficiency." *Journal of Cleaner Production* 123 (2016): 155–66. https://doi.org/10.1016/j.jclepro.2015.12.035

De Jesus, Ana, and Sandro Mendonça. "Lost in Transition? Drivers and Barriers in the Eco-Innovation Road to the Circular Economy." *Ecological Economics* 145 (2018): 75–89. https://doi.org/10.1016/j.ecolecon.2017.08.001

Del Rosario, Reyes-Santiago María, Sánchez-Medina Patricia S., and René Díaz-Pichardo. "Eco-Innovation and Organizational Culture in the Hotel Industry." *International Journal of Hospitality Management* 65 (2017): 71–80. https://doi.org/10.1016/j.ijhm.2017.06.001

Durif, Fabien, Catherine Boivin, and Claire Julien. "In Search of a Green Product Definition." *Innovative Marketing* 6, no. 1 (2010): 25–33.

EC. *Eco-Innovation at the Heart of European Policies*. European Commission, Directorate-General for Environment, 2022.

Ekins, P., Hughes, N., et al. *Resource Efficiency: Potential and Economic Implications. A report of the International Resource Panel*. UNEP, 2017.

El-Kassar, Abdul-Nasser, and Sanjay Kumar Singh. "Green Innovation and Organizational Performance: The Influence of Big Data and the Moderating Role of Management Commitment and HR Practices." *Technological Forecasting & Social Change/ Technological Forecasting and Social Change* 144 (2019): 483–98. https://doi.org/10.1016/j.techfore.2017.12.016

European Environment Agency (EEA). *More from Less – Material Resource Efficiency in Europe*. Luxembourg: Publications Office of the European Union, 2016. https://doi.org/10.2800/240736

European Environment Agency (EEA). *Eco-Innovation Scoreboard*. Publications Office of the European Union, 2023.

European Commission. *Roadmap to a Resource Efficient Europe*. COM (2011) 571 Final, 2011.

European Commission. *Resource Efficiency Scoreboard*. Brussels, 2016.

European Commission. *European Semester Thematic Factsheet – Resource Efficiency*. Brussels, 2017.

European Commission. *Monitoring Framework for the Circular Economy*. COM/2018/029 Final, 2018.

European Commission. *Entrepreneurship and SMEs, SMEs, Green Markets and Resource Efficiency*. Report, Publications Office of the European Union, 2018. https://data.europa.eu/doi/10.2873/490067

European Commission. *Science for Environment Policy. Eco-Innovation in SMEs*. Future Brief 22, Brief Produced for the European Commission DG Environment by the Science Communication Unit, UWE, Bristol, 2020.

European Commission. *Flash Eurobarometer 498 Report: SMEs, Green Markets and Resource Efficiency*. Publications Office of the European Union, 2022. https://doi.org/10.2873/490067

European Commission. *Circular Economy – New Tool for Measuring Progress.* Publications Office of the European Union, 2023. https://data.europa.eu/doi/10.2779/1618

European Innovation Observatory (EIO). *Europe in Transition: Paving the Way to a Green Economy through Eco-Innovation.* Brussels, Eco-Innovation Observatory. Funded by the European Commission, DG Environment, 2013.

Faisal, Saba, and Mohammed Naushad. "An Overview of Green HRM Practices among SMEs in Saudi Arabia." *Entrepreneurship and Sustainability Issues* 8, no. 2 (2020): 1228–44.

Faulkner, William, and Fazleena Badurdeen. "Sustainable Value Stream Mapping (Sus-VSM): Methodology to Visualize and Assess Manufacturing Sustainability Performance." *Journal of Cleaner Production* 85 (2014): 8–18. https://doi.org/10.1016/j.jclepro.2014.05.042

García-Quevedo, Jose, Elisenda Jové-Llopis, and Ester Martínez-Ros. "Barriers to the Circular Economy in European Small and Medium-Sized Firms." *Business Strategy and the Environment* 29, no. 6 (2020): 2450–64. https://doi.org/10.1002/bse.2513

Gasparatos, Alexandros, Christopher N. H. Doll, Miguel Esteban, Abubakari Ahmed, and Tabitha A. Olang. "Renewable Energy and Biodiversity: Implications for Transitioning to a Green Economy." *Renewable & Sustainable Energy Reviews* 70 (2017): 161–84. https://doi.org/10.1016/j.rser.2016.08.030

Gharfalkar, Mangesh, Zulfiqur Ali, and Graham Hillier. "Measuring Resource Efficiency and Resource Effectiveness in Manufacturing." *International Journal of Productivity and Performance Management* 67, no. 9 (2018): 1854–81. https://doi.org/10.1108/ijppm-11-2017-0282

Ghisetti, Claudia, Susanna Mancinelli, Massimiliano Mazzanti, and Mariangela Zoli. "Financial Barriers and Environmental Innovations: Evidence From EU Manufacturing Firms." *Climate Policy* 17, no. 1 (2016): 131–47. https://doi.org/10.1080/14693062.2016.1242057

Green Win. *Enabling Environments for Green SMEs: Recommendations for Action.* Policy Brief, 2018.

Caporale, Guglielmo Maria, Cristiana Donati, Nicola Spagnolo. *European SMEs and Resource Efficiency Measures: Firm Characteristics and Contextual Factors.* Working Paper, No. 10799. Munich Society for the Promotion of Economic Research – CESifo GmbH, 2023.

Gupta, Himanshu, and Mukesh Kumar Barua. "A Framework to Overcome Barriers to Green Innovation in SMEs Using BWM and Fuzzy TOPSIS." *Science of the Total Environment* 633 (2018): 122–39. https://doi.org/10.1016/j.scitotenv.2018.03.173

Hasan, Md Jahid, and Muhammad Shajib Rahman. "Determinants of Eco-Innovation Initiatives Toward Sustainability in Manufacturing SMEs: Evidence from Bangladesh." *Heliyon* 9, no. 7 (2023): e18102. https://doi.org/10.1016/j.heliyon.2023.e18102

Hojnik, Jana, and Mitja Ruzzier. "What Drives Eco-Innovation? A Review of an Emerging Literature." *Environmental Innovation and Societal Transitions* 19 (2016): 31–41. https://doi.org/10.1016/j.eist.2015.09.006

Horbach, Jens. "Empirical Determinants of Eco-Innovation in European Countries Using the Community Innovation Survey." *Environmental Innovation and Societal Transitions* 19 (2016): 1–14. https://doi.org/10.1016/j.eist.2015.09.005

Ikotun, Abiodun M., Absalom E. Ezugwu, Laith Abualigah, Belal Abuhaija, and Jia Heming. "K-Means Clustering Algorithms: A Comprehensive Review, Variants Analysis, and Advances in the Era of Big Data." *Information Sciences* 622 (2023): 178–210. https://doi.org/10.1016/j.ins.2022.11.139

ISSA. *Climate Change and Natural Resource Scarcity.* International Social Security Association Geneva, 2014.

Ivanović-Đukić, Maja, Tamara Rađenović and Miljana Talić. "Innovative Entrepreneurship and Economic Growth in Emerging Markets." *Ekonomika* 67, no 4 (2021): 55–64. https://doi.org/10.5937/ekonomika2104065I

Janjić, Ivana, and Tamara Rađenović. "The Importance of Managing Innovation in Modern Enterprises." *Ekonomika* 65, no. 3 (2019): 45–54. https://doi.org/10.5937/ekonomika1903045j

Janjić, Ivana, BojanKrstić, and Sandra Milanović, "The Impact of R&D Activity on the Business Performance of High-Technology Companies." *FactaUniversitatis. Series: Economics and Organization* 19, no. 4 (2022): 253–71. https://doi.org/10.22190/fueo220707018j

Janjić, Ivana, Milica Jovanović, and Zoran Simonović. "The Importance of Research and Development for Innovative Activity: The Overview of the Top Countries in Europe and Worldwide." *Economics of Sustainable Development* 5, no. 2 (2021): 19–28. https://doi.org/10.5937/esd2102019j

Kalantzis, Fotios, and Revoltella, Debora. "Do Energy Audits Help SMEs to Realize Energy-Efficiency Opportunities?" *Energy Economics* 83 (2019): 229–39. https://doi.org/10.1016/j.eneco.2019.07.005

Kaoungku, Nuntawut, Keerachart Suksut, Ratiporn Chanklan, Kittisak Kerdprasop, and Nittaya Kerdprasop. "The Silhouette Width Criterion for Clustering and Association Mining to Select Image Features." *International Journal of Machine Learning and Computing* 8, no. 1 (2018): 69–73. https://doi.org/10.18178/ijmlc.2018.8.1.665

Kar, Sanjay Kumar, and Sidhartha Harichandan. "Green Marketing Innovation and Sustainable Consumption: A Bibliometric Analysis." *Journal of Cleaner Production* 361 (2022): 132290. https://doi.org/10.1016/j.jclepro.2022.132290

Kemp, René. "Eco-Innovation: Definition, Measurement and Open Research Issues." *EconomiaPolitica*, no. 3 (2010): 397–420. https://doi.org/10.1428/33131

Khan, Rizwan Ullah, Hina Arif, Noor E. Sahar, Arif Ali, and Munir A. Abbasi. "The Role of Financial Resources in SMEs' Financial and Environmental Performance; The Mediating Role of Green Innovation." *Green Finance* 4, no. 1 (2021): 36–53. https://doi.org/10.3934/gf.2022002

Konar, Shameek, and Mark A. Cohen. "Does the Market Value Environmental Performance?" *Review of Economics and Statistics* 83, no. 2 (2001): 281–89. https://doi.org/10.1162/003465301511143815

Krstić, Bojan, Ivana Janjić, Milica Jovanović, and Sandra Milanović. "The Significance of R&D Activities for Managing Intellectual Capital of Enteprises" *BH Ekonomski Forum* 14, no. 1 (2021): 117–32. https://doi.org/10.5937/bhekofor2101117K

Kurapatskie, Brent, and Nicole Darnall. "Which Corporate Sustainability Activities Are Associated with Greater Financial Payoffs?" *Business Strategy and the Environment* 22, no. 1 (2012): 49–61. https://doi.org/10.1002/bse.1735

Li, Dayuan, Mi Zheng, Cuicui Cao, Xiaohong Chen, Shenggang Ren, and Min Huang. "The Impact of Legitimacy Pressure and Corporate Profitability on Green Innovation: Evidence from China Top 100." *Journal of Cleaner Production* 141 (January 1, 2017): 41–49. https://doi.org/10.1016/j.jclepro.2016.08.123

Lutfi, Abdalwali, Hamza Alqudah, Mahmaod Alrawad, Ahmad Farhan Alshira'h, Malek Hamed Alshirah, Mohammed Amin Almaiah, Adi Alsyouf, and Mohammed Faisal Hassan. "Green Environmental Management System to Support Environmental Performance: What Factors Influence SMEs to Adopt Green Innovations?" *Sustainability* 15, no. 13 (2023): 10645. https://doi.org/10.3390/su151310645

Marin, Giovanni, Alberto Marzucchi, and Roberto Zoboli. "SMEs and Barriers to Eco-Innovation in the EU: Exploring Different Firm Profiles." *Journal of Evolutionary Economics* 25, no. 3 (2015): 671–705. https://doi.org/10.1007/s00191-015-0407-7

Mathiyazhagan, K., Kannan Govindan, and A. Noorul Haq. "Pressure Analysis for Green Supply Chain Management Implementation in Indian Industries Using Analytic Hierarchy Process." *International Journal of Production Research* 52, no. 1 (2013): 188–202. https://doi.org/10.1080/00207543.2013.831190

Milanović, Sandra, Miljana Talić, and Žarko Rađenović. "Social Entrepreneurial Intention Factors of Tourism Management Students." In *Association of Economists and Managers of the Balkans; Faculty of Tourism and Hospitality eBooks*, 567–83, 2022. https://doi.org/10.31410/tmt.2021-2022.567

Miller, K., A. Neubauer, A. Varma, and E. Williams. *First Assessment of the Environmental Assistance Programme for SMEs (ECAP): Final Report Prepared by AEA Technology Plc for the European Commission.* London: DG Environmental and Climate Action, 2011.

Misuraca, Michelangelo, Maria Spano, and Simona Balbi, "BMS: An Improved Dunn Index for Document Clustering Validation." *Communications in Statistics. Theory and Methods/Communications in Statistics, Theory and Methods* 48, no. 20 (2018): 5036–49. https://doi.org/10.1080/03610926.2018.1504968

Mohiuddin, Ahmed, Seraj Raihan, and Islam Syed Mohammed Shamsul. "The K-Means Algorithm: A Comprehensive Survey and Performance Evaluation". *Electronics* 9, no. 8 (2020): 1295. https://doi.org/10.3390/electronics9081295

Moric, Ilija, Jelena Šaković Jovanović, Rajka Đoković, Sanja Peković, and Đurđica Perović. "The Effect of Phases of the Adoption of the Circular Economy on Firm Performance: Evidence From 28 EU Countries." *Sustainability* 12, no. 6 (2020): 2557. https://doi.org/10.3390/su12062557

O Almalki Sultan Musaad, Zhang Zhuo, Almalki Otaibi Musaad O, Zafar Ali Siyal, Hammad Hashmi, and Syed Ahsan Ali Shah. "A Fuzzy Multi-Criteria Analysis of Barriers and Policy Strategies for Small and Medium Enterprises to Adopt Green Innovation." *Symmetry* 12, no. 1 (2020): 116. https://doi.org/10.3390/sym12010116

OECD. *Sustainable Manufacturing and Eco-Innovation: Framework, Practices and Measurement-Synthesis.* Paris: OECD Publishing, 2009.

OECD. *Working Party on SMEs and Entrepreneurship (WPSMEE) – Green Entrepreneurship, Eco-Innovation and SMEs.* Final Report, CFE/SME(2011)9/FINAL. 2011.

OECD. *No Net Zero without SMEs: Exploring the Key Issues for Greening SMEs and Green Entrepreneurship.* OECD SME and Entrepreneurship. Papers No. 30. 2021. https://dx.doi.org/10.1787/bab63915-en

Oti, Eric U., Michael O. Olusola, Francis C. Eze, and Samuel U. Enogwe. "Comprehensive Review of K-Means Clustering Algorithms." *International Journal of Advances in Scientific Research and Engineering* 7, no. 8 (2021): 64–9. https://doi.org/10.31695/ijasre.2021.34050

Özbuğday, Fatih Cemil, Derya Fındık, Kıvılcım Metin Özcan, and Sıdıka Başçı. "Resource Efficiency Investments and Firm Performance: Evidence From European SMEs." *Journal of Cleaner Production* 252 (2020): 119824. https://doi.org/10.1016/j.jclepro.2019.119824

Panthai, Pimchanok, and Kanokporn Kungwalsong. "Resource Efficiency and Environmental Impact Assessment Method for Small-Scale Producers: A Case Study of Pond and In-Pond Raceway System Production for Growing Nile Tilapia." *Sustainability* 16, no. 3 (2024): 1237. https://doi.org/10.3390/su16031237

Patricio, Joao, Lovisa Axelsson, Simon Blomé, and Leonardo Rosado. "Enabling Industrial Symbiosis Collaborations Between SMEs From a Regional Perspective." *Journal of Cleaner Production* 202 (2018): 1120–30. https://doi.org/10.1016/j.jclepro.2018.07.230

Peiró-Signes, Ángel, and Marival Segarra-Oña. "How Past Decisions Affect Future Behavior on Eco-Innovation: An Empirical Study." *Business Strategy and the Environment* 27, no. 8 (May 3, 2018): 1233–44. https://doi.org/10.1002/bse.2071

Pinget, Amandine, Rachel Bocquet, and Caroline Mothe. "Barriers to Environmental Innovation in SMEs: Empirical Evidence From French Firms." *M@N@Gement* 18, no. 2 (2015): 132. https://doi.org/10.3917/mana.182.0132

Purwandani, Junia A., and Gilbert Michaud. "What Are the Drivers and Barriers for Green Business Practice Adoption for SMEs?" *Environment Systems & Decisions* 41, no. 4 (2021): 577–93. https://doi.org/10.1007/s10669-021-09821-3

Rađenović, Žarko, Ivana Janjić, and Miljana Talić. "Targeting Eco-Innovation Performance among EU countries: Cluster Outcome." *Science International Journal* 3, no. 1 (2024): 189–94. https://doi.org/10.35120/sciencej0301189r

Rađenović, Tamara, Bojan Krstić, Ivana Janjić, and Milica Jovanović-Vujatović. "The Effects of R&D Performance on the Profitability of Highly Innovative Companies." *Strategic Management* 28, no. 3 (2023): 34–45. https://doi.org/10.5937/straman2200034r

Rađenović, Žarko, Bojan Krstić, and Milan Marković. "Economic Performance of Agriculture in the European Union Countries." *Zagadnienia Ekonomiki Rolnej* 370, no. 1 (2022): 5–21. https://doi.org/10.30858/zer/145686.

Rađenović, Žarko, and Milena Rajić. "Assessing Circular Economy Performances of Selected Countries Using Mcdm Methods: Promethee vs. CoCoSo." Navigating the Circular Age of a Sustainable Digital Revolution. *IGI Global* (2024): 59–91. https://doi.org/10.4018/979-8-3693-2827-9

Rizos, Vasileios, Arno Behrens, Wytze Van Der Gaast, Erwin Hofman, Anastasia Ioannou, Terri Kafyeke, Alexandros Flamos, et al. "Implementation of Circular Economy Business Models by Small and Medium-Sized Enterprises (SMEs): Barriers and Enablers." *Sustainability* 8, no. 11 (2016): 1212. https://doi.org/10.3390/su8111212

Rodrigues, Margarida, and Mário Franco. "Green Innovation in Small and Medium-Sized Enterprises (SMEs): A Qualitative Approach." *Sustainability* 15, no. 5 (2023): 4510. https://doi.org/10.3390/su15054510

Rodríguez-Rebés, Laura, Raquel Ibar-Alonso, Luis Manuel Ruíz Gómez, and Julio Navío-Marco. "The Use and Drivers of Organisational Eco-Innovation in European SMEs." *Research in International Business and Finance* 70 (2024): 102297. https://doi.org/10.1016/j.ribaf.2024.102297

Rosa, Bruna Oliveira, and Fábio De Oliveira Paula. "Circular Economy Adoption by European Small and Medium-Sized Enterprises: Influence on Firm Performance." *Revista Brasileira De Gestão De Negócios* 25, no. 3 (2023): 421–38. https://doi.org/10.7819/rbgn.v25i3.4232

Scikit-Learn. n.d. "Selecting the Number of Clusters with Silhouette Analysis on KMeans Clustering," Scikit-Learn-Scikit Learn 0.21.2 Documentatnio, 2019. https://scikit-learn.org/stable/auto_examples/cluster/plot_kmeans_silhouette_analysis.html

Shahzad, Mohsin, Ying Qu, Abaid Ullah Zafar, Saif Ur Rehman, and Tahir Islam. "Exploring the Influence of Knowledge Management Process on Corporate Sustainable Performance through Green Innovation." *Journal of Knowledge Management* 24, no. 9 (2020): 2079–2106. https://doi.org/10.1108/jkm-11-2019-0624

Shao, Jing, and Enes Ünal. "What Do Consumers Value More in Green Purchasing? Assessing the Sustainability Practices From Demand Side of Business." *Journal of Cleaner Production* 209 (2019): 1473–83. https://doi.org/10.1016/j.jclepro.2018.11.022

Simms, Christopher, Paul Trott, Ellis Van Den Hende, and Erik Jan Hultink. "Barriers to the Adoption of Waste-Reducing Eco-Innovations in the Packaged Food Sector: A Study in the UK and the Netherlands." *Journal of Cleaner Production* 244 (2020): 118792. https://doi.org/10.1016/j.jclepro.2019.118792

Singh, Manvendra Pratap, Arpita Chakraborty, and Mousumi Roy. "The Link Among Innovation Drivers, Green Innovation and Business Performance: Empirical Evidence from a Developing Economy." *World Review of Science, Technology and Sustainable Development* 12, no. 4 (2016): 316. https://doi.org/10.1504/wrstsd.2016.10003088

Song, Zhiting, Jianhua Zhu, and Jianfeng Shi. "Evolution Analysis of Green Innovation in Small and Medium-Sized Manufacturing Enterprises." *Systems* 11, no. 2 (2023): 95. https://doi.org/10.3390/systems11020095

Talić, Miljana, Maja Ivanović-Đukić, and Tamara Rađenović. "Sustainable Entrepreneurship: Creating Opportunities for Green Products Development." *Economics of Sustainable Development* 4, no. 2 (2020): 1–13. https://doi.org/10.5937/esd2002001t

Talić, Miljana, Suzana Stefanović, and Maja Ivanović-Đukić. "Barriers to Sustainable Business Practices of Sustainability-Oriented SMEs in Southeast Europe." Conference proceedings 10th REDETE Conference 2023. *Researching Economic Development and Entrepreneurship in Transition Economies* (pp. 73–84) Trebinje, RepublikaSrpska, BiH. October 26–27, 2023a.

Talić, Miljana, Suzana Stefanović, and Anđela Milenković. "Eko-inovacije u malimisrednjimpreduzećima: analizaregulatornogokvira u Republici Srbiji." International Scientific Conference: *Smart and Sustainable Economy: Trends and Perspectives*. Faculty of Economics, University of Niš. 13 October, 2023b.

Taranic, Igor, Arno Behrens, and Corrado Topi. *Understanding the Circular Economy in Europe, from Resource Efficiency to Sharing Platforms: The CEPS Framework*. CEPS: Brussels, Belgium, No. 143, 1–24, 2016. https://doi.org/10.13140/RG.2.2.14272.94728

Terzić, Lejla. "The Importance of Resource Efficiency Indicators in Assessing Eco-Innovation in the European Union Countries." *Journal of Theoretical and Applied Economics* XXXI (2024): 15–28.

Trushkina, Nataliia. "Green Economy in the Conditions of Modern Challenges: Conceptual Frameworks." *International Science Journal of Management, Economics & Finance* 1, no. 1 (2022): 1–6. https://doi.org/10.46299/j.isjmef.20220101.1

Umargono, Edy, Jatmiko Endro Suseno, and S. K. Vincensius Gunawan. "K-Means Clustering Optimization Using the Elbow Method and Early Centroid Determination Based on Mean and Median Formula," *Proceedings of the 2nd International Seminar on Science and Technology (ISSTEC 2019)*, 2020. https://doi.org/10.2991/assehr.k.201010.019

UNEP. *Towards a Green Economy: Pathways to Sustainable Development and Poverty Eradication - A Synthesis for Policy Makers*. United Nations Environment Programme, 2011.

Van Ewijk, Stijn. *Resource Efficiency and the Circular Economy: Concepts, Economic Benefits, Barriers, and Policies*. A Report Prepared for the Department for Environment, Food & Rural Affairs (Defra), 2018.

Wang, Haijun, Muhammad Aamir Shafique Khan, Farooq Anwar, Fakhar Shahzad, Daniel Adu, and Majid Murad. "Green Innovation Practices and Its Impacts on Environmental and Organizational Performance." *Frontiers in Psychology* 11 (2021). https://doi.org/10.3389/fpsyg.2020.553625

Wang, Mingyue, Yingming Li, Junqiang Li, and Zitong Wang. "Green Process Innovation, Green Product Innovation and Its Economic Performance Improvement Paths: A Survey and Structural Model." *Journal of Environmental Management* 297 (2021): 113282. https://doi.org/10.1016/j.jenvman.2021.113282

Warrens, Matthijs J., and Hanneke Van Der Hoef. "Understanding the Adjusted Rand Index and Other Partition Comparison Indices Based on Counting Object Pairs." *Journal of Classification* 39, no. 3 (2022): 487–509. https://doi.org/10.1007/s00357-022-09413-z

Wong, Stanley Kam Sing. "Environmental Requirements, Knowledge Sharing and Green Innovation: Empirical Evidence from the Electronics Industry in China." *Business Strategy and the Environment* 22, no. 5 (2012): 321–38. https://doi.org/10.1002/bse.1746

Wijkman, A., and K. Skånberg. *The Circular Economy and Benefits for Society: Swedish Case Study Shows Jobs and Climate as Clear Winners*. Club of Rome Report; Club of Rome: New York, 2015.

3 AI-Driven Green HR Management and Business Performance Insights

A Comprehensive Literature Review

Lakshmikanth Raja, Geetha Manoharan, and M Nivedha

3.1 INTRODUCTION

The emergence of the genuine environmental movement in the mid-1960s was driven by the ongoing utilization and depletion of non-renewable resources, as well as the significant rise in consumption rates, waste generation, and environmental contamination. This movement gained momentum during the following decades. As society increasingly attributed many of the world's environmental problems to corporations, these companies were expected to take responsibility and provide answers. Consequently, firms had little choice but to include environmentally friendly management techniques in their operating procedures. Over time, the concept of green management gained global recognition in the 2000s, and leaders (Manoharan & Ashtikar, 2024) in management realized that commercial and environmental goals should align. Implementing environmentally sustainable (Alagpuria et al., 2024) practices allows organizations to fulfil their social responsibility and act ethically towards the environment. Additionally, it benefits businesses by enhancing their competitiveness in the market, improving their financial performance (P. Kumar et al., 2024; Manoharan et al., 2023), increasing their firm value, fostering product innovation, and ensuring long-term sustainability.

The worldwide community is currently experiencing the repercussions of global climate change. Consequently, it is increasingly crucial for firms to embrace sustainable practices. Green human resource management has garnered considerable interest in recent years as a means for organizations to integrate environmental considerations into their operations while also preserving profitability. Green human resource management is the practice of incorporating environmental factors into an

DOI: 10.1201/9781003626527-3

organization's decision-making process (Chaudhary et al., 2024) decision-making process. It entails implementing sustainable (Shiva et al., 2024) business strategies that strive to minimize adverse environmental effects, save resources, and foster ecological sustainability (Shiva et al., 2024; A. Singh et al., 2024).

Corporations that prioritize environmental and social concerns generally adopt more progressive practices compared to other corporations. To be classified as "green", a corporation must meet the following four criteria: The company's dedication to environmental stewardship influences every decision made in the business (S. Kumar et al., 2024). Ecologically friendly alternatives are substituted for non-green goods and services. Conventional competition has a lower environmental impact. As a corporation, it has exhibited a persistent dedication to environmental principles. A corporation can achieve profitability while also tackling existing environmental issues by adopting environmentally friendly or sustainable practices, examining the utilization of renewable resources in a company's product manufacturing process, as well as determining the most effective strategies for designing new products to optimize their environmental impact.

3.1.1 Emergence of GHRM

The inception of the green movement and green policies can be traced back to the late 1970s with the establishment of the first Green Party in Germany. The phrase "green" is the English equivalent of the German word *Grün*. Green politics promotes the concerns of ecology, environment, feminism, conservation, and peace. The concept of green politics is considered to be influenced by the ideas of Mahatma Gandhi, Spinoza, and Uexkull. These thinkers emphasized the need for personal responsibility in making morally correct decisions, which forms the foundation of green politics. The green management movement may appear to be a recent phenomenon (Purushottam Ashtikar, & Manoharan, 2024), but it has been in progress for several decades. Understanding corporate greening's history is essential to understanding its current state. In the 1960s, "pollution", which was seen as filthy and harmful to our lifestyle, became a growing concern. Therefore, pollution control was created to prohibit illegal operations and smokestack and drainpipe emissions. Air and water pollution regulation and control in the US and abroad began with the creation of the Environmental Protection Agency (EPA) in 1970. In the 1980s, a few corporations realized that preventing pollution might minimize the need for regulations and cleanup. As part of their pollution control efforts, firms reorganized their operations and management systems to decrease waste and costs, focusing on energy efficiency and waste reduction. Corporate demand for environmental management systems and ISO 14001 increased in the 1990s. ISO 14001 was introduced by the International Organization for Standardization to develop a fundamental set of guidelines for businesses to structure their operations in an environmentally responsible manner and evaluate the environmental effects linked to their products. Environmental managers were employing emerging concepts such as life-cycle evaluations, design for the environment, dematerialization, demanufacturing, and remanufacturing. Companies effectively assessed and controlled the throughput of raw materials by quantifying the number of completed goods that could be manufactured from each unit of raw material consumed. The message was finally sent when William McDonough and Swiss scientist Michael Braungart collaborated to caution us against pursuing cradle-to-grave thinking. They emphasized the necessity of establishing mechanisms to create closed loops, products, and processes that were cradle-to-cradle referred to as “recycling”. Upon analyzing their products and procedures, businesses started to recognize the significant influence their suppliers, contractors, and business partners had on their environmental effects. Consequently, the world was transformed into environmental supply chain management, where corporations aim to extend the clean-and-green mindset to earlier stages of production. Under the mentorship of scientific author Janine Benyus, biologists, engineers, and industrial designers cooperated to develop novel products and procedures that drew inspiration from a diverse range of species, such as insects, fungi, mammals, birds, and other creatures. The researchers asked, “How would nature create this?” and found a comprehensive set of tools that had evolved over more than three billion years. As a result, the 3R strategy—reduce, reuse, and recycle—became prominent. Businesses must understand and embrace the concept of sustainability, which involves three key elements: people, profit, and the environment. This approach is sometimes referred to as “green human resource management”.

3.1.2 Concept of Reduce, Reuse, Recycle

The fundamental concept of green management is the traditional reduce, reuse, recycle methodology. To effectively minimize our company’s ecological footprint, we should actively decrease and regulate the quantity of trash produced during our activities, make use of resources again whenever feasible, and resort to recycling when necessary.

Energy efficiency means using energy efficiently to maximize production and decrease waste. Energy efficiency is another crucial component of green management. By embracing energy-efficient measures, such as employing LED lighting, upgrading to energy-efficient equipment, and installing intelligent building systems, we may substantially diminish our energy usage and the accompanying expenses.

Ecologically responsible and long-lasting framework for managing commodities and services from origin to consumption, sustainable supply chain management (Manoharan et al., 2023) is a vital element of green human resource management. This entails implementing sustainable practices at every stage of supply chain management (Almahairah et al., 2023), starting with the procurement of raw materials to the final delivery of the finished product to clients. By collaborating with suppliers who adhere to sustainable practices and advocating for responsible procurement, we may effectively minimize our ecological footprint.

Corporate social responsibility (CSR)/green human resource management encompasses a dedication to business social responsibility. This refers to actively participating in endeavours that promote the well-being of our local community and society, such as volunteering, making charity donations, and advocating for social concerns (Ananthan et al., 2024; Saranya et al., 2024).

3.1.3 Drives Favouring Green Human Resource Management

Government and business sector initiatives have strengthened environmental management systems. Here are some notable cases according to public news reports:

- The Ministry of Environment and Forests (India) announced the Wetlands Conservation and Management Rules, 2010. To ensure improved preservation and control and prevent additional deterioration of current wetlands in India, these regulations have been established to create legally binding procedures for the protection and management of wetlands.
- India hosted a collaborative workshop with the United States on February 18, 2011, in Delhi, focusing on hydrofluorocarbons (HFCs). Indian Minister Jairam Ramesh stated that his country's involvement in the Montreal Protocol had been accountable. Regarding ozone protection, HFCs might be considered a "solution", but they also pose a "problem" to climate change. India aims to address this issue innovatively. A collaborative Indo-US Task Force has been formed to thoroughly evaluate the several options for gradually eliminating HFCs.
- The World Bank allocated $220 million on June 30, 2010, for an Integrated Coastal Zone Management Project in India. The project aims to enhance the management and conservation of India's coastal zones while simultaneously safeguarding the livelihoods of the people dependent on them. The Integrated Coastal Zone Management Project is facilitating the establishment of a national coordinating structure for coastal zone management across Gujarat, Orissa, and West Bengal.
- The World Bank allocated $65 million to the Industrial Pollution Management Project to restore specific sites affected by industrial pollution and establish a comprehensive plan for the remediation of contaminated sites that have been abandoned or neglected. Furthermore, there is a $20 million energy-saving effort currently in progress that simultaneously advocates for the removal of ozone-depleting chemicals. India's government is executing its ambitious initiative to cleanse and conserve the Ganga River basin with the assistance of the World Bank. This involves the establishment of the National Ganga River Basin Authority and other organizations at the state level. Some of the funds will be allocated for sewage collection and treatment, as well as the management of municipal solid waste.
- Pesticides and herbicides are recognized as enduring environmental contaminants that have detrimental effects on human and animal health once they enter the food chain. The US Environment Protection Act (EPA) is a significant policy designed to promote sustainable and environmentally friendly growth. According to the EPA, to effectively control weeds, it is recommended to water the grass deeply but seldom, allowing the soil to partially dry between waterings. Additionally, it is advised not to cut the grass too short, as taller grass is more capable of dealing with weeds.

3.1.4 AI-Driven Insights Related to GHRM

Artificial Intelligence (AI) is progressively impacting GHRM practices, optimizing sustainability and operational efficiency within organizations. AI-driven tools and analytics are revolutionizing conventional HR processes like recruitment, training, and performance management through the integration of sustainable practices. These technologies enable organizations to analyze employee behaviour patterns, forecast job performance, and synchronize human resource strategies with sustainability objectives. For example, artificial intelligence can enhance energy efficiency and minimize carbon emissions by analyzing workplace behaviours, including resource usage and commuting trends, thus facilitating GHRM efforts. Moreover, AI enhances decision-making based on data, allowing organizations to evaluate the environmental effects of HR policies and practices with increased accuracy. It additionally facilitates the automation of tasks such as overseeing adherence to environmental standards, administering remote work protocols, and promoting sustainable employee involvement.

The integration of AI in GHRM significantly advances environmental sustainability while simultaneously enhancing business performance. Organizations can enhance employee well-being, improve corporate reputation, and boost productivity by promoting environmentally sustainable practices, which ultimately result in increased job satisfaction and retention rates. The capability of AI to process extensive datasets and deliver immediate insights guarantees that GHRM practices are both sustainable and capable of scaling and adapting, serving the interests of businesses and the environment alike.

3.2 THEORETICAL BACKGROUND OF THE STUDY

Elshaer, Azazz, and Fayyad (2023) claimed that green management practices (GMPs) improve enterprises' environmental, economic, and social performance and give them a competitive edge. The study solved environmental difficulties and met tourist, government, and community environmental expectations. Many scholars believe that green management requires employees' pro-environmental behaviours to ensure business sustainability, but few have researched such behaviours in developing nations' small and medium-sized enterprises (SMEs). Partial least squares structural equation modeling (PLS-SEM) demonstrated that GMPs increase environmental, economic, and social performance. Environmentally conscious employees can increase GMP performance. The paper recommends environmental stewardship for small and medium-sized hospitality companies. These firms should involve employees in green projects to improve sustainability. This study examines how SME hospitality employees' pro-environmental behaviour impacts green management and sustainability.

Kanan et al. (2023) investigated the growth of GHRM research in recent decades. GHRM practices improve Palestinian manufacturing enterprises' sustainable performance through green innovation. A randomly selected sample of 58 high-level executives in Palestinian manufacturing enterprises provided responses that were collected via a self-administered questionnaire. The study was analyzed using Smart-PLS and PLS-SEM. Green innovation, sustainable performance, and mild GHRM were identified. The findings showed that GHRM and green innovation boost sustainability. Innovation benefits from GHRM. Sustainability and GHRM are mediated by green innovation. This study provides manufacturing business policy-makers with a conceptual framework and roadmap for GHRM initiatives to promote employee commitment to environmental sustainability and performance.

Awan et al. (2022) noted that modern manufacturing businesses are actively adopting new manufacturing process techniques. Corporations aim for optimum production efficiency with minimal expenditure. This study examines how lean manufacturing tactics affect organizational sustainability and how green supply chain management mediates this effect. Surveys of 250 Pakistani industrial businesses were examined using AMOS 25. Processes, equipment, product design, supplier partnerships, and customer interactions strongly impact sustainability.

Naseer et al. (2023) used resource-based perspective theory to analyze green human resource and supply chain management. This study randomly sampled 212 chemical, pharmaceutical, automotive, textile, and food industries. GHRM initially benefited from the triple bottom line. Internal green supply chain management mediates GHRM and sustainable performance. External green supply chain management methods bridge GHRM initiatives and environmental and social sustainability concerns.

Jasim (2023) seeks to examine the correlation between green human resource practices and sustainable performance, as well as the magnitude of this correlation. This includes evaluating green HR practices inside and outside the company. The Environment Ministry prioritized sustainable development. Questionnaire validity and reliability were tested. Of the 245 surveys, 150 were returned. Descriptive analysis was utilized. The study surveyed perspectives and gave results. The Iraqi Ministry of Environment's use of GHRM approaches, particularly job analysis and design, planning, and performance evaluation, to improve sustainable performance was the most noteworthy finding. The text links green work to worker engagement in society. It also emphasizes the importance of these two variables and sustainable performance in GHRM.

Hazim Al-Abady and Saadoon Alsamman (2023) want to achieve consistent performance by implementing environmentally friendly techniques. A crucial question arises: How does global supply chain management impact the attainment of consistent performance through the use of green technology as an intermediary factor? The effects of these variables will be studied on Baghdad's State Company for Electrical and Electronic Industries. The research community was comprised of personnel from the organization under study, and a purposeful sample of 221 persons was selected, specifically targeting administrative leaders. To assess the accuracy of their proposed hypotheses, data were gathered data on the attitudes of the participants using a questionnaire. The model yielded diverse conclusions, with the most notable being the significant direct and indirect effects among the research variables. The findings prioritize the management of the Global Supply Chain Management Program (GSCMP) and the use of environmentally friendly technologies since these factors play a large role in attaining consistent and enduring performance.

Allam and Mansour (2024) studied how GHRM improves private hospital sustainability in Egypt. Egyptian private hospitals have incorporated GHRM techniques, but not as well in green compensations, job analysis, HR planning, or green recruiting and selection from 398 personnel respondents. A minor association between GHRM practices and sustainable performance was found. Egyptian private hospitals are obligated to adhere to GHRM principles to foster environmental consciousness among their personnel, encourage environmentally friendly conduct, and promote a green culture that enhances sustainable performance.

Shayegan, Bazrkar, and Yadegari (2023) proposed that the adoption of an environmental responsibility culture is not only a means of branding but also a crucial element in the advancement of businesses. This study analyzes how environmentally friendly human management might improve environmental performance in Iranian car manufacturers. The research uses survey and correlation-analytical methods to show that environmentally friendly recruitment, training, and performance management improve firms' environmental performance. Learning management systems, cloud computing, and AI improve these activities. Environmentally friendly people management fosters a sustainable development strategy and increases employee career opportunities. Enhancing its corporate image, fulfilling rising product demand, boosting production safety, and improving the internal organizational environment benefit the company.

Shahzad, Du, and Junaid (2023) discussed sustainability in HRM, specifically GHRM, which has changed significantly. Companies need to value their employees. This study examines HRM's sustainability roles and analyzes how GHRM affects Chinese business performance through green innovation (GI), green employee behaviour (GEB), and organizational culture. More than 300 Chinese industrial company HR specialists met the study's goals. The statistics reveal that GI and GEB affect firm sustainability. Academics, HR experts, Chinese security exchange commissioners, and firm senior management are impacted by the research theory, methodology, and applications.

Faizan, Tanveer, and Haq (2023) explore how GHRM practices affect operational sustainability in manufacturing SMEs in industrial cities in Pakistan. Positive and quantitative studies were utilized to explore how primary HRM practices like green recruitment selection and green training and development affect green innovation and sustainable firm performance. Researchers surveyed 350 managers and utilized SmartPLS and SEM to find relationships. The findings indicate that green recruitment and selection and green training and development promote green innovation and sustainable firm performance. The mediation study showed that green innovation links GHRM practices to sustainable results. This work illuminates the complicated relationships between these variables. The results have practical implications for HR professionals and organizational leaders, highlighting the strategic significance of incorporating environmentally friendly HRM practices to promote innovation and improve overall business sustainability in the context of manufacturing SMEs.

3.3 RESEARCH METHODOLOGY

This study intends to fill a gap in the literature by reviewing the relationships between sustainable performance, green human resource management practices, and academic research in sustainable management (Deviprasadet al., 2023).

3.3.1 Relationship between GHRM and Sustainable Performance

Environmentally friendly policies and practices are the main emphasis of green human resource management, which aims to reduce harmful effects on the environment (refer to Table 3.1). The goal is to promote environmentally friendly products and services while reducing waste and conserving resources through the integration of sustainable techniques into business operations. This method improves sustainable performance, which is defined as attaining long-term financial success while preserving ecological harmony, and it also helps the environment.

In many cases, operational efficiency, cost savings, and the company's reputation all increase after implementing green management practices. Carbon footprint reduction, energy efficiency, and renewable resource utilization are indicators of sustainable performance. To drive innovation, satisfy regulatory obligations, and attract environmentally conscious consumers, firms should integrate their commercial goals with environmental stewardship.

In the end, there is a connection between green human resource management and sustainable performance; this connection promotes a more complete view of business that helps with growth, sustainability, and social responsibility. Because of this complementary relationship, businesses can succeed in today's market while still making a positive impact on society and the environment.

TABLE 3.1
Goals of Green Human Resource Management

Goals of Green Human Resource Management
Reducing energy and resource consumption
Minimizing waste and pollution
Promoting sustainable sourcing and procurement
Designing products with minimal environmental impact
Implementing sustainable practices
Long-term sustainability
Balancing economic, social, and environmental performance
Socially responsible and environmental sensitive

3.4 CHALLENGES

Implementing ethical and ideal standards might potentially affect profits and the overall financial performance (D. Singh et al., 2024; Winster et al., 2024) of a company to establish long-term and environmentally responsible business practices. Numerous chief executive officers recoil at the idea of this predicament. Aside from the typical challenges related to the situation, it is important to acknowledge the change in identity and the potential for achievement. The likelihood of employees (Suresh et al., 2024; Megala et al., 2024; Shaikh et al., 2023) and other stakeholders embracing the idea of implementing green initiatives increases when there is a widespread focus on environmental awareness. The business sector addresses the significant negative correlation between environmental performance (Joyce et al., 2024) and economic success.

3.5 FUTURE TRENDS AND SUGGESTIONS FOR GREEN HUMAN RESOURCE MANAGEMENT

The role of the marketer is crucial, as they are responsible for enlightening clients about the significance and benefits of environmentally friendly products compared to their non-environmentally friendly counterparts. Green business is highly pertinent in rising nations such as India. The government should actively collaborate with private sector initiatives to ensure the welfare of future generations. Companies should acknowledge and incentivize environmentally friendly business practices. Therefore, the government must allocate additional funds towards the study and development of environmentally sustainable technology (Rukmini et al., 2024; Ponduri et al., 2024), while also implementing subsidies for individuals who adopt such practices. There exists a robust association between environmentally conscious business operations and achieving success in the market (Dattatreyulu et al., 2024). Therefore, firms should prioritize the development of eco-friendly products and adapt their product design to gain a larger market share and a competitive advantage.

- Enhancing the capacity of pollution control boards: To enhance the effectiveness of the pollution control boards, a planned initiative will be implemented to construct new monitoring and data management methods, as well as develop and maintain skills and infrastructure.
- Biodiversity conservation: In this study on biodiversity, various habitats are analyzed to assess the potential effectiveness of conservation initiatives. As part of this endeavour, the diverse range of plant and animal species in India's hotspots will be systematically documented, and the economic well-being of the local communities residing in these areas will be enhanced, enabling them to enjoy the advantages of investments made in these regions.

3.6 CONCLUSION

Sustainability requires critical self-analysis, honesty, creativity, and risk over time. In other words, a company must be prepared to critically assess its own operations and environmental impact, as well as to be transparent about it, before pursuing sustainability. Additionally, it should be prepared to branch out of its typical comfort zone and experiment with new ideas, taking risks. In the context of green business sustainability, it is imperative to consider the operations of a company. Companies that implement only a handful of enhancements are incapable of asserting sustainability. All business processes, activities, and functions must be prepared to implement the aforementioned critical self-analysis, honesty, creativity, and risk. The term "sustainability" denotes a comprehensive transformation in the organization's values, practices, attitudes, and outlook that influences its operations. Finally, green business uses the triple bottom line (often known as "planet, people, and profit") to assess a company's impact on the environment, society, and economics. The Indian economy is making strides towards becoming more environmentally sustainable; however, it still has a significant amount of work to do to compete with the developed world. The necessity of green marketing in ensuring sustainability is demonstrated by the global concern for environment management systems and the evident degradation in environmental norms. Restoring the natural heritage that has been lost has been a commendable endeavour by both the government and corporations, which became more eminent after the COVID-19 pandemic (Sadasivan et al., 2024). In parallel, numerous exhaustive studies and reports have been conducted to examine the issue of green marketing, which has become increasingly prominent in recent decades. It is imperative to devise constructive and practical solutions for environmental management, rather than merely discussing the subject and disseminating the results. It is possible that we must still rediscover our understanding of the term "green" or confront our environmental responsibilities to attempt to restore our ecosystem and its lost legacy. Only then can we ensure that our planet remains habitable for future generations and accomplish sustainable development.

REFERENCES

Alagpuria, M., Dhandapani, K., Manoharan, G., Kannan, R., Suchith, C., & Selvaraj, F. J. (2024, June). Survey of Intelligent Human Society Sustainable Communities for Agri-Food Production Industries Using Artificial Neural Network. In *AIP Conference Proceedings* (Vol. 2971, No. 1). AIP Publishing.

Allam, N. M., & Mansour, M. M. (2024). Do Green Human Resource Management Practices Improve Sustainable Performance? *International Journal of Customer Relationship Marketing and Management*, 15(1), 1–20. https://doi.org/10.4018/ijcrmm.336915

Almahairah, M. S. Z., Goswami, S., Karri, P. N., Krishna, I. M., Aarif, M., & Manoharan, G. (2023, December). Application of Internet of Things and Big Data in Iimproving Supply Chain Financial Risk Management System. In *2023 10th IEEE Uttar Pradesh Section International Conference on Electrical, Electronics and Computer Engineering* (UPCON) (Vol. 10, pp. 276–280). IEEE.

Ananthan, M., Shanmugam, V., Udayanan, M., Selvaraj, F., & Manoharan, G. (2024, June). Corporate Social Responsibility Is Used as a Marketing Strategy During Covid-19 Pandemic. In *AIP Conference Proceedings* (Vol. 2971, No. 1). AIP Publishing.

Ashtikar, S. P., Manoharan, G., & Billa, P. K. (2022). COVID-19: A Special Review of MSMEs for Sustaining Entrepreneurship. *Challenges and Recovery Measures*, 152.

Awan, F.H., Liu, D., Jamil, K., Mustafa, S., Atif, M., Gul, R.F., & Qin, G. (2022). Mediating Role of Green Supply Chain Management Between Lean Manufacturing Practices and Sustainable Performance. *Frontiers in Psychology*, 12.

Chaudhary, V., Vemuri, V. P., Cavaliere, L. P. L., Verma, V., Manoharan, G., & Bharti, A. (2024, March). A Comparative Analysis of Job Satisfaction and Motivational Factors of Employees in Public Versus Private Organizations. In *AIP Conference Proceedings* (Vol. 2816, No. 1). AIP Publishing.

Dattatreyulu, C., Krishna, C. B., & Geetha, M. (2024, June). Assessment of the Effect of Covid–19 Situation on Digital Marketing. In *AIP Conference Proceedings* (Vol. 2971, No. 1). AIP Publishing.

Deviprasad, S., Madhumithaa, N., Vikas, I. W., Yadav, A., & Manoharan, G. (2023). The Machine Learning-Based Task Automation Framework for Human Resource Management in MNC Companies. *Engineering Proceedings*, 59(1), 63.

Durai, S., Krishnaveni, R., Manoharan, G., & Ashtikar, S. P. (2024). Leveraging HR Metrics and Workforce Analytics: The Road to Organizational Efficacy. In *Practical Approaches to Agile Project Management* (pp. 346–368). IGI Global.

Durai, S., Krishnaveni, R., Manoharan, G., & Nivedha, M. (2024). Leveraging HR Metrics for an Effective Recruitment and Selection Process with Special Reference to the Manufacturing Sector in Coimbatore. In *Convergence of Human Resources Technologies and Industry 5.0* (pp. 188–210). IGI Global.

Durai, S., Manoharan, G., & Ashtikar, S. P. (2024). Harnessing Artificial Intelligence: Pioneering Sustainable Solutions for a Greener Future. In *Social and Ethical Implications of AI in Finance for Sustainability* (pp. 89–117). IGI Global.

Elshaer, I. A., Azazz, A. M., & Fayyad, S. (2023). Green Management and Sustainable Performance of Small-and Medium-Sized Hospitality Businesses: Moderating the Role of an Employee's Pro-Environmental Behaviour. *International Journal of Environmental Research and Public Health*, 20(3), 2244.

Faizan, S., Tanveer, N., & Haq, A. U. (2023). Green Human Resource Management Practices: Driving Sustainable Innovation and Performance in Manufacturing SMEs. *Contemporary Issues in Social Sciences and Management Practices*. https://doi.org/10.61503/cissmp.v2i4.106

García-Tadeo, D. A., Peram, D. R., Kumar, K. S., Vives, L., Sharma, T., & Manoharan, G. (2022). Comparing the Impact of Internet of Things and Cloud Computing on Organisational Behavior: A Survey. *Materials Today: Proceedings*, 51, 2281–2285.

Geetha, M., & Sunitha, P. A. (2022). *Work-Life Balance among Women in the Private Higher Education Industry during COVID-19: A Path to Organisational Sustainability.* Building Resilient Organizations.

Gopinathan, R., & Manoharan, G. (2022). Work-Life Balance Practices and Organizational Performance for Achieving Superior Performance. *The Changing Role of Human Resource Management in the Global Competitive Environment*, 217.

Gulati, N., Sethi, A., Mahesh, D., Makkar, S., Manoharan, G., Megaladevi, M., & Palanivel, R. (2022, May). Implementation of Blockchain in Supply Chain. In *AIP Conference Proceedings* (Vol. 2418, No. 1). AIP Publishing.

Hazim Al-Abady, A.W., & Saadoon Alsamman, T.A. (2023). *Green Supply Chain Management Practices and ITS Impacts on the Sustainable Performance / Testing the Mediating Role of Green Technology.* NexoRevistaCientífica.

Jasim, W. T. (2023). The Effect of Some Green Human Resource Management Practices in Sustainable Performance: Analytical Research in the Ministry of Environment. *International Journal of Research in Social Sciences & Humanities*. https://doi.org/10.37648/ijrssh.v13i02.032

Joyce, P. R., Selvaraj, F. J., Manoharan, G., Priya, C., Vijayalakshmi, R., Dwivedi, P. K., & Veerakumar, K. (2024). To Study The Role of Marketing In Human Resource Management. *Migration Letters*, 21(S2), 1191–1196.

Kanan, M., Taha, B., Saleh, Y., Alsayed, M., Assaf, R., Ben Hassen, M. F., Alshaibani, E., Bakir, A., & Tunsi, W. H. (2023). Green Innovation as a Mediator Between Green Human Resource Management Practices and Sustainable Performance in Palestinian Manufacturing Industries. *Sustainability*. https://doi.org/10.3390/su15021077

Keserwani, H., PT, R., PR, J., Manoharan, G., Mane, P., & Gupta, S. K. (2021). Effect of Employee Empowerment on Job Satisfaction in Manufacturing Industry. *Turkish Online Journal of Qualitative Inquiry*, 12(3).

Krishna, S. H., & Manoharan, G. (2022). Making the Link between Work-Life Balance Practices and Organizational Performance in the Hospitality Industry. *The Changing Role of Human Resource Management in the Global Competitive Environment*, 201.

Krishnan, S. G., & Manoharan, G. (2022). An Investigation in Analysing the Key Aspects of High-Performance Work Practice for Enhancing Productivity in the Organisation. *The Changing Role of Human Resource Management in the Global Competitive Environment*, 31.

Kumar, P., Maurya, N., Keerthiraj, Krishna, S. H., Manoharan, G., & Bharti, A. (2024). Evaluating the Impact of Healthcare 4.0 on the Performance of Hospitals. *Human Cancer Diagnosis and Detection Using Exascale Computing*, 1–17.

Kumar, S., Kumar, P., Ahmad, S. S., Jayasaradadevi, P., Rajeyyagari, S., Manoharan, G., & Radhakrishnan, V. (2024). A Novel Approach for IoT-Based Cloud Computing Technology and its Impact on Business Entrepreneurship. *International Journal of Intelligent Systems and Applications in Engineering*, 12(10s), 624–628.

Makkar, S., Bajpai, T., Bhola, M., Mahesh, D., & Manoharan, G. (2022, May). Blockchain Disruption in Banking Sector. In *AIP Conference Proceedings* (Vol. 2418, No. 1). AIP Publishing.

Manoharan, G., Agarwal, S., Agarwalla, S., Dutt, A., Vani, V. D., & Sandhya, G. (2023, December). Blockchain Technology in Supply Chain Management: Prospects and Challenges. In *2023 10th IEEE Uttar Pradesh Section International Conference on Electrical, Electronics and Computer Engineering (UPCON)* (Vol. 10, pp. 557–560). IEEE.

Manoharan, G., & Ashtikar, S. P. (2024). A Theoretical Framework for Emotional Intelligence in Academic Leadership in Higher Education. In *Building Organizational Resilience with Neuroleadership* (pp. 96–112). IGI Global.

Manoharan, G., & Ashtikar, S. P. (2024). Nexus Between Leadership and Effective Communication: Implications for Educational Institutions. In *Neuroleadership Development and Effective Communication in Modern Business* (pp. 274–291). IGI Global.

Manoharan, G., Ashtikar, S. P., Rao, C. G., Natarajan, S., & Nivedha, M. (2024). Masterminds and Machines: Harnessing AI in Strategic Leadership. In *Impact of New Technology on Next-Generation Leadership* (pp. 107–119). IGI Global.

Manoharan, G., Ashtikar, S. P., Smitha, V., Sundaramoorthi, S., & Krishna, I. M. (2023). Work-Life Balance Perceptions of Women in the IT and ITeS Sectors in Kerala: A Research Study. *Journal of Pharmaceutical Negative Results*, 3363–3375.

Manoharan, G., Durai, S., & Rajesh, G. A. (2022, May). Emotional Intelligence: A Comparison of Male and Female Doctors in the Workplace. In *AIP Conference Proceedings* (Vol. 2418, No. 1). AIP Publishing.

Manoharan, G., Durai, S., & Rajesh, G. A. (2022, May). Identifying Performance Indicators and Metrics for Performance Measurement of the Workforce Is the Need of the Hour: A Case of a Retail Garment Store in Coimbatore. In *AIP Conference Proceedings* (Vol. 2418, No. 1). AIP Publishing.

Manoharan, G., Durai, S., Rajesh, G. A., & Ashtikar, S. P. (2023). A Study on the Application of Natural Language Processing Used in Business Analytics for Better Management Decisions: A Literature Review. *Artificial Intelligence and Knowledge Processing*, 249–261.

Manoharan, G., Durai, S., Rajesh, G. A., Razak, A., Rao, C. B., & Ashtikar, S. P. (2023). An Investigation into the Effectiveness of Smart City Projects by Identifying the Framework for Measuring Performance. In *Artificial Intelligence and Machine Learning in Smart City Planning* (pp. 71–84). Elsevier.

Manoharan, G., Jalaja, V., Sathe, M. A., Lourens, M., & Suresh, K. (2023, December). Machine Learning and Data Privacy in Human Resource Management. In *2023 4th International Conference on Computation, Automation and Knowledge Management (ICCAKM)* (pp. 1–6). IEEE.

Manoharan, G., & Narayanan, S. (2021). A Research Study to Investigate the Feasibility of Digital Marketing Strategies in Advertising. *PalArch's Journal of Archaeology of Egypt/Egyptology*, 18(9), 450–456.

Manoharan, G., Nithya, G., Razak, A., Durai, S., Sharma, S., & Ashtikar, S. P. (2024). Enhancing Financial Literacy: Strategies, Challenges, and Global Initiatives. In *Emerging Perspectives on Financial Well-Being* (pp. 111–128). IGI Global.

Manoharan, G., Nithya, G., Razak, A., Sharma, S., &Ashtikkar, S. P. (2024). Challenges and Opportunities in Green Finance: Overcoming Barriers and Scaling up Sustainable Investments. *Practical Approaches to Agile Project Management*, 398–413.

Manoharan, G., Rao, B. S., Ashtikar, S. P., Razak, A., Durai, S., & Kukreja, J. (2024). A Study of Influence of Leadership with Effective Communication in Educational Institutions. In *Neuroleadership Development and Effective Communication in Modern Business* (pp. 214–227). IGI Global.

Manoharan, G., Razak, A., Rajchandar, K., Nithya, G., Durai, S., & Ashtikar, S. P. (2024). Digital Learning for Professional Development in Varied Fields of Service Sectors: Embracing Technological Advancements. In *Embracing Technological Advancements for Lifelong Learning* (pp. 111–137). IGI Global.

Manoharan, G., Sharma, V. K., Lourens, M., Kumar, A., Nayak, B. B., & Hinge, P. (2023, September). Design and Empirical Analysis of a Artificial Intelligence-Based Human Resource Management Processing Systems for Detecting Personal Stress. In *2023 6th International Conference on Contemporary Computing and Informatics (IC3I)* (Vol. 6, pp. 2421–2426). IEEE.

Megala, A., Patil, R. D., Pushparathi, V. G., Manoharan, G., Hemambika, S., & Khan, B. (2024, March). Experimental Analysis of Employee Depression Identification System Using Modified Learning Procedure with AI Assistance. In *2024 5th International Conference on Intelligent Communication Technologies and Virtual Mobile Networks (ICICV)* (pp. 282–289). IEEE.

Naseer, S., Song, H., Adu-Gyamfi, G., Abbass, K., & Naseer, S. (2023). Impact of Green Supply Chain Management and Green Human Resource Management Practices on the Sustainable Performance of Manufacturing Firms in Pakistan. *Environmental Science and Pollution Research*, 30, 48021–48035.

Ponduri, S. B., Ahmad, S. S., Ravisankar, P., Thakur, D. J., Chawla, K., Chary, D. T., & Sharma, S. (2024). A Study on Recent Trends of Technology and its Impact on Business and Hotel Industry. *Migration Letters*, 21(S1), 801–806.

Purushottam Ashtikar, S., & Manoharan, G. (2024, June). Emotional Intelligence on Higher Education Educators: A Systematic Literature Review. In *AIP Conference Proceedings* (Vol. 2971, No. 1). AIP Publishing.

Purushottam Ashtikar, S., & Manoharan, G. (2024, June). Studying the Level of Work-Life Balance and Emotional Intelligence: University Educators. In *AIP Conference Proceedings* (Vol. 2971, No. 1). AIP Publishing.

Ramachandran, K. K., Mary, S. S. C., Painoli, A. K., Satyala, H., Singh, B., & Manoharan, G. (2022). Assessing the Full Impact of Technological Advances on Business Management Techniques. *The British Journal of Administrative Management.* https://tbjam. org/vol58-special-issue-06

Rukmini, S., Sreenivasulu, Y., Raghu, T., Nagatulasi, K., & Manoharan, G. (2024, June). English Language Teaching in Technological and Management Studies in India: An Edu-Techno Approach. In *AIP Conference Proceedings* (Vol. 2971, No. 1). AIP Publishing.

Sadasivan, A., Udayanan, M., Alexrajesh, G., Manoharan, G., Razak, A., &Purushottamashtikar, S. (2024, June). An Analysis of Consumer's Online Shopping Behaviors Before and During the COVID-19 Pandemic. In *AIP Conference Proceedings* (Vol. 2971, No. 1). AIP Publishing.

Saranya, M., Durai, S., Manoharan, G., Ashtikar, S. P., Rajesh, G. A., & Razak, A. (2024, June). Social Value Influencing Entrepreneurial Intention Among Postgraduate Students. In *AIP Conference Proceedings* (Vol. 2971, No. 1). AIP Publishing.

Satpathy, A., Samal, A., Gupta, S., Kumar, S., Sharma, S., Manoharan, G., & Sharma, S. (2024). To Study the Sustainable Development Practices in Business and Food Industry. *Migration Letters*, 21(S1), 743–747.

Shahzad, M.A., Du, J., & Junaid, M. (2023). Impact of Green HRM Practices on Sustainable Performance: The Mediating Role of Green Innovation, Green Culture, and Green Employees' Behavior. *Environmental Science and Pollution Research*, 30, 88524–88547.

Shaikh, I. A. K., Kumar, C. N. S., Rohini, P., Jafersadhiq, A., Manoharan, G., & Suryanarayana, V. (2023, August). AST-Graph Convolution Network and LSTM Based Employees Behavioral and Emotional Reactions to Corporate Social Irresponsibility. In *2023 Second International Conference on Augmented Intelligence and Sustainable Systems (ICAISS)* (pp. 966–971). IEEE.

Shayegan, S., Bazrkar, A., & Yadegari, R. J. (2023). *Realization of Sustainable Organizational Performance Using New Technologies and Green Human Resource Management Practices.* Foresight and STI Governance.

Shiva, C. K., Vedik, B., & Manoharan, G. (2024). Sustainable Smart Transportation: Technologies, Benefits, Challenges to Urbanisation Concept. In *Secure and Intelligent IoT-Enabled Smart Cities* (pp. 355–370). IGI Global.

Singh, A., Virmani, D., Kulkarni, N., Tiwari, M., Manoharan, G., & Alanya-Beltran, J. (2024). The Emerging Role and Application of Adopting Revolutionary Internet of Things (IoT) for Future Technology Sustainability. *Computer Science Engineering and Emerging Technologies: Proceedings of ICCS 2022*, 103.

Singh, D., Makkar, S., Durai, S., Manoharan, G., Purushottamashtikar, S., & Alexrajesh, G. (2024, June). Human Resource Information System Applications in Small and Medium Enterprises: A Study (SMEs). In *AIP Conference Proceedings* (Vol. 2971, No. 1). AIP Publishing.

Singh, D., Makkar, S., Manoharan, G., Durai, S., Purushottamashtikar, S., &Alexrajesh, G. (2024, June). The Impact of Information Technology Enabler and Organizational Learning on Performance: A Systematic Literature Review. In *AIP Conference Proceedings* (Vol. 2971, No. 1). AIP Publishing.

Suresh, E., Ravi, B., Udayanan, M., Manoharan, G., Ashtikar, S. P., & Rajesh, G. A. (2024, June). Prospects and Determinants Towards Work from Home Associated Perception of IT Employees: With Special Reference to Chennai. In *AIP Conference Proceedings* (Vol. 2971, No. 1). AIP Publishing.

Winster, D., Ramakrishnan, S., Veluthakkal, J., Manoharan, G., Purushottamashtikar, S., & Selvaraj, F. (2024, June). Interactions Between Emotional and Spiritual Intelligence and Their Effects on Employee Performance. In *AIP Conference Proceedings* (Vol. 2971, No. 1). AIP Publishing.

4 Harnessing the Sun
Role of State Action Plan on Climate Change and Tech-Driven Solutions for Solar Energy Sector in the Indian States of Uttar Pradesh, Kerala, and Assam

Ashish Saraswat, Garima Tiwari, and Viktoriia Hamaiunova

4.1 INTRODUCTION

India, with a population of 1.42 billion (O'Neill 2024) and the fifth-largest economy (Forbes India 2024;Prasad 2020;IEA 2020), needs to strike a balance in its economic development and carbon emissions. India ranks among the top five countries in terms of climate change performance. According to the Climate Change Performance Index (CCPI 2023), India ranks highest among the G-20 nations. India advanced two positions on the CCPI 2023, an assessment of climate change policies and actions involving 59 countries and the European Union, which was released at COP 27 (Government of India, Ministry of Power 2024). In 2024, it is ranked at seventh in the climate change performance index (Bosse 2023). In spite of this, India continues to meet its expanding energy demands through its substantial dependence on coal, oil, and gas. This dependence contributes significantly to the emission of greenhouse gases and exacerbates air pollution, particularly in urban areas (Bosse 2023).

The major contributor to CO_2 emission in India is coal (see Figure 4.1) (Dubash 2019, 169;IEA 2020; Sharma 2018). In order to address the issue, there is an emergent need to shift toward renewable energy, which would enables the reduction in emissions. This is because renewable energy is also considered sustainable energy, which will eventually lead to sustainable development (J and Majid 2020). At global level as well, India has shown interest in a transition toward renewable energy by signing a non-binding declaration (Major Economies Forum Declaration 2009) that

DOI: 10.1201/9781003626527-4

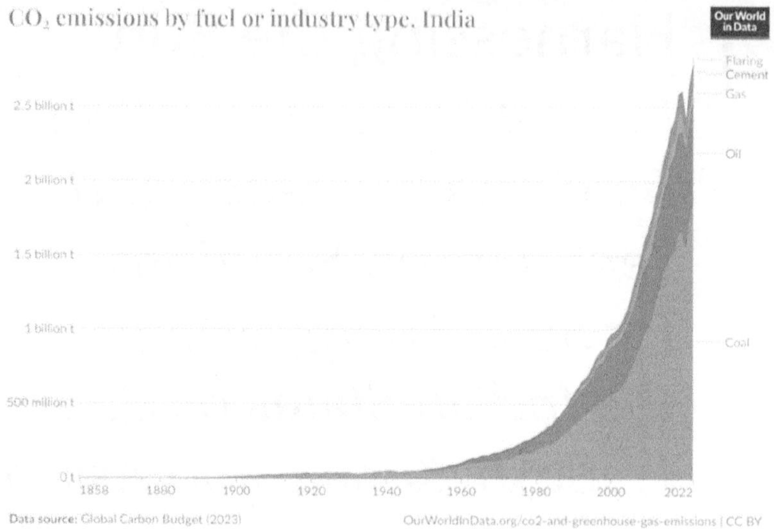

FIGURE 4.1 CO_2 emissions by fuel or industry type in India (Source: Global Carbon Budget, 2023).

recognized for the first time that global temperatures should not exceed 2 degrees Celsius and that the member countries must work toward a "global goal" to reduce "global emissions" by 2050 (Sengupta 2019; Choudhary 2022; Sharma and Rao 2021).

Pursuant to this declaration, India presented its Nationally Determined Contributions (NDCs) to decrease the emission intensity of its gross domestic product (GDP) by 33–35% compared to 2005 levels by the year 2030. The goal is to attain around 40% of the total installed capacity of electric power from energy sources that are not dependent on fossil fuels by 2030 (UNFCCC 2015, 29).

Solar energy is considered the cleanest form of renewable energy (Shinn 2024). India possesses a vast solar capacity of over 5,000 trillion kWh/year of energy, which is received across the country's land area. The majority of regions get an average of 4–7 kWh/m^2. each day (Government of India, Ministry of New and Renewable Energy 2024). In order to harness this untapped capacity, solar energy has been prioritized in the National Action Plan on Climate Change (NAPCC), which was introduced in 2008 and consists of eight missions (Government of India, Ministry of New and Renewable Energy 2024). The National Solar Mission and National Mission on Enhanced Energy Efficiency emphasize the utilization of solar energy to tackle climate change and minimize emissions.

The NAPCC focuses on the principles of sustainable development and outlines strategy by way of various missions to lead India on the path of development while addressing the concerns of climate change (Chandel et al. 2016, 867). These objectives can only be achieved when they are pursued by states as per their emissions, vulnerability, and climate consciousness (Gogoi 2017, 4). Consequently, states have developed their State Action Plan on Climate Change (SAPCC) as an expansion of the NAPCC. The SAPCC is a strategic framework designed to address the impacts of climate change at the local and regional level (Jogesh and Dubash 2015, 248).

With this background, this chapter takes into account the role of solar energy in meeting the emission reduction objectives enshrined under the SAPCCs for the states of Uttar Pradesh, Kerala, and Assam. In the first part, it discusses the purpose of the SAPCC and its relation to the NAPCC, followed by tracing solar energy in the SAPCC. Thirdly, it traces the climate profile, vulnerabilities, role of solar energy, and progress in the aforesaid states. Fourthly, the chapter shows how tech-driven solutions like artificial intelligence, machine learning, data analytics, and automation in business management could enhance solar energy transition in India. Finally, it attempts to highlight major obstacles and challenges for the way forward.

4.2 PURPOSE OF SAPCC AND RELATION TO NAPCC

In line with the NAPCC implemented in 2008, state governments initiated the SAPCC to coordinate efforts across the national and state level (Chandel et al. 2016, 867). The diversity of the terrain, geography, demography, economic growth rates, and development across states and within states requires state-specific efforts to tackle climate change (Gogoi 2017, 4). Hence, the SAPCC for various states should prioritize subnational planning while adhering to the eight national missions outlined in the NAPCC. These missions include the National Solar Mission, National Mission for Enhanced Energy Efficiency, National Mission on Sustainable Habitat, National Water Mission, National Mission for Sustaining the Himalayan Ecosystem, National Mission for Green India, National Mission for Sustainable Agriculture, and National Mission on Strategic Knowledge on Climate Change (Prime Minister's Council on Climate Change 2008, 4–7).

India's NAPCC, released in 2008, is guided by the principles of sustainable development that seek to balance economic growth with environmental concerns (Chandel et al. 2016, 868). It lays out a national strategy to guide the economy on an inclusive, sustainable, climate-resistant developmental path (Chaudhary, Krishna, and Sagar 2014, 78). It emphasizes the significance of maintaining a trajectory of rapid economic expansion in order to uplift millions of people from poverty and mitigate their susceptibility to the catastrophic consequences of climate change disasters (Prime Minister's Council on Climate Change 2008, 3, 9, 12, 46). It also seeks to create awareness of the threat posed by climate change among the public, government agencies, industry professionals, and communities at large. Technology assumes a central role in adaptation and mitigation efforts (Government of India, Department of Science & Technology 2024). The primary focus is on developing innovative market, regulatory, and voluntary methods to promote sustainable development. Additionally, there is a strong emphasis on international collaboration for research and development in order to address climate change and facilitate technology transfer (Prime Minister's Council on Climate Change 2008, 4).

Following the announcement of the NAPCC, states were given the task of preparing their respective SAPCCs in consonance with the eight missions enumerated under the NAPCC (Gogoi 2017, 3–4). In essence, the NAPCC intended to serve as a guiding document for each state to design its respective action plans suiting their development levels, resource constraints, demography, ecology, and target economic growth rate. Thus, each state differs in its targets related to emissions per capita, GDP intensity of emissions, forest cover, energy efficiency, etc. No two states have the same targets with the same deliverables and deadlines. Also, the policy mechanisms related to formulation and implementation vary across states, and the NAPCC only acts as a guiding document.

Keeping the uniqueness of each state in consideration, each SAPCC focuses on assessing the local climate vulnerability and identifying the major opportunities to contribute toward the national goal of tackling climate change (UNDP 2024). Major objectives of SAPCCs are to ensure sustainable development of the state so that the interests of present and future generations are not threatened due to climate change; to provide an adaptation framework to deal with climate change while ensuring a high rate of economic growth; to establish a financing mechanism for adaptation and mitigation purposes; and to formulate innovative policies to deal with climate change. It involves mapping regional climate vulnerabilities, developing future projections, understanding sectoral implications, and devising actionable strategies (UNDP 2024).

States are required to cooperate with multiple stakeholders in the participatory process toward the creation of SAPCCs, establish institutional structures to carry out the plans, form new committees, organize workshops, conduct research plans on existing policy programs, conduct vulnerability and impact assessments, and identify appropriate adaptation and mitigation strategies. Pilots and demonstration projects under SAPCCs are financed either by the designated agencies or the union government under the National Adaptation Fund on Climate Change (Patra 2016, 27, 34, 76).

4.3 SOLAR ENERGY AND SAPCC

The importance of shifting toward renewable energy became more imperative after the Paris Climate Agreement, 2015 (COP 21) (Gielen et al. 2019). India has implemented many domestic measures through its NAPCC and numerous other policy efforts in alignment with its NDCs. India has established a goal to decrease the carbon intensity of its economy by a minimum of 45% by the end of the next ten years. Additionally, India aims to generate 50% of its total electricity from renewable sources by 2030 and attain a state of net zero carbon emissions by 2070. India's objective is to achieve a total installed capacity of 500 gigawatts (GW) for renewable energy by the year 2030.

India's objective is to have a production capacity of 5 million tonnes of green hydrogen by the year 2030. This will be backed by a renewable energy capacity of 125 GW. India has granted approval for the establishment of 50 solar parks, which collectively have a capacity of 37.49 GW. India aims to achieve a capacity of 30

GW from offshore wind farms by the year 2030. Potential locations for these wind farms have been identified (Government of India, Press Information Bureau 2024). However, this is possible when there is synchronization between India's international commitments and domestic actions coupled with uninterrupted financial flow. To this end, India's NAPCC has two out of eight missions specifically focusing on harnessing the solar potential of the country and shifting to a non-fossil-based economy, i.e., National Solar Mission and National Mission for Enhanced Energy Efficiency (Chandel et al. 2016, 867).

The National Solar Mission clearly emphasizes the necessity of promoting the use of renewable energy to enhance the proportion of solar energy in the energy mix, while also advancing the development of innovative solar technologies. In addition, the implementation of renewable energy may be attributed to the goal of promoting the installation of 2000 MW of off-grid solar applications, which includes the distribution of 20 million solar lights by the year 2022. The mission's objective is to attain a total of 100 GW of solar power over a span of seven years, commencing from 2014–2015. This will be accomplished by the implementation of 40 GW of grid-connected rooftop projects and 60 GW of large and medium land-based solar power projects (Prime Minister's Council on Climate Change 2008, 3). The country achieved that milestone and is working further to harness the solar potential. This national mission is going to play a significant role in increasing the share of renewable energy in the energy mix of India if it is able to achieve its desired targets. The use of solar energy will increase with a reduction in its cost and ease of storage and transfer, and the share of renewable consumption in India will also increase by 13% by the year 2040 (J and Majid 2020).

The National Mission for Enhanced Energy Efficiency promotes the deployment of renewable energy by strengthening the market for energy efficiency and fostering innovative and sustainable business models (Prime Minister's Council on Climate Change 2008, 3–4). The mission indicates the use of energy-efficient and alternative energy systems. Though renewable energy is not specifically mentioned in this mission, the aims and objectives are aligned toward the incorporation and shift toward renewable energy with the goal of CO_2 emission mitigation of 98 million tons per year and the issuance of renewable energy certificates (Government of India, Ministry of Power 2024).

As already mentioned, after the advent of NAPCC, various state governments, along with the assistance of the Government of India, are working to align their state action plans on climate change with the national objectives for the promotion of solar power in industry, household, agriculture, transportation, etc. (Chaudhary, Krishna, and Sagar 2014, 72). States have formulated their respective targets based on their capacity, infrastructure, technological support, and vulnerability to climate change under their SAPCCs. The common feature across most of the state action plans related to solar is cutting their emissions by deploying solar-based technology in solar parks, ultra-mega solar power projects, various schemes under the Jawaharlal Nehru National Solar Mission (JNNSM), solar rooftop programs, Atal Jyoti Yojana (AJAY), Pradhan Mantri Kisan Urja Suraksha evam Utthan Mahaabhiyan (KUSUM), PM Suryodaya Yojana, and PM Suryaghar (Government of India, Ministry of New and

Renewable Energy 2024). All these policies are functional in coordination with the Government of India and state governments through their respective SAPCCs. In these policies, the Government of India provides technical and financial assistance in coordination with states for the formulation and implementation of various initiatives in the solar power sector under the umbrella of the NAPCC.

Despite all the initiatives in an SAPCC, the focus is only on adaptation, whereas the NAPCC focuses both on adaptation and mitigation. The reason for this is that the non-Annex 1 countries to the United Nations Framework Convention on Climate Change (UNFCCC) need not adopt mitigation strategies (Jogesh and Dubash 2015, 252). However, there is a lack of coherence between the national plan and the state action plans regarding these two aspects of climate change. To tackle the climate change effectively, both strategies must be allowed to be practiced by states in letter and spirit. Especially, the use of solar energy to reduce emissions and tackle climate change requires both mitigation and adaptation to go hand in hand.

4.4 PERFORMANCE BY STATES

For the purpose of this research, states from three different regions of India, i.e., east, north, and south, have been taken to reflect upon the diverse energy requirements and to trace the developments of state action plans as per their respective capacity and capabilities. The analysis considers the following states: Uttar Pradesh, Kerala, and Assam.

4.4.1 Uttar Pradesh

Being the fifth biggest and most populated state in India, Uttar Pradesh has diverse energy requirements ranging from transport, household, industry, and agriculture. Rapid urbanization, floods, and heat waves particularly affecting its main source of livelihood, i.e., agriculture, make it more climate sensitive and vulnerable (Government of Uttar Pradesh, Department of Environment 2014, 13). In its SAPCC 2014, which has been majorly aligned with missions under NAPCC, there is a predicted increase in maximum temperature of 1.8°C to 2.1°C by 2080 and rainfall by 15% to 20% in the 2050s (Government of Uttar Pradesh, Department of Environment 2014, 17). If emissions in the state are not proactively reduced, the future prospects for the state are bleak. It now has the highest emissions in India, contributing around 14% of the country's total greenhouse gases (GHGs) (Government of Uttar Pradesh, Department of Environment 2014, 13).

Power is considered a driver of the economy and catalyst for infrastructure development. Uttar Pradesh was shown to have a 15% power deficit as per its state action plan (Government of Uttar Pradesh, Department of Environment 2014, 68), but remarkable progress has been made as per the Load Generation Balance Report 2023–2024 by the Central Electricity Authority. The anticipated power deficit of energy is 4.5%, and during peak hours, it is –8.0% (Government of India, Ministry of Power, Central Electricity Authority 2023). The state has been able fill this gap with its robust implementation of policies related to power generation.

In 2005–2006, there was an installed electricity capacity of 8181.7 MW; by 2009–2010, it had increased to 9983.7 MW (Government of Uttar Pradesh, Department of Environment 2014, 27). At the end of the fiscal year 2018, the installed power capacity was around 24,909 MW (Jaganmohan 2024). The state implemented an energy policy in 2009 with the objective of ensuring that its residents, industry, and agricultural sector have access to dependable, reasonably priced, and high-quality electricity. In 2018, the state implemented a solar power policy with the aim of achieving a renewable installed capacity of 10.7 GW by 2020. Out of this total, 4.3 GW was to be generated from rooftop solar photovoltaic projects, in accordance with the solar rooftop policy, and 6.4 GW from utility-scale sources (International Energy Agency 2021).

Although significant advancements have been achieved, Uttar Pradesh's electricity transmission and distribution loss stands at over 31%, above the national average of around 25% in FY2009 (Government of Uttar Pradesh, Department of Environment 2014, 68). To overcome all the challenges related to the power sector, the state has shifted its focus to generating power through solar rather than fuel while also focusing on energy efficiency. Much progress has been made after the localization of Sustainable Development Goals (SDGs) into the state policies by way of a state action plan in line with the national plan. Consequently, it still has the largest number of "energy poor" populations, and the rapid climate change happening across makes it more detrimental to aspects of health, housing, education, industry, household, agriculture, etc.

Uttar Pradesh's solar policy provides incentives for grid-connected solar power projects with a capacity of 5 MW and above, as well as solar power plants with capacities above 5 MW for captive use. The state has established certain time frames for the initiation of solar photovoltaic projects, which is set at 13 months, and solar thermal projects, which is set at 28 months, starting from the date of signing the power purchase agreement. In order to address its electricity shortage and combat the effects of climate change, the state additionally offers specific incentives on a case-by-case basis for the establishment of solar farms with investments above INR 500 crores. To make the use of solar more accessible and reachable, the Uttar Pradesh New & Renewable Energy Development Agency (UPNEDA) acts as a single-window clearance agency. Additionally, all incentives of the Infrastructure and Industrial Investment Policy 2012 of Uttar Pradesh are also made applicable to power plants based on solar energy (Government of Uttar Pradesh, Department of Environment 2014, 69). The National Solar Mission grants a 30% subsidy in the state for the construction of solar water heating systems and solar street lighting systems (Government of Uttar Pradesh, Department of Environment 2014, 70). With budgetary allocation of Rs. 25 crores for grid-connected rooftop solar power plants and Rs. 30 crores for the Pt. Deen Dayal Upadhyay Solar Light Scheme, the state aims to produce 10,700 MW of power utilizing solar energy (PRS Legislative Research 2024).

By the year 2030, the Indian government has committed to establishing a power generation capacity comprised of energy resources that do not rely on fossil fuels. This capacity will constitute 50% of the total capacity. The state government of Uttar

Pradesh recognizes the adverse effects of climate change, both now and in the future, and is committed to promoting the development of renewable energy sources. In line with this, the state has launched the Uttar Pradesh Solar Energy Policy 2022.

Uttar Pradesh is in possession of a significant quantity of solar energy resources and has the capability to generate a tremendous amount of solar power. It is the goal of the state to establish a holistic ecosystem in order to make full use of the solar energy potential that the region possesses. In order to provide assistance for India's planned ambition to increase its capacity for solar energy, the state plans to speed up the installation of solar power. This goal will be accomplished by means of distributed solar systems, large-scale solar parks and projects, the building of enormous solar parks, and rooftop solar photovoltaic projects. It will be very beneficial to the accomplishment of these goals to set up innovative systems that will support clients, businesses, and innovators working in the industry. With this, achieving these goals will be a lot easier. Furthermore, promoted by the strategy would be the construction of energy storage facilities and the integration of solar projects with other sources of production. The policy would also promote energy storage. This will result in the assurance of an uninterrupted electricity supply and a long-term improvement of grid stability (Government of Uttar Pradesh, Uttar Pradesh New & Renewable Energy Development Agency 2022, 3).

The vision and objectives of the Uttar Pradesh Solar Energy Policy 2022, are as follows (Government of Uttar Pradesh, Uttar Pradesh New & Renewable Energy Development Agency 2022, 4):

- The goal is to guarantee that the residents of Uttar Pradesh have access to reliable, ecologically friendly solar energy.
- To achieve the goals of lowering reliance on fossil fuels and ensuring energy security in the state, an "optimal energy mix" made up of both traditional and renewable energy sources needs to be achieved.
- To foster an environment that will stimulate private sector investment in the solar energy generation and storage technology industry.
- To promote involvement from the business sector and create opportunities for investment in the solar energy sector.
- Human resource development aims at improving current abilities and opening new career opportunities in the renewable energy sector.
- To educate the general public on the various technologies that are utilized for solar energy.

The policy will remain in effect for five years or until the government declares a new policy, whichever occurs earlier (Government of Uttar Pradesh, Uttar Pradesh New & Renewable Energy Development Agency 2022, 4) and aspires to take the following initiatives (Government of Uttar Pradesh 2024):

- This initiative aims to decrease reliance on non-renewable energy sources, such as fossil fuels, while promoting the use of sustainable and renewable energy alternatives. The major purpose of the Uttar Pradesh Solar Energy

Policy 2022 is to guarantee that citizens of the state of Uttar Pradesh have access to power that is both inexpensive and reliable.

- A target of 22,000 MW of solar power projects is established by the policy for the period of 2026–2027.
- Solar power projects will be established in Uttar Pradesh for the purpose of self-use as well as for the purpose of selling power to organizations other than Uttar Pradesh Power Corporation Ltd., not only within the state but also outside its borders.
- Creation of solar parks as a solution.
- The Bundelkhand Region of the state is now in the process of establishing a green energy corridor, which will allow for the transmission of power generated by solar systems that have a capacity of 4000 MW. The corridor has the potential to be enlarged in order to handle more capacity, provided that private investment is made.
- Stamp duty is completely waived for land used specifically for the construction of a solar power plant or solar park. Additionally, there is a complete exemption from electricity duty for a period of ten years with this exemption.
- Over the course of the next five years, a total of 30,000 young people will be instructed as Surya Mitra at the several training facilities that are owned and operated by UPNEDA.
- In order to encourage the installation of utility-scale solar systems that are accompanied by battery storage systems that can store energy for four hours, the state is providing subsidies of Rs 2.50 Cr/MW.
- Increased emphasis will be placed on residential solar rooftop installations. It is estimated that during the policy period, about 1.35 million residential properties will have solar rooftop power plants constructed.
- The deployment of solar energy in the city will be modeled after Ayodhya, which will serve as a prototype for the project. There will be 16 Nagar Nigam cities that will be developed into solar cities.
- Solarizing agricultural feeders that are separated from one another and connecting private tube wells to the grid will be done as an example of solarization.
- Solar rooftop plants will be installed on many buildings in the state, including government offices, government public sector undertaking buildings, schools, government colleges, technical institutions, and universities progressively.

The policy aims to achieve a target of 22,000 MW solar power projects by 2026–2027 in Uttar Pradesh (see Table 4.1) (Government of Uttar Pradesh 2024).

The SAPCC has shown integration with other departments through the UPNEDA, working in line with the state's solar policy of 2018 with defined targets and budgetary allocation. It was the first state to have a Climate Change Authority under the chairmanship of the Chief Minister for the implementation of its objectives under the SAPCC with an initial corpus of 20 crores (Shukla 2017). However, it lacks a

TABLE 4.1
Targets for Uttar Pradesh Solar Energy Policy 2022

Particulars	Capacity
Utility scale solar projects/parks	14000 megawatts
Solar rooftop projects (residential)	4500 megawatts
Solar rooftop projects (non-residential institutions)	1500 megawatts
Distributed solar generation (PM KUSUM component C1 and C2)	2000 megawatts
Employment generation/skill development	30,000 people

Source: UP Solar Energy Policy (2022, p. 10).

decentralized governance mechanism regarding solar energy as there are no specific plans for various districts and cities where emission rates are high. Providing incentives and budgetary allocations is not sufficient enough to solve the energy crisis and climate change issues faced by the state unless certain aspects related to solar energy use are made mandatory by virtue of bylaws, rules, and regulations.

4.4.2 Kerala

Kerala, an Indian coastal state, ranks as the 13th most populous state in India. Kerala, with a population of 33.3 million and a population density of 859 individuals per square kilometer, has come up with a new state action plan for 2023–2030 (The Directorate of Environment and Climate Change, Department of Environment, Government of Kerala 2022, 7). The Department of Environment has thoroughly reviewed the SAPCC via extensive discussions and comprehensive evaluations, in accordance with the guidelines provided by the Ministry of Environment, Forest and Climate Change (MoEFCC), in alignment with the NDCs and SDGs (The Directorate of Environment and Climate Change, Department of Environment, Government of Kerala 2022, ix).

Having 3.01% of India's population (The Directorate of Environment and Climate Change, Department of Environment, Government of Kerala 2022, ix) indicates the substantial energy requirements for the inhabitants of the state, especially considering that the economy heavily depends on climate-sensitive industries such as agriculture, fisheries, and forestry. The state is often exposed to climate-induced floods and droughts. The impact of climate change is clearly evident in the statistics provided by the Indian Meteorological Department. The average maximum temperature in Kerala has increased by around 0.8 degrees Celsius, while the lowest temperature has climbed by about 0.2 degrees Celsius. Additionally, the average temperature has gone up by 0.6 degrees Celsius, resulting in a range of 27.3 to 27.9 degrees Celsius (The Directorate of Environment and Climate Change, Department of Environment, Government of Kerala 2022, 22). This rise in temperature and the effects of climate change has led to higher rates of evaporation and transpiration, shifting patterns of the monsoon, soil erosion, groundwater depletion, shrinking of water bodies, rise in droughts, floods, landslides, and coastal erosion (The Directorate of Environment and Climate Change, Department of Environment, Government of Kerala 2022, 23).

The new SAPCC 2023–2030 has implemented the adoption of a risk framework from the Intergovernmental Panel on Climate Change and monitoring mechanisms (The Directorate of Environment and Climate Change, Department of Environment, Government of Kerala 2022, 4). The effects of climate change can be controlled to some extent if the state action plan works in sound consonance with the NAPCC. As per the projected climate scenario, the number of rainy days is expected to decline, the temperature is expected to rise by 2°C by 2050 in the state (The Directorate of Environment and Climate Change, Department of Environment, Government of Kerala 2022, 38), and the sea level is likely to rise by 100 to 200 mm over the next 100 years if the state fails to meet its stated goals under its state action plan (The Directorate of Environment and Climate Change, Department of Environment, Government of Kerala 2022, 39).

To overcome such a situation, there is an urgent need to shift toward alternative sources of energy to decrease GHG emissions. By 2030, cumulative solar projects would reach 3 GW (The Directorate of Environment and Climate Change, Department of Environment, Government of Kerala 2022, 83). There is a geographical advantage associated with the state as it is near the equator and has vast potential for power generation through solar energy. Vacant lands can be used effectively for the same purpose. The state has been proactive in deciphering the coming situation and became the first state of India to implement a comprehensive renewable energy policy in 2002. It was also the first state to implement the Bachat Lamp Yojana in the entire state, where 1,27,93,540 CFLs were distributed to domestic consumers, resulting in a reduction of 300 MW of electricity during peak hours (Government of India, Ministry of Power, Bureau of Energy Efficiency 2019, 74).

Efforts to mitigate the effects of climate change may be broken down into two primary categories: reducing the emission of GHGs and increasing the amount of carbon that is stored in natural systems. The state of Kerala is the fifth least contributor to the emission of greenhouse gases among all of India's territories. The government's efforts to promote energy conservation and enhance energy efficiency, as well as the huge forest cover that the country possesses, are primarily responsible for the relatively low amount of emissions that are produced across the country. Over the course of the last 20 years, there has been a significant increase in the amount of emissions that originate from the energy sectors themselves. These sectors include power generation, transportation, industry, agriculture (energy use), and structures. Each of these industries is responsible for 80% of the total emissions that are produced in the state. The transportation sector was responsible for the greatest amount of GHG emissions. This sector was responsible for 54% of the total emissions in the energy sector. Twenty-one percent of greenhouse gas emissions were attributed to the building industry, making it the second largest contributor (The Directorate of Environment and Climate Change, Department of Environment, Government of Kerala 2022, xvi).

According to projections made by the Central Energy Authority, the demand for energy in Kerala is anticipated to increase by 50% percent by the year 2030 compared to the levels that were observed in 2015. To forestall a potential increase in greenhouse gas emissions due to increased energy consumption, the action plan specifies conservative emission reduction goals. Because of Kerala's significant reliance on energy imports from other states, those other states may also experience a reduction in their production of emissions (The Directorate of Environment and Climate Change, Department of Environment, Government of Kerala 2022, xvi).

The suggested mitigation measures to reduce negative impacts in 2030 include: augmenting the installed capacity of renewable energy to 3.46 GW, minimizing transmission and distribution loss to 8.8%, equipping 53% of residential lighting points with energy-efficient lighting, encouraging a higher adoption of electric vehicles in public and intermediate transport, expanding the utilization of solar-based and energy-efficient agricultural supports, and improving the energy efficiency of non-PAT (Perform, Achieve, and Trade) industries and other enterprises (The Directorate of Environment and Climate Change, Department of Environment, Government of Kerala 2022, xvi).

The Chief Minister's Governing Council on Climate Change, which is backed by the State Level Steering Committee, serves as the highest-level mechanism. The State Climate Change Cell and the Working Groups will persist in offering assistance to the State Level Steering Committee, which will remain the supreme governing body. In addition, it is recommended that a prominent advisory board be established, which would include members from the State Planning Board, academic and research institutes that are highly regarded, and experts working in a variety of fields at both the national and worldwide scopes. To ensure that any required modifications are made for continued implementation, it is vital to regularly convene the Strategic Leadership and Strategy Committee, in addition to the independent advisory committee, to concentrate on monitoring and assessing various strategies (The Directorate of Environment and Climate Change, Department of Environment, Government of Kerala 2022, xviii).

For the purpose of financing the activities that are scheduled to be carried out in accordance with the SAPCC, it is essential to have an ideal combination of money coming from the central government, state governments, and other sources. Nevertheless, it is the responsibility of the state government to allocate the resources that are required to carry out the SAPCC within the constraints of the available financial space. It is anticipated that the central government will allocate more funds to the state in order to provide assistance to the state in meeting the increased costs that are associated with mitigation and adaptation activities.

In 2018–2019, Kerala implemented several changes in the electricity industry. The total installed capacity was 2995.88 MW, with the state sector contributing 2237.60 MW (74.69%), the central sector contributing 359.60 MW (12.00%), and the private sector contributing 398.68 MW (13.31%). However, the danger of climate change-induced catastrophe is still in front of the state, as the projected energy consumption is likely to increase by 74% in the commercial sector and domestic consumption by 60% in the state by 2026–2027 (Kerenvis 2024).

The Solar Energy Policy 2013 of the state of Kerala aims to increase the installed capacity to 2500 MW by 2023. The total installed capacity of solar in March 2019 was around 134 MW. To achieve the target, the Kerala State Regulatory Commission (KSERC) issued the Kerala State Electricity Regulatory Commission (Renewable Energy and Net Metering) Regulations 2020. The state has realized that in order to reduce its carbon emissions, solar is the best way to go about it, as can be seen with the comprehensive initiatives taken by the state to upgrade its solar capacity. A harmonious balance between state objectives in its action plan of 2023–2030 and the NAPCC needs to be implemented, and the rich solar potential of the state must be utilized.

4.4.3 Assam

Every state across India has been impacted by the ill effects of climate change and is eagerly looking forward to transitioning to renewable energy, with a significant portion of solar energy in its energy mix. Similarly, Assam's State Action Plan of 2015 acknowledges that climate change might present issues that may affect its developmental goals. In its Green Budget, the state has acknowledged that the Assam State Action Plan on Climate Change 2.0 (2021–2030) is under review by the union government (Government of Assam 2023, 2), wherein the state has tried to address the problem of climate change in line with India's international commitments. Assam is located in the northeastern Himalayas with the Brahmaputra flowing through it, and it is faced with recurrent floods (Government of Assam, Department of Environment and Forest 2015, 40). Increased warming of temperatures has changed precipitation patterns in the state, impacting its unique biodiversity, water resources, agriculture, and livelihoods. By mid-century, the temperature is expected to rise by 1.7–2.0 degrees Celsius tin the state (Government of Assam, Department of Environment and Forest 2015, 51).

Extreme rainfall events are also projected to increase from 5% to 38%, drought weeks may increase by 75%, and floods by 25% in southern parts of Assam (Government of Assam, Department of Environment and Forest 2015, 51). The vulnerability projected by the state action plan clearly indicates that if the effects of climate change are not controlled, they will impact the infrastructure and economy of the state to a great extent. Road transport and inland waterways transport are the primary contributors to GHG emissions in the state. These emissions are expected to increase due to population growth and a higher number of personal automobiles per person (Government of Assam, Department of Environment and Forest 2015, 96–98). Failure to implement proactive steps to mitigate emissions would result in a projected increase in the economic impact of climate change on infrastructure, estimated to range between 0.3% and 0.9% of GDP by 2050 (Government of Assam, Department of Environment and Forest 2015, 55).

The state action plan is aligned with the national action plan and prioritizes the identification of adaptation strategies to combat the effects of climate change in the state and maintain its developmental goals and economic equilibrium. Regarding energy efficiency, the focus is more on a shift toward renewable forms of energy, technological innovations in the energy sector, and energy conservation. The major sources of power generation in the state are thermal (75%), hydro (19%), and biomass and solar

(6%) (Government of Assam, Department of Environment and Forest 2015, 103–107). As a flood-prone state, Assam is also concerned about its degrading soil health and would consider solar electrical pump-operated drip/sprinkler irrigation of harvested rainwater as a last resort (Government of Assam, Department of Environment and Forest 2015, 65). There are various initiatives taken in the state for the promotion of solar power to replace the conventional sources for energy generation under the solar mission, such as 2000 solar pumping systems, solar lighting, and ten solar-powered passenger vehicles (Government of Assam, Department of Environment and Forest 2015, 9). The state also has the State Renewable Energy Policy, which aims to meet the energy requirements of the state through alternate sources of energy.

In compliance with the provisions of section 108 of the Electricity Act 2003, the state of Assam brought into effect its Renewable Energy Policy on October 13, 2022 (Government of Assam 2022, 1). In the event that it is not superseded or replaced by another policy, the policy will continue to be in force until the March 31, 2027 (Government of Assam 2022, 6). Through the implementation of programs at both the central and state levels, the plan aims to significantly increase the capacity of the state to generate renewable energy, with a special emphasis on solar energy. Increasing the overall capacity by 1200 MW is one of the goals of the Assam Renewable Energy Policy 2022 strategy. This comprises 620 MW from solar projects that are connected to the grid, 300 MW from rooftop solar projects that are connected to the grid, 80 MW from off-grid solar projects, and 200 MW from a variety of other sources including small hydro, pump storage, biomass, and solid waste. In addition to the Mukhya Mantri Souro Shakti Prokolpo project, which is scheduled to be unveiled in 2022–2023, this objective is intended to expand the capacity of renewable energy sources to a total of 1000 MW (Government of Assam 2022, 7–8).

More recently, the government of Assam has been making concerted efforts to increase its capacity to generate energy from renewable sources. The most recent development is the state budget for the fiscal year 2024–2025, which proposes the mandatory installation of rooftop solar systems in all newly constructed buildings, whether they are government or private. In addition to that, it allots 100 million rupees to the process of performing energy audits of government buildings. The Green Growth for a Greener Assam project that is included in the budget gives electric automobiles and renewable energy the highest priority (Das 2024).

In accordance with the Assam Renewable Energy Policy 2022, the state intends to realize its goal of increasing its installed capacity for renewable energy to 1200 MW by the year 2027. This purpose is driven by the state's commitment to the Sustainable Development Goals as well as the necessity of preserving energy security. Further, as part of the Mukhya Mantri Sauro Shakti Prokolpo plan, the state government is in the process of putting into action a solar project with a capacity of 1000 MW (International Forum for Environment, Sustainability and Technology 2024).

While the anticipated development is certainly worthy of commendation, the state of Assam must significantly accelerate the rate at which it increases its capacity for renewable energy. The Ministry of Power has released a revised renewable purchase obligation (RPO) that demands a gradual rise in the requirement, beginning with 29.91% in 2024–2025 and increasing to 43.3% in 2029–2030 when it is fully implemented. This upgraded RPO also contains specific requirements for increased capacity for renewable energy sources such as wind, hydro, and distributed renewable energy (Government of India, Ministry of Power 2022).

The Assam Renewable Energy Policy 2022 was an effort developed at the right moment with the intention of reducing carbon emissions in the energy sector. Nevertheless, the majority of investments are still driven by state and central government bodies, with a particular concentration on big solar ventures. The technological improvement of smaller solar systems and other types of renewable energy is now moving at a snail's pace, and the corporate sector is not showing much excitement for the subject (Das 2024).

The central government is also supporting the state to augment its solar power by way of technical and financial assistance in areas such as (Government of Assam, Department of Environment and Forest 2015, 105–107):

a. Installation of a grid interactive rooftop/ground-mounted solar power plant (with battery backup) at important government buildings in Assam (90% MNRE [Ministry of New and Renewable Energy] and GOI [Government of India]; 10% GOA [Government of Assam]) (Government of Assam, Department of Environment and Forest 2015, 40).
b. Installation and promotion of 150 solar water pumping systems for irrigation purposes (30% MNRE and GOI; 70% GOA).
c. Electrification of 75 remote villages through solar photovoltaic-based power plants under decentralized distribution generation (90% GOI; 10% GOA).

However, the state government is showing incredible interest in a shift toward solar energy generation with 100% investment by the government of Assam in areas such as the installation and promotion of 25 MW aggregate capacity of grid-connected solar power plants (Government of Assam, Department of Environment and Forest 2015, 105–107) at different vacant lands and illumination of important towns/historical locations of Assam through a solar street lighting system. State governments and union territories across India have realized the potential of solar energy and are inclined toward the generation of energy by harnessing solar energy through their action plans on climate change in line with the national action plan. The SAPCC on climate is relatively new and tried to address various loopholes prevalent in the action plans of other states. However, it lacks an implementation mechanism for its desired objectives laid down under the plan and requires robust governance and decentralization to make it a success.

4.5 HARNESSING ARTIFICIAL INTELLIGENCE, MACHINE LEARNING, DATA ANALYTICS, AND AUTOMATION IN BUSINESS MANAGEMENT FOR SOLAR ENERGY TRANSITION IN INDIA

The chapter has highlighted that solar energy has played an important role in the states of Uttar Pradesh, Kerala, and Assam in aiming to achieve their emission reduction goals as outlined in their SAPCCs. It is also seen that as one traces the progress made, several obstacles and challenges in the implementation of these policies also emerge. There have been discussions on how to integrate technological advancements, particularly machine learning (ML), artificial intelligence (AI), data analytics, and automation, among others, into the field of business management whereby these could drive India's sustainable energy transition in the solar energy sector.

As discussed, India is one of the largest emitters of carbon, and it is also vulnerable to climate change. This highlights the pressing need to find means to reduce carbon emissions and move toward renewable sources of energy through efficient use of emerging technologies. ML, AI, data analytics, and automation would be helpful for businesses, regulators, and policymakers in developing a nuanced understanding of the requirement of energy, optimization of energy generation as well as improvement in operational efficiency. This would certainly enable public and private businesses to make informed decisions. Advanced predictive analysis could be one of the primary ways ML and AI can contribute to the renewable energy sector. ML algorithms could be used to examine huge amounts of data pertaining to climate, weather patterns, and energy consumption, among others, to boost the production of solar energy and make solar power facilities efficient. Tools that use AI could predict the production of solar energy based on real-time weather data. This would in turn ensure that energy companies respond to fluctuations in solar output. Such informed forecasts become useful in integrating solar energy into the national grid in a stable and dependable manner. Further, data analytics does play an important role in tracing the performance of solar power projects and recognizing important aspects for energy optimization and cost reduction. Such automated data collection and its analytics would help in generating detailed reports that allow careful monitoring of the progress of state-level solar initiatives, likc the ones mentioned in the SAPCCs for Uttar Pradesh, Kerala, and Assam. These learnings would be helpful for business leaders and policymakers in making relevant refinements to the project plans, supervising resources economically, and meeting climate goals in line with the NAPCC. Additionally, business management could also could be assisted by AI and automation in the areas of supply chain and resource management in the solar energy sector. If companies employ automation in different segments like manufacturing, installation, and maintenance, they can cut costs and errors, allowing for making solar energy projects more feasible and scalable. An illustration of this could be automatic drone investigations that allow for monitoring large solar farms, detecting potential issues, and reducing the time spent on maintenance. Automation, therefore, allows for cost-saving and rapid expansion of solar energy projects in different areas. Policy formulation and implementation along with the social impact of using solar energy could be studied through the use of ML and AI-driven data analytics. The results of data analytics could help in revealing demographic patterns in energy access and consumption, aiding businesses in ensuring equitable distribution of renewable energy and contributing to social welfare aims alongside environmental goals.

Therefore, while the SAPCC and NAPCC could function as important plans, the use of technological advancements in the solar energy sector, its policy planning, and business management practices could significantly enhance India's transition to solar energy, particularly in states such as Uttar Pradesh, Kerala, and Assam. This is in line with the business and policy objectives under the sustainable development goals, designing a data-informed route toward a low-carbon future.

4.6 CONCLUDING REMARKS

Solar energy has been gaining momentum in various SAPCCs in line with national as well as their respective objectives and obligations and is still in the formative stage. To bring the solar energy aspect into the main agenda of SAPCCs requires coherence, integration, and coordination with the NAPCC in accordance with the respective capacities, vulnerabilities, and capacities of states. In policy, both mitigation and adaptation strategies must be incorporated in SAPCCs in line with international objectives and commitments of India, and such integration must be reflected in letter and in action.

In most states, the SAPCC has been implemented as a standalone document. The plans lack integration with other line departments like environment, rural development, and agriculture. This integration, if brought about, can contribute immensely toward combating climate change. Mainstreaming policy objectives as per national commitments in the solar energy sector requires adequate financial flow, technical assistance, and clarity of objectives to develop smart responses to tap solar potential in the states to address climate change and also to cut emissions. As observed after the analysis of the three states, there is no single institution dealing with the deployment of solar energy except for the policies in the state of Uttar Pradesh and Kerala, and it is scattered across various ministries and departments. However, Uttar Pradesh has an agency dealing with renewable energy matters and, in the new state action plan, Kerala has also included a State Level Steering Committee as a top-level executive body. Therefore, to achieve the desired target, a single nodal agency must be established in every state for the deployment and implementation of solar energy initiatives, with their own progress indicators and bi-annual publication of achievements.

Another hurdle faced by states in the implementation of SAPCCs is the lack of adequate budgetary allocation from both state funds as well as central funds. It has been observed from the preceding analysis of the three states that the central government is trying to augment the solar capacities of states with the allocation of funds according to their vulnerabilities and capacities. However, a strategic and smart plan needs to be there for the yearly allocation of funds to states, and a proper audit report must be published annually for the utilization of funds. It is a matter of concern that, despite having funds for some solar projects, there is no clearly defined implementation mechanism in state action plans. From the analysis, it can be observed that the scope of SAPCCs is primarily restricted to respective state jurisdictions, with no clear vision for decentralized governance systems. The plans lack a climate action focus on cities and districts. Therefore, states must clearly define their objectives and budgetary allocation for every district for the augmentation of solar energy to comprehensively achieve state objectives.

For the effective implementation of solar initiatives in states, a state must make it mandatory to implement solar initiatives by virtue of bylaws, municipal rules, and regulations in line with their SAPCC. This will not only achieve the obligatory mandate to tap the solar potential of the state but will also make the implementation process more seamless, effective, and result oriented.

REFERENCES

Bosse, Jolana. 2023. "India – Climate Performance Ranking 2024 | Climate Change Performance Index." *Climate Change Performance Index (CCPI) Is a Scoring System Designed to Enhance Transparency in International Climate Politics.* https://ccpi.org/country/ind/.

Chandel, S.S., Rajnish Shrivastva, Vikrant Sharma, and P. Ramasamy. 2016. "Overview of the Initiatives in Renewable Energy Sector under the National Action Plan on Climate Change in India." *Renewable & Sustainable Energy Reviews* 54 (February): 866–873. https://doi.org/10.1016/j.rser.2015.10.057.

Chaudhary, Ankur, Chetan Krishna, and Ambuj Sagar. 2014. "Policy Making for Renewable Energy in India: Lessons from Wind and Solar Power Sectors." *Climate Policy* 15 (1): 58–87. https://doi.org/10.1080/14693062.2014.941318.

Choudhary, Kamya. 2022. "How Is India Tackling Climate Change? Grantham Research Institute on Climate Change and the Environment." *Grantham Research Institute on Climate Change and the Environment.* https://www.lse.ac.uk/granthaminstitute/explainers/how-is-india-tackling-climate-change/.

Das, Barasha. 2024. "Assam Has More Renewable Potential than the Government Estimates, Says Study." *Mongabay-India.* https://india.mongabay.com/2024/06/assam-has-more-renewable-potential-than-the-government-estimates-claims-a-study/#:~:text=The%20Ministry%20of%20New%20and,waste%2Dto%2Denergy%20sources.

Dubash, Navroz K. 2019. *India in a Warming World.* Oxford University Press. http://books.google.ie/books?id=CtjODwAAQBAJ&printsec=frontcover&dq=India+in+the+warming+world&hl=&cd=2&source=gbs_api.

Forbes India. 2024. "The Top 10 Largest Economies in the World in 2024." *Forbes India.* https://www.forbesindia.com/article/explainers/top-10-largest-economies-in-the-world/86159/1.

Gielen, Dolf, Francisco Boshell, Deger Saygin, Morgan D. Bazilian, Nicholas Wagner, and Ricardo Gorini. 2019. "The Role of Renewable Energy in the Global Energy Transformation." *Energy Strategy Reviews* 24 (April): 38–50. https://doi.org/10.1016/j.esr.2019.01.006.

Gogoi, Elizabeth. 2017. *In Depth: India's State Action Plans on Climate Change: Towards Meaningful Action.* ID17-005v1. United Kingdom: Oxford Policy Management. https://www.opml.co.uk/sites/default/files/migrated_bolt_files/id-state-action-plan-climate-india.pdf.

Government of Assam. 2022. *Assam Renewable Energy Policy, 2022.* PEL.230/2021/138. India: Government of Assam. https://cdnbbsr.s3waas.gov.in/s3716e1b8c6cd17b771da77391355749f3/uploads/2023/10/2023101780042599.pdf.

Government of Assam. 2023. *Green Budgeting in Assam for FY 2023-24.* India: Government of Assam. https://finance.assam.gov.in/sites/default/files/swf_utility_folder/departments/agriculture_com_oid_2/portlet/level_1/files/goa_green_budget_2023-24.pdf.

Government of Assam, Department of Environment and Forest. 2015. *Assam State Action Plan on Climate Change 2015-20.* Government of Assam, Department of Environment and Forest. https://moef.gov.in/wp-content/uploads/2017/08/ASSAM-SAPCC.pdf.

Government of India, Department of Science & Technology. 2024. "Climate Change Programme." Department of Science & Technology. Accessed June 19. https://dst.gov.in/climate-change-programme.

Government of India, Ministry of Environment, Forest and Climate Change. 2024. "India Achieves Two Targets of Nationally Determined Contribution Well Ahead of the Time." Accessed June 22. https://pib.gov.in/PressReleaseIframePage.aspx?PRID=1987752#:~:text=In%20August%202022%2C%20India%20updated,enhanced%20to%2050%25%20by%202030.

Government of India, Ministry of New and Renewable Energy. 2024a. "Solar Overview | Ministry of New and Renewable Energy | India." Accessed June 22. https://mnre.gov.in/solar-overview/.

Government of India, Ministry of New and Renewable Energy. 2024b. "Schemes | Ministry of New and Renewable Energy | India." *Ministry of New and Renewable Energy.* Accessed June 19. https://mnre.gov.in/solar/schemes.

Government of India, Ministry of Power. 2022. *Renewable Purchase Obligation (RPO) and Energy Storage Obligation Trajectory till 2029-30.* F.NO. 09/13/21-RCM. India: Ministry of Power, Government of India. https://powermin.gov.in/sites/default/files/Renewable_Purchase_Obligation_and_Energy_Storage_Obligation_Trajectory_till_2029_30.pdf.

Government of India, Ministry of Power. 2024a. "India Jumps 2 Spots Higher, and Now Ranks 8th as per Climate Change Performance Index (CCPI, 2023)." Accessed June 22. https://www.pib.gov.in/PressReleasePage.aspx?PRID=1878023.

Government of India, Ministry of Power. 2024b. "NMEEE | Bureau of Energy Efficiency, Government of India, Ministry of Power." *Bureau of Energy Efficiency.* Accessed June 19. https://beeindia.gov.in/en/nmeee-0.

Government of India, Ministry of Power, Bureau of Energy Efficiency. 2019. *Unlocking National Energy Efficiency Potential.* India: Government of India, Ministry of Power, Bureau of Energy Efficiency. https://beeindia.gov.in/sites/default/files/press_releases/UNNATEE%20Report.pdf.

Government of India, Ministry of Power, Central Electricity Authority. 2023. *Load Generation Balance Report 2023-24.* File No.CEA-GO-12-31/2/2022-GM Division. *Central Electricity Authority.* India: Government of India, Ministry of Power, Central Electricity Authority. https://cea.nic.in/l-g-b-r-report/?lang=en.

Government of India, Press Information Bureau. 2024. "Renewable Energy in India – Indian Power Industry Investment." *Invest India.* Accessed June 19. https://www.investindia.gov.in/sector/renewable-energy#:~:text=India%20aims%20for%20500%20GW,have%20been%20approved%20in%20India.

Government of Uttar Pradesh. 2024. "Solar Energy Policy 2022." *Invest UP.* Accessed June 19. https://invest.up.gov.in/solar-energy-policy-2022/.

Government of Uttar Pradesh, Department of Environment. 2014. *Uttar Pradesh State Action Plan on Climate Change.* CNTR NO 83181079. India: Government of Uttar Pradesh, Ministry of Environment, Forest and Climate Change. https://moef.gov.in/wp-content/uploads/2017/09/SAPCC_UP_final_version_0.pdf.

Government of Uttar Pradesh, Uttar Pradesh New & Renewable Energy Development Agency. 2022. *Uttar Pradesh Solar Energy Policy 2022. UPNEDA.* India: Government of Uttar Pradesh, Uttar Pradesh New & Renewable Energy Development Agency. https://www.upneda.org.in/MediaGallery/Uttar_Pradesh_Solar_Energy_Policy2022_English_.pdf.

International Energy Agency. 2020. "India 2020 – Analysis – IEA." *IEA.* https://www.iea.org/reports/india-2020.

International Energy Agency. 2021. "The Uttar Pradesh Solar Power Policy 2017 – Policies – IEA." *IEA.* https://www.iea.org/policies/6465-the-uttar-pradesh-solar-power-policy-2017.

International Forum for Environment, Sustainability and Technology. 2024. *Assam Renewable Energy Potential Re-Assesment: Focus of Solar, Wind and Biomass. Saurenergy.* India: International Forum for Environment, Sustainability and Technology. https://img.saurenergy.com/2024/03/assam-re-potential-assessment.pdf.

Jaganmohan, Madhumitha. 2024. "Installed Power Capacity in Uttar Pradesh India FY 2005-2022." *Statista.* https://www.statista.com/statistics/1077923/india-uttar-pradesh-installed-power-capacity/.

Jogesh, Anu, and Navroz K. Dubash. 2015. "State-Led Experimentation or Centrally-Motivated Replication? A Study of State Action Plans on Climate Change in India." *Journal of Integrative Environmental Sciences* 12 (4): 247–266. https://doi.org/10.1080/1943815x.2015.1077869.

Kerenvis. 2024. "Economy: Energy– Status of Environment Related Issues: Kerala ENVIS Centre, Ministry of Environment and Forests, Govt. of India." Kerala State Council for Science, Technology and Environment. Accessed June 19. http://www.kerenvis.nic.in/Database/ENERGY_811.aspx.

Kumar. J, Charles Rajesh, and M. A. Majid. 2020. "Renewable Energy for Sustainable Development in India: Current Status, Future Prospects, Challenges, Employment, and Investment Opportunities." *Energy, Sustainability and Society* 10 (1). https://doi.org/10.1186/s13705-019-0232-1.

O'Neill, Aaron. 2024. "Total Population of India 2029." *Statista.* https://www.statista.com/statistics/263766/total-population-of-india.

Patra, Jyotiraj. 2016. *Review of Current and Planned Adaptation Action in India.* CARIAA Working Paper #10. Canada: International Development Research Centre. https://www.iisd.org/system/files/publications/idl-55866-india.pdf.

Prasad, Sangeeta. 2020. "Why the World Should Be Watching India's Fast-Growing Cities." *World Economic Forum.* https://www.weforum.org/agenda/2019/01/india-urbanization-why-the-world-should-watch/.

Prime Minister's Council on Climate Change. 2008. *National Action Plan on Climate Change. Ministry of Environment, Forest and Climate Change.* India: Prime Minister's Council on Climate Change. https://moef.gov.in/wp-content/uploads/2018/04/Pg0152.pdf.

PRS Legislative Research. 2024. "Uttar Pradesh Budget Analysis 2018-19." *PRS Legislative Research.* Accessed June 19. https://prsindia.org/budgets/states/uttar-pradesh-budget-analysis-2018-19.

Rajendran, K. 2022. "This Indian Solar Energy Scheme Will Turn Homes into Power Stations." *World Economic Forum.* https://www.weforum.org/agenda/2021/03/india-kerala-benefit-new-solar-power-scheme/.

Sengupta, Sandeep. 2019. "India's Engagement in Global Climate Negotiations from Rio to Paris." In *Navroz* India in a Warming World: Integrating Climate Change and Development, edited by K. Dubash, 114–141. Oxford University Press, Delhi, 2019. https://doi.org/10.1093/oso/9780199498734.003.0007.

Sharma, Prakash. 2018. "Climate Change, Technology Transfer and Access to Clean Energy: The Role of Intellectual Property Rights in the Transfer of Environmentally Sustainable Technologies." English. In *Socio-Legal Dimensions of Climate Change*, edited by T.V. Subba Rao, V.S. Mallar, and S. Anuja, 1st ed., 1:51–71. India: ENVIS Centre on Environmental Law and Policy, National Law School of India University.

Sharma, Prakash, and R. Venkata Rao. 2021. "Linking Climate Change and Sustainable Development Goals: India's Responsive Efforts." *Indian Bar Review* 48 (3 & 4): 22–24. https://papers.ssrn.com/sol3/papers.cfm?abstract_id=4046365.

Shinn, Lora. 2024. "Renewable Energy: The Clean Facts." *NRDC.* https://www.nrdc.org/stories/renewable-energy-clean-facts.

Shukla, Neha. 2017. "Uttar Pradesh First State to Have Climate Change Authority in Country." *The Times of India*, October 10. https://timesofindia.indiatimes.com/city/lucknow/uttar-pradesh-first-to-have-climate-change-authority/articleshow/61023177.cms.

The Directorate of Environment and Climate Change, Department of Environment, Government of Kerala. 2022. *Kerala State Action Plan on Climate Change 2023-2030*. India: The Directorate of Environment and Climate Change, Department of Environment, Government of Kerala. https://envt.kerala.gov.in/wp-content/uploads/2022/12/Kerala-State-Action-Plan-on-Climate-Change-2.0.pdf.

UNDP. 2024. "India Case Study: Developing the State Level Action Plan on Climate Change for Madhya Pradesh | Climate Change Adaptation." Accessed June 19. https://www.adaptation-undp.org/resources/case-study/india-case-study-sapcc.

UNFCCC. 2015. *India's Intended Nationally Determined Contribution: Working Towards Climate Justice*. United Nations Framework Convention on Climate Change. https://unfccc.int/sites/default/files/NDC/2022-06/INDIA%20INDC%20TO%20UNFCCC.pdf.

5 Exploring the Intersection

Technological Innovations and Green Management in Enhancing Sustainable Organizational Performance

Shivakami Rajan and L R Niranjan

5.1 INTRODUCTION

In today's competitive world, companies' pursuit of profit has no limits in the way they operate. Operational benefits are the sole objective and sometimes serve as a competitive advantage (Hsu et al. 2015). Companies create current value, but it ends up creating disvalue when the current value harms ecological balances and externalities (Beckmann and Schaltegger, 2021). Manufacturing companies increasingly take advantage of artificial intelligence (AI), machine learning (ML), and automation, finding it necessary to balance environmental preservation and sustainable manufacturing (Ghobakhloo et al.,2021). These technologies aid them in effectively optimizing their consumption of resources, minimizing the ecological footprint, and offering sustainable products (Cioffi et al., 2020). The global concerns of global warming, emissions, and environmental degradation are often in the background of organizations' strategic planning and implementation. But many times, there exist trade-offs in many areas to keep up the operational benefits (Wang and Juo 2021). In the past couple of decades, the world has been under severe pressure from environmental degradation and social tensions as a result of it. Political and geopolitical pressures led to the creation in 2015 of the United Nations Sustainable Development Goals (SDGs) for 2030. The SDGs give business organizations 17 main themes, 169 specific areas, and thousands of indicators as opportunities to contribute toward society, ecology, and the environment. SDGs are thus a business organization's "crystal ball" (Pedersen 2018) to look into their future and strategize their activities to achieve their desired goals without compromising their economic profits. Banking and financial institutions and services firms are governed by the Equator Principles, serving as an environmental and social risk benchmark for loans, lending, and other project management collaboration. Financial institutions, in addition to adhering to

 DOI: 10.1201/9781003626527-5

the Equator Principles, are aligned with the UN's SDGs and to the global goal of net zero by 2050 (Equator Principles, 2024).

Today, investors and financial institutions value organizations that not only exhibit social and environmental behavior but also give importance to ethical and socially responsible values in their operations. With the increased call from stakeholders to reduce, reuse, and recycle resources (Thanki et al., 2016), organizations have been impelled to comply with restrictions and regulations of environmental requirements. Furthermore, the growth of social media and globalization, coupled with increased adoption of Industry 4.0 practices, has propelled organizations to adopt sustainable practices. Firms can adopt innovative business models that are ethically sound and economically viable with growing investors' expectations and the demands of financial institutions (Moghrabi et al., 2023). These practices are essential to preserve their competitive stance in domestic and international markets as increased public awareness of corporate social responsibility (CSR) and environmental concerns demands transparent operations (Smith and Lee, 2021). This has led organizations to "go green"—exploring options that can reduce their negative impact on the environment and society through responsible behavior. Greening all business functions—internally and externally—has become a necessarily important strategic initiative to bring about green development. Raut et al. (2019) states that green management can be a proactive strategy that can reduce an organization's environmental effects on its supply chain, while simultaneously enhancing its economic efficiency. Loiseau et al. (2016) pinpoints that a clear pathway to sustainability is to focus on green-economy-related strategies. Additionally, Sroufe (2018) contends that every organization's top agenda should be sustainability and that integration is the key theme—an organization should integrate all its business functions, behavior, strategy formulation, information, and environmental issues. Further, Sroufe emphasizes integrated management that provides organizations opportunities toward environmental, social, and governance (ESG) performances. As governments are providing various financial incentives for organizations that adopt green initiatives, there has been a growing emphasis on integrating sustainability measures in management practices and corporate philosophy across various industries (Jamwal et al., 2021). These organizations are increasingly implementing green initiatives in their operational frameworks. Efficient and effective green management strategies help organizations reduce expenses and cut costs (Yu and Huo, 2021). The most effective long-term strategy to achieve sustainable cost reduction is by adopting the elimination and minimization of waste and wasteful processes, monitoring resource consumption, and energy-efficient operations and practices. Sustainability not only offers environment advantages but also increases an organization's reputation and fulfills customer expectations. It further increases customer loyalty and customer satisfaction. Galbreath (2019) pinpoints that an organization can achieve sustainable development through the effective and efficient management of its strategic internal and external resources.

Wernerfelt, Prahalad, Hamel, and Barney are the key proponents of the resource-based view (RBV) theory that emerged during the 1980s and 1990s for achieving competitive advantage by organizations. The organizational resources—internal and external, including tangible and intangible resources (Wernerfelt, 1984)—when used effectively, create a sustainable competitive advantage. Organizational resources are of paramount and key importance in the pursuit of enhanced performance. They can attain competitive advantage by leveraging their distinct, varied, and unique resources in diverse strategies, enabling them to outperform their competitors. The resources and capabilities of an organization significantly impact its performance and long-term success (Lusthaus, 1999). In addition, Lusthaus (1999) stress that the foundation of organizational performance fundamentally rests upon its capacity in the areas of strategic leadership, human resources, finance, infrastructure, process and project management, and its various inter- and intrainstitutional linkages. It is an established fact that environmental issues—such as degradation, pollution, and resource exploitation—are precipitated by and serve as a consequence of human actions (Anwar et al., 2020). In this context, efficient deployment of resources is crucial for success. In the perspective of sustainability, green management should be construed as one vital resource that organizations require to improve their performance. A significant positive connection between ecological practice and organizational performance was identified. Quality green management initiatives are necessary for effective business solutions. The critical focus on sustainability through green management of organizational resources—tangible and intangible—has been explored by numerous firms (Ansari and Kant, 2017).

5.2 LITERATURE REVIEW

Integrating environmental aspects into business strategies has gained traction in the recent past. An attempt has been made to review literature and explore the current status of research pertaining to green business strategies, their implementation, and impact on various aspects of organizational performance. Sustainable manufacturing, corporate governance, supply chain management, and the role of innovation in promoting environmental performance are some of the themes covered in this work.

5.2.1 Sustainable Operations and Manufacturing

Researchers have shown interest in understanding the implementation of green practices in manufacturing and operations. The integration of green, lean, and six sigma operations has been used to promote sustainable manufacturing, highlighting synergies among these approaches. Generative AI, coupled with ML and large language models, enables cloud computing technology applications in manufacturing, resulting in minimal ecological footprints while maximizing economic and social benefits. These can be achieved by calibrating efficient model design, data management, energy-efficient hardware, and efficient life cycle management. These integrations not only lead to improved efficiency but also reduce the impact on the environment, thereby aligning with the growing demand for sustainable production methods. A model is in place that links lean and green supply chain performance, suggesting how this combination can drive innovation and improve overall supply chain sustainability.

5.2.2 Green Supply Chain Management (GSCM)

Several studies have focused on the importance of green supply chains, investigating the impact of lean, agile, and green practices on competitiveness in business. Blockchain technologies and distributed ledger applications are widely applied across the supply chain to ensure transparency and sustainability. Data sharing, collaboration, and coordination between SCM partners enable sustainable goals for all concerned. Automation through smart contracts ensures immutability and enforcement of requirements for each partner's sustainable goals. A study on using GSCM as a competitive tool in the fast-moving consumer goods (FMCG) manufacturing sector focused on the potential of GSCM implementation for cost-effective power consumption, sustainable green initiatives in manufacturing, waste management, leadership support, effective collaboration between stakeholders, compliance, and digital transformation. The convergence of these technologies with AI further provides innumerable opportunities to SCM partners for predictions, circularity, and interconnectedness.

5.2.3 Sustainability and Corporate Governance

Scholars have examined the role of corporate governance in promulgating sustainability, investigating the relationship between total quality management (TQM) and sustainability, with special reference to the public service sector. The mediating role of corporate governance in the relationship between TQM and sustainability was examined and it was found that effective governance structures are essential for translating quality management practices into sustainable outcomes. Another study indicated how strong corporate governance can drive sustainability practices, especially when mediated through effective operations management.

5.2.4 Environmental Performance and Innovation

The impact of innovation on the performance of the firm from an environmental perspective has been examined in the past. Green R&D, firm performance, and eco-innovation were investigated, revealing that investments in green R&D can lead to improved environmental performance and, in a few cases, enhanced financial performance too. Further to this finding, another study investigated the role of environmental ethics and environmental performance in driving competitive advantage. It was found that firms with strong environmental ethics are more likely to engage in eco-friendly innovations, leading to better environmental performance and resulting in a competitive advantage.

5.2.5 Green Business Strategies and Resource-Based View

The adoption and outcomes of green business strategies have been explained using a resource-based view. The internal drivers and performance consequences of green business strategies of small firms were examined and it was found that firms with unique environmental capabilities can achieve better performance by implementing green strategies. Another study applied the resource-based view and identified boundary conditions that influence the relationship between green business strategy and export performance.

5.2.6 Leveraging Digital Transformations for Green Sustainability in Organizational Performance Management

Digital transformations by integrating green technologies such as AI, ML, and cloud computing have been evolving the business landscape and play a pivotal role in enhancing business management. Green sustainability, being the focal point for firms to align with global environmental goals and compliance, necessitates integrating green technologies to enhance their sustainability efforts (Harkawat, 2024). Firms optimize operations, decision-making, and improve customer experiences by using these technologies. Digital transformation allows businesses to anticipate consumer expectations well in advance, decipher market trends, and enables them to take strategic decisions and actions alongside efficient green operations (Singh, 2023).

5.3 OBJECTIVES AND RESEARCH QUESTIONS

The purpose of this study is to analyze current trends and topics, key regions, and countries in green management and sustainability research by examining the metadata of all papers indexed in the Scopus database that deal with the topic of green management and sustainability, keeping digital transformation as the lens. This study provides a novel contribution to the review literature on the state of research on the environmental perspective of sustainable organizational performance through green management. Publication patterns, countries of influence, prominent journals, and pertinent topics of interest keywords are explored using bibliometric insights. Social and conceptual network analyses were conducted, and the findings enhanced our understanding of researchers' outputs. The following questions are put forth:

RQ1. How is the scientific growth of sustainable business performance shown through green management publications?

RQ2. What are the top countries, journals, and most frequent words and keywords co-occurrence?

5.4 RESEARCH METHODS

This review maps research studies from 2005 to 2024 at the global level on the basis of sustainable business performance from the Scopus database. The Scopus database was selected to mine the documents published in the defined areas of research. In June 2024, we collected the data from the Scopus database using keyword search queries for the following keywords: "Green Management" or "Green techniques" or "Green initiatives" as the first line of search. Then, we associated these with the keywords "Sustainability" or "Sustainable performance" or "Sustainable development"

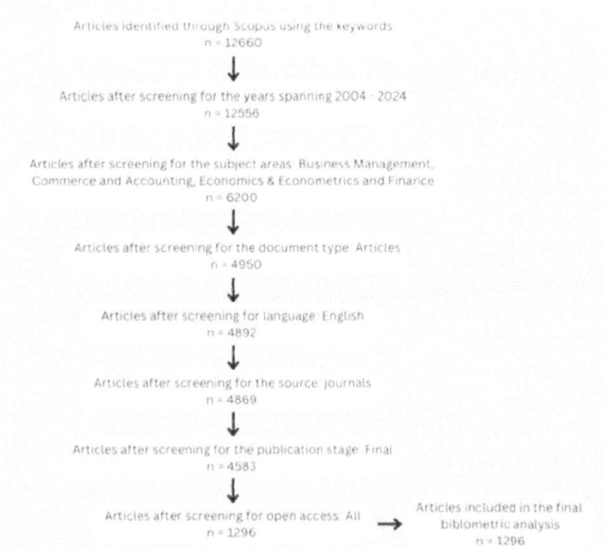

FIGURE 5.1 The search and selection criteria of articles on Scopus for bibliometric analysis.

or "Organizational performance" or "Productivity" or "Growth" in "title-abstract-keywords" in the second line of search.

The details of the document results and the subsequent narrowing using the filtering options in Scopus are shown in Figure 5.1. The inclusion and exclusion criteria resulted in 1296 documents. These 1296 documents were subjected to manual screening based on titles and abstracts to identify relevant articles to the subject areas of research. This resulted in 1107 articles, which were used for further analysis.

We then used the "bibliometrix" R package with Biblioshiny, an online application package of R language (Aria and Cuccurullo, 2017) for bibliometric analysis.

5.5 DATA SYNTHESIS, VISUALIZATIONS, AND DISCUSSIONS

The details of the research scientific production, including the number of publications over the years, are presented in Table 5.1 and Figure 5.2. The contributions in the number of documents—research articles from 2005 to 2024—show that the contributions have had steady growth since 2013. The keywords related to the topics—sustainability, green management, and organizational performance—from research articles are 3723 from 96 single-author documents and 3606 from collaborated works.

Further, there were 418 articles published up to 2019, whereas 979 articles were published from 2019 until 2024 in the area of research. The growth of articles has increased more than twofold post-2019, signaling increased awareness, participation, and involvement of all stakeholders in the topics of sustainability and green management initiatives that focus on organizational performance. Figure 5.2 incorporates the number of articles published per year, and we notice there is increased traction in the

TABLE 5.1
Details of Publications over the Years 2005 to 2024

Main Information About Data	Results
Timespan	2005–2024
Sources (journals, books, etc.)	439
Documents	1296
Annual growth rate	30.04%
Document average age	3.68
Average citations per document	45.81
References	0
Document Contents	**Results**
Keywords plus (ID)	2041
Author's keywords (DE)	3723
Authors	**Results**
Authors	3606
Authors of single-authored documents	96
Author Collaborations	**Results**
Single-authored document	102
Co-authors per document	3.34
International co-authorships	42.44%
Document Type	**Results**
Article	1296

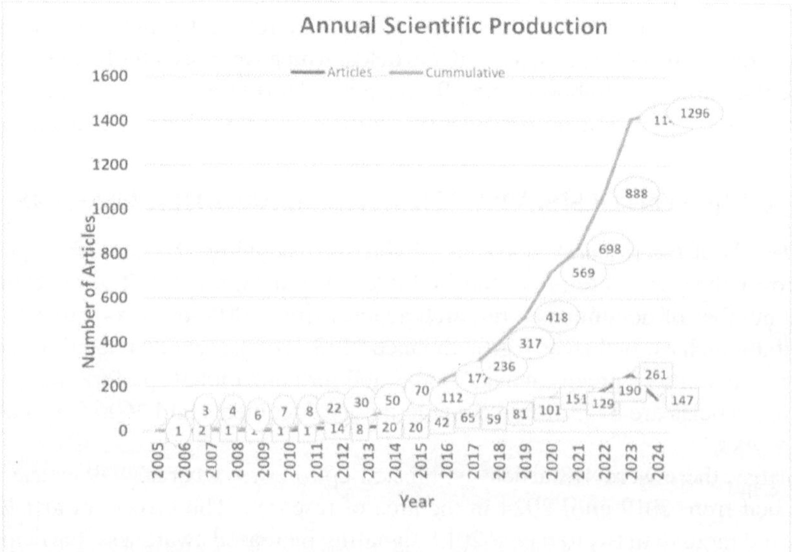

FIGURE 5.2 The annual production of documents over the years 2005 to 2024.

area of sustainability over time, which has affected the impact metrics. It is observed that the year 2023 saw the highest publication count of 261 articles, indicating that post the year of COVID-19, the study of sustainability saw a significant contribution. Growing control measures by various international bodies and governments have sparked the stakeholder's interest and further initiatives to tighten investments in projects that are environmentally viable have probably contributed to this growing interest. Fewer resource usages and production of less waste are the compelling calls for a rewarding future of economic welfare and environmental health (Thiele, 2024). The sources of document production, as seen in Figure 5.3 and Table 5.2, show the increase in document production post-2015. The journal *Business Strategy and the Environment* produced 88 documents from 2005 to 2014, followed by production from the *Journal of Cleaner* with 79 documents, followed by *Corporate Social Responsibility and Environmental Management* with 48 documents.

Bradford's law of scattering, shown in Figure 5.4, depicts the top ten journals referred in Table 5.2 as the core sources that contributed about 60% of relevant documents in the study area. The nuclear zone (Zone 1) contains a small number of 14 journals contributing 60% to the subject area, followed by Zone 2 with 91 moderately productive journals, and Zone 3 with a large number of 334 journals of low productivity. Figure 5.4 shows the core sources shaded in gray with high productivity contributions to sustainability. Notably in this core are the journals *Business Strategy and the Environment*; *Journal of Cleaner Production*; *Corporate Social Responsibility and Environmental Management*; and *Cogent Business & Management.*

The average citations per year—a common indicator of the impact of a research article published—are shown in Table 5.3. The year 2007 has the highest average citation (mean total citations per article), followed by the year 2011, with the most cited country being the United Kingdom (Figure 5.5).

The most relevant authors having at least two or more documents are shown in Table 5.4 for the study period.

The most cited author during this period is shown in Table 5.5 and Figure 5.6. The first three authors in Table 5.5 with high citations of 1474, 1294, and 1127, respectively, are authors who have reviewed previous research on the topics of CSR and its implications for the firm, governance, and democracy; ESG and financial performance; green human resource management; and sustainability-oriented innovation. The author El-Kassar was the first to explore the relationship of green innovation with its drivers, factors, and technological challenges on a firm's performance, triggering further research on green concepts.

The themes that evolved during the study period show that in the initial years, the major themes were "exploitations" and "exploration" (frequencies of 7 times) immediately following the year of the SDGs announcement 2016. The theme "sustainable operations" evolved in 2017 (with frequencies of 5), followed by the term "corporate sustainability" (with frequencies of 17) from the year 2019. In 2021, the theme of "sustainable development" exponentially developed (with frequencies of 71), and the term "sustainability" also showed high frequency (134) in the same year. The theme "circular economy" gained traction in 2021 (with frequencies of 31).

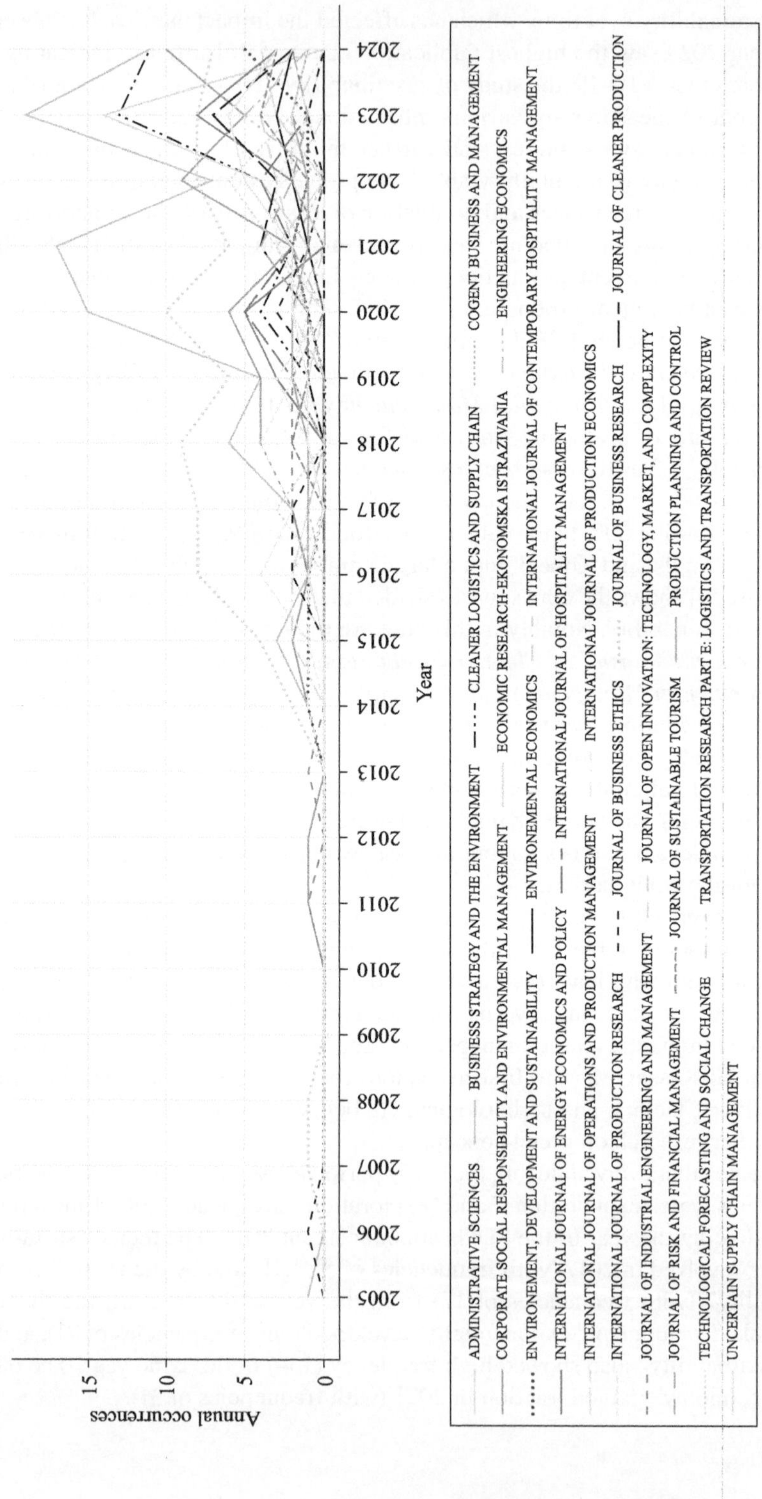

FIGURE 5.3 Top 25 sources (2005–2024).

TABLE 5.2
Journals with Production of More than Ten Documents

Journal	Number of Documents
Business Strategy and the Environment	88
Journal of Cleaner Production	79
Corporate Social Responsibility and Environmental Management	48
Cogent Business & Management	38
International Journal of Production Economics	22
International Journal of Energy Economics and Policy	21
Uncertain Supply Chain Management	21
Economic Research–Ekonomska Istrazivanja	19
Journal of Business Research	19
Technological Forecasting and Social Change	18
International Journal of Hospitality Management	17
Journal of Sustainable Tourism	14
Journal of Open Innovation: Technology, Market, and Complexity	13
Administrative Sciences	12
Journal of Industrial Engineering and Management	12
Journal of Risk and Financial Management	12
Production Planning and Control	11
International Journal of Operations and Production Management	10

Figure 5.7 shows the theme of "green" evolved in 2020 in documents stressing its usage in business functions—human resource, supply chain management, and transformational leadership—and has been frequently explored, including the current year of study (2024). The concept of "organizational performance" in terms of the environment began in 2019 (33 times) and it evolved over the years to be called "environmental performance" (2020, 80 times), "firm performance" (2022, 32 times) and "sustainable performance" (2022, 32 times).

The words in the abstracts for the study are grouped into basic, emerging/declining, motor, and niche themes based on the conjunction of two common themes or words (bigrams). Figure 5.8 shows the details of themes based on centrality and density in four groups. The themes of supply chain and its various related themes, and sustainability development have merged in the motor theme, explaining that these themes are well established and developed for further research. Meanwhile, green concepts reflecting both in the niche group and motor group, bring out that the concept of green is very specialized, as can be seen from its applications in all functional areas of study, and is of peripheral nature. Supply chain management themes are almost central to all themes.

The basic themes that transverse between different areas of research on sustainability are "environmental management," "social responsibility," "corporate social responsibility," and "financial performance." Furthermore, the terms "human

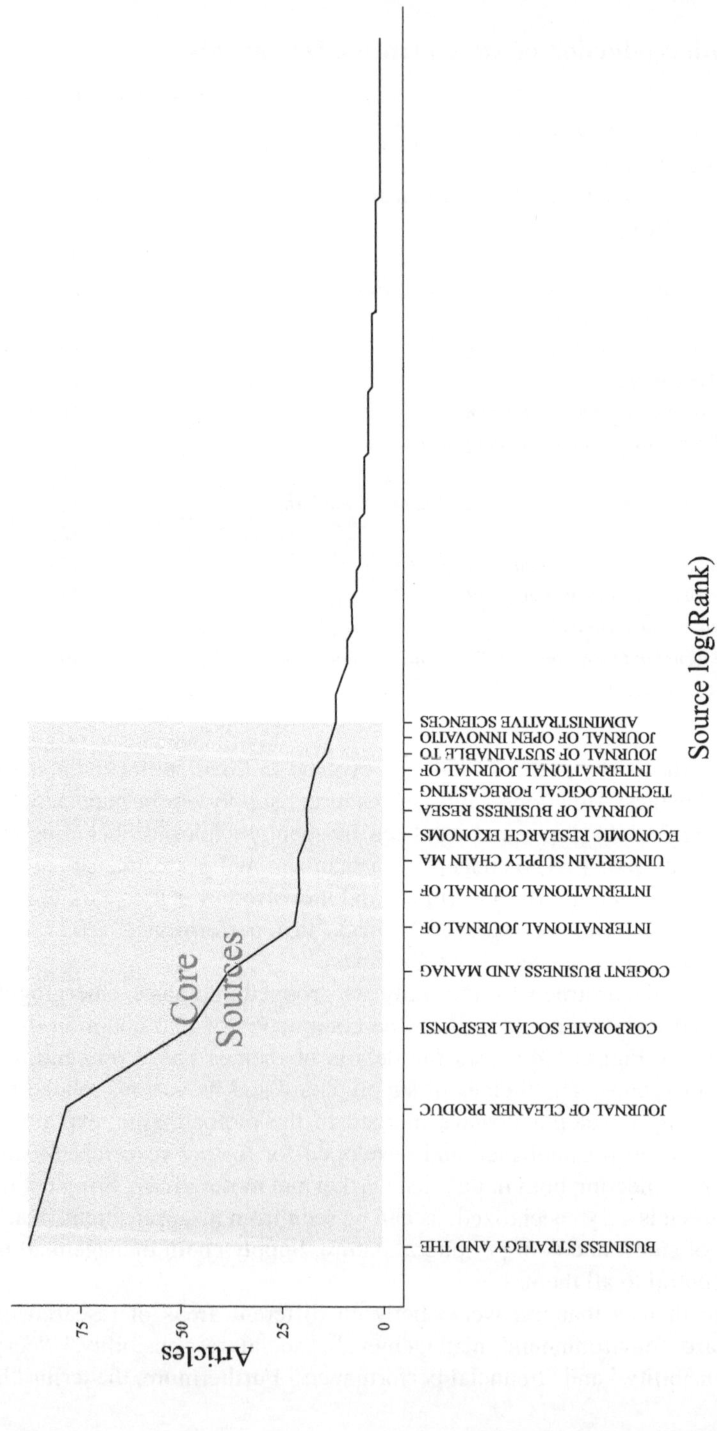

FIGURE 5.4 Core sources by Bradford's law.

TABLE 5.3
Average Citations per Year for the Study Period

Year	MeanTCperArt	MeanTCperYear	N
2005	93.00	4.65	1
2006	116.50	6.13	2
2007	380.00	21.11	1
2008	94.50	5.56	2
2009	179.00	11.19	1
2010	220.00	14.67	1
2011	257.36	18.38	14
2012	74.88	5.76	8
2013	170.20	14.18	20
2014	132.25	12.02	20
2015	134.33	13.43	42
2016	118.40	13.16	65
2017	78.88	9.86	59
2018	72.93	10.42	81
2019	72.80	12.13	101
2020	47.08	9.42	151
2021	27.33	6.83	129
2022	19.13	6.38	190
2023	7.68	3.84	261
2024	2.05	2.05	147

Note: MeanTCperArt, average total citations per article; MeantTCperYear, average total citations per year.

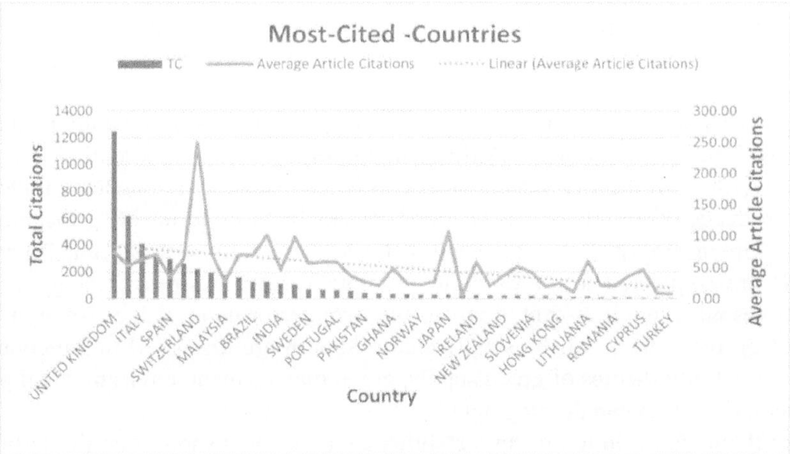

FIGURE 5.5 The most cited countries for the years 2005 to 2024.

TABLE 5.4
Most Relevant Authors

Author	Articles Fractionalized	Articles
Delmas MA	2.17	5
Lee K-H	1.50	3
Leonidou CN	0.83	3
Cheng TCE	0.50	2
Dahlgaard JJ	0.40	2
Dahlgaard-Park SM	0.40	2
Haleem A	0.58	2
Kumar V	0.75	2
Lai K-H	0.50	2
Luthra S	0.58	2
Maletič D	0.40	2
Maletič M	0.40	2
Nairn-Birch N	0.67	2
Palazzo G	0.83	2
Pekovic S	1.00	2
Perotti S	0.58	2
Scherer AG	0.83	2
Venus Lun YH	0.50	2
Wong CWY	0.50	2
Zhang J	0.38	2
Others (1 document each)	0.33	292
Total		**337**

resource" and "competitive advantage" are emerging trends. The themes of "green management," "firm performance," "management practices," and "business strategies" in the niche segment group highlights highly developed but isolated themes with low centrality.

Figure 5.9 shows the clusters of co-occurrence networks of the author's abstract bigrams. The visualized network diagram shows the nodes representing individual theme words with their co-occurrence with other related and connected topics that were produced during the study period. The sustainability terms of "green management," "green supply," "management practices," "green innovation," and "corporate social" are connected to the domains of supply chain, business strategy, trading, human resource, environment, finance, and firm performance. Likewise, notice that themes on the concept of the supply chain domain are discussed in the context of all sustainability themes of green supply, green management, environmental performance, and sustainable development.

The thematic evolution using a stability index of 0.50 shows that the term "corporate social responsibility" figures as the highest weighted inclusion and has been the most impactful theme of this period, followed by "environmental sustainability"

TABLE 5.5
Most Influential Cited Authors

Author	Total Citations per Year	Total Citations
Scherer AG, 2011, *J Manage Stud*	105.29	1474
Friede G, 2015, *Journal of Sustain Finance Invest*	129.40	1294
Renwick DW, 2013, *Int J Manage Rev*	93.92	1127
Adams R, 2016, *Int J Manage Rev*	96.78	871
El-Kassar A-N, 2019, *Technol Forecast Soc Change*	116.33	698
Lee K-H, 2015, *J Clean Prod*	65.90	659
Kim YJ, 2019, *Int J Hosp Manage*	102.83	617
Babiak K, 2011, *Corp Soc Responsib Environ Manage*	43.86	614
Michelon G, 2015, *Crit Perspect Account*	57.80	578
Kraus S, 2020, *Technol Forecast Soc Change*	113.00	565
Qiu Y, 2016, *Br Account Rev*	56.89	512
Mellahi K, 2016, *J Manage*	56.56	509
Jabbour CJC, 2016, *J Clean Prod*	54.22	488
Lozano R, 2015, *Corp Soc Responsib Environ Manage*	47.30	473
Roscoe S, 2019, *Bus Strategy Environ*	77.67	466
Juvan E, 2014, *Ann Tour Res*	42.27	465
Tang G, 2018, *Asia Pac J Hum Resour*	65.29	457
Dangelico RM, 2017, *Bus Strategy Environ*	56.63	453
Baumgartner RJ, 2014, *Corp Soc Responsib Environ Manage*	41.00	451
Tang M, 2018, *Bus Strategy Environ*	62.71	439
Luthra S, 2011, *J Ind Eng Manage*	31.36	439
Bohnsack R, 2014, *Res Policy*	38.36	422
Geng R, 2017, *Int J Prod Econ*	51.88	415
Li D, 2017, *J Clean Prod*	50.75	406
Scherer AG, 2013, *J Manage Stud*	32.17	386

and "environmental management" based on the stability index (Table 5.6). The term "sustainability" with the second highest weighted inclusion index has emerged as the most frequently used theme in the research articles. The thematic word "corporate social responsibility" has evolved over this period to "sustainability" if the weighted inclusion index is considered.

Figure 5.10 depicting the thematic evolution using the cutting period of five years highlights the themes that evolved during the study period. The size of the rectangle highlights the inflow and outflow count of the theme, and the thickness of the gray lines highlight the inflow to the subsequent evolving theme.

Figure 5.11 depicts a three-field plot, also called the Sankey plot, showing the three interrelated elements of article author, keywords of authors, and authors' countries

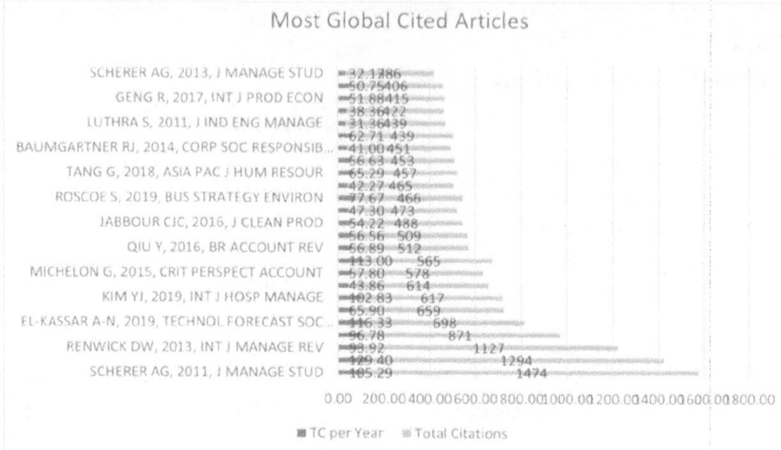

FIGURE 5.6 The most cited authors for the years 2005 to 2024.

of sustainability-related publications. The authors and their associated keywords display the major countries that have contributed toward sustainability areas. The gray lines establish the relationship between the three concepts under study.

The size of each rectangle represents the number of publications of each associated element (Figure 5.11). The keyword "sustainable development" is the largest focused term with 16 inflows of author articles, spread with an outflow of 18 countries, having major contributions from the UK and China. This is then followed by the keyword "sustainability" with 14 inflows and 15 outflows from the UK and China. The term "environmental management" has 5 inflows with 17 outflows. Italy and Spain have 20 and 18 inflows, respectively, while the term "supply chain management" has 13 inflows and 15 outflows, and the UK has major contribution. Authors from the UK, China, Italy, and Spain have the most influential topics related to sustainability research. The gray lines' thickness represents significant relationships among the three elements under consideration—authors, keywords, and author's countries.

The most frequent words are shown in the word cloud (Figure 5.12). The larger keyword reflects the high occurrences it has among the publications under study, as seen in the word cloud. In the period of study from 2005 to 2024, note that there are a total of 2043 keywords used by authors in the domain of sustainability research, in which the highest frequency word is "sustainable development" (139 occurrences), followed by the term "sustainability" (82 occurrences). Each rectangle reflects the word's occurrences with frequency percentage, where the rectangle area is proportional to the value of the frequency.

In Figure 5.13 note that during the study period, the list of terms having a frequency of 25 and above in the publications under study is shown in Table 5.2. The term "environmental management" is closely followed by the term "supply chain management."

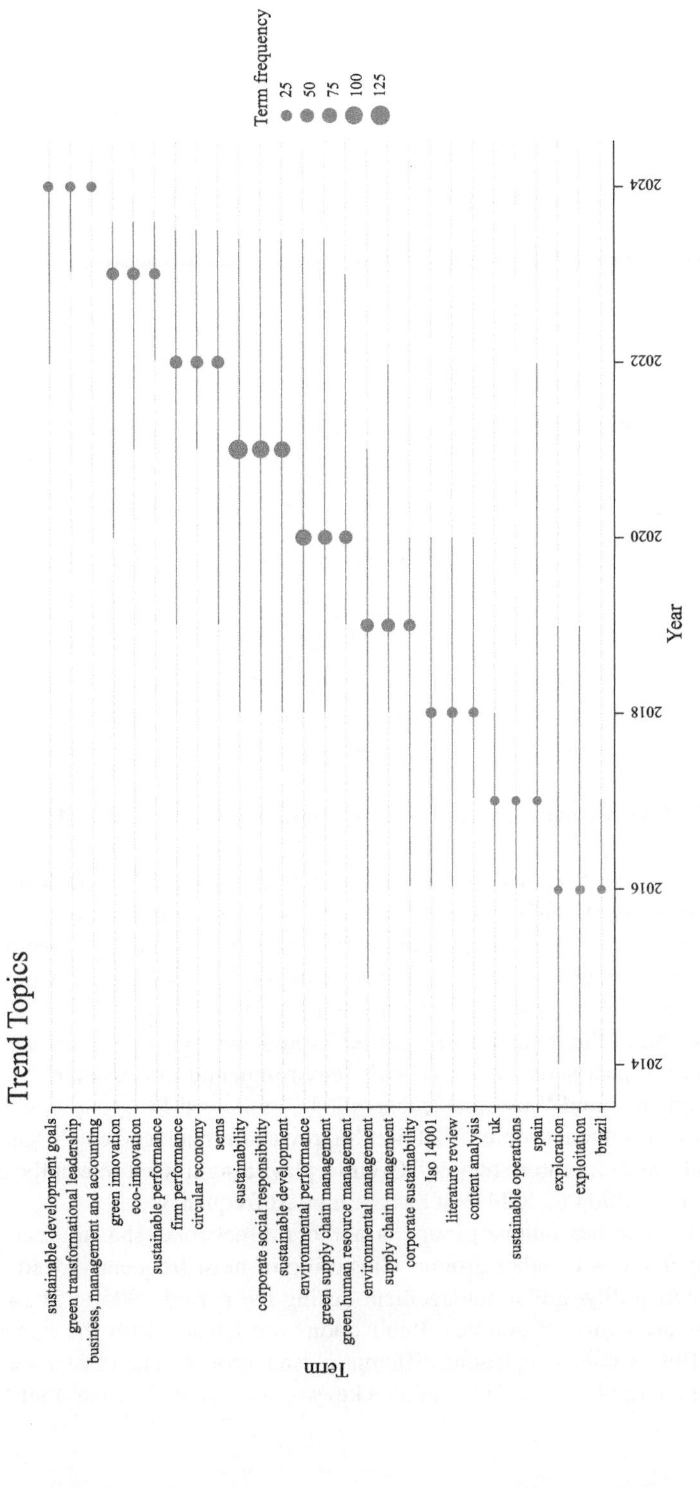

FIGURE 5.7 Theme development over the years 2005 to 2024.

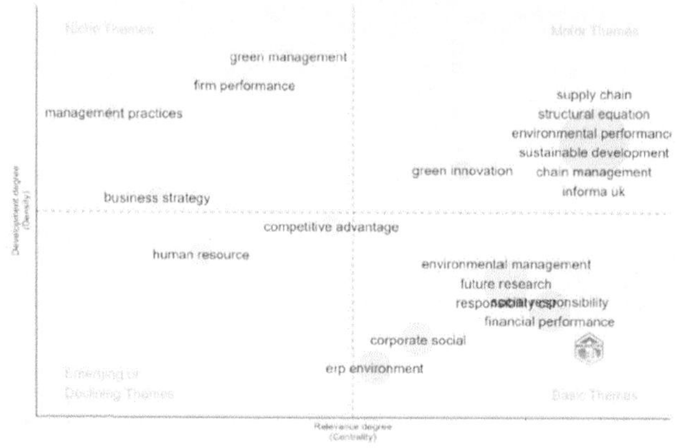

FIGURE 5.8 Abstract bigrams.

FIGURE 5.9 Co-occurrence, abstract bigrams, Louvain method (2005–2024).

Table 5.7 depicts the terms with a frequency mention of 25 and above in the sustainability research during the study period. This confirms that research on "sustainable development" draws keen interest from researchers. For example, the term "environmental management" was the most frequently used term used by most of the authors, followed by "supply chain management" and "innovation."

Table 5.8 shows the details of the cluster groups where group 5 has the highest centrality and impact score of 2.064 with "environmental management," "sustainable development," and "sustainable operations," followed by group 7 with highest frequency and impact score of 2.054: "supply chain management," "sustainable development," and "literature review." Group 5 plays a significant role in the publication dataset from 2004 to 2024 as it has the highest frequency.

Clustering identifies unique groups of authors in networks sharing features and values comparable with other groups. Clustering of most frequently cited publications in sustainability–green management during the period 2005 to 2024. Figure 5.14 is the cluster map of sources. Publications are labeled with the name of the journal. Different colors represent different cluster groups. The cluster map using authors as the unit of analysis and author's keywords as a coupling parameter, shows

TABLE 5.6
Thematic Evolution of Keywords

From	To	Words	Weighted Inclusion Index	Inclusion Index	Occurrences	Stability Index
corporate social responsibility (2005–2014)	corporate social responsibility (2015)	corporate social responsibility	1.00	1.00	4	0.50
corporate social responsibility (2005–2014)	financial performance (2015)	environmental sustainability	0.29	0.50	3	0.17
firm performance (2005–2014)	environmental management (2015)	environmental management	0.18	0.33	2	0.17
environmental performance (2005–2014)	financial performance (2015)	environmental performance	0.44	0.33	4	0.14
sustainability (2005–2014)	sustainability (2015)	sustainability	0.78	0.50	8	0.11
sustainability (2005–2014)	financial performance (2015)	financial performance	0.24	0.20	4	0.08

FIGURE 5.10 Thematic evolution.

FIGURE 5.11 The three-field plots (2005–2024).

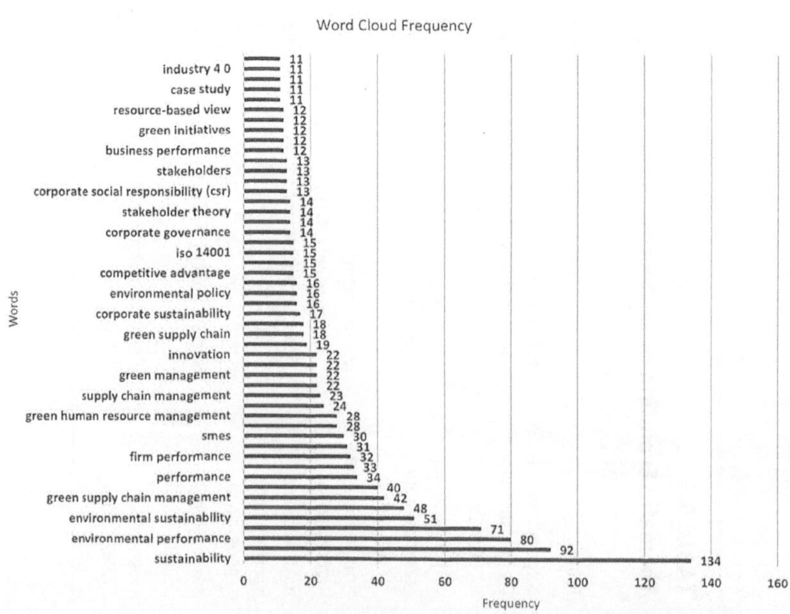

FIGURE 5.12 The word cloud of the most frequent author keywords (2005–2024).

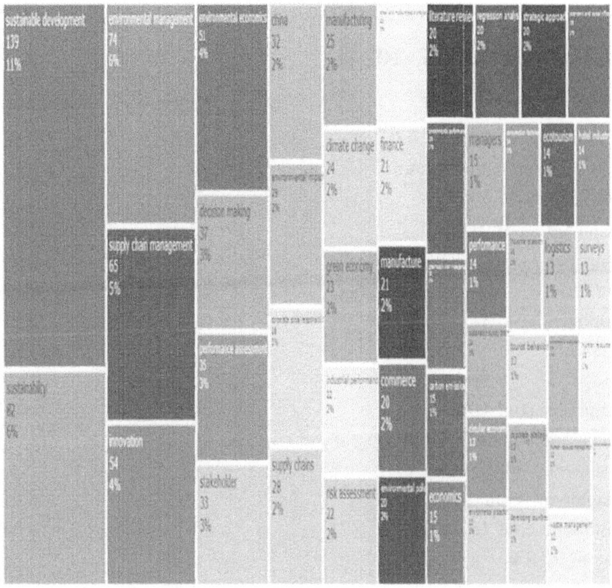

FIGURE 5.13 Tree map displaying words with the most frequent occurrences (2005–2024).

TABLE 5.7
Terms with Frequency Mentions of 25 and Above (2005–2024)

Term	Occurrences
sustainable development	139
sustainability	82
environmental management	74
supply chain management	65
innovation	54
environmental economics	51
decision making	37
performance assessment	35
stakeholder	33
china	32
environmental impact	29
corporate social responsibility	28
supply chains	28
manufacturing	25

TABLE 5.8
Coupling Bibliometric (2005–2024)

Label	Group	Frequency	Centrality	Impact
sustainable development, conf 27.5% environmental management, conf 23.7% sustainability, conf 20%	3	22	0.822	1.849
sustainable development, conf 21.2% supply chain management, conf 28% sustainability, conf 35%	6	26	0.694	2.327
sustainable development, conf 11.2% sustainability, conf 12.5% india, conf 28.6%	4	11	0.429	2.179
sustainable development, conf 10% environmental management, conf 18.4% performance assessment, conf 23.8%	2	11	0.493	1.713
supply chain management, conf 36% sustainable development, conf 18.8% literature review, conf 90.9%	7	49	0.636	2.054
environmental management, conf 34.2% sustainable development, conf 11.2% sustainable operations, conf 43.8%	5	17	0.874	2.064

Note: conf = co-occurrence frequency.

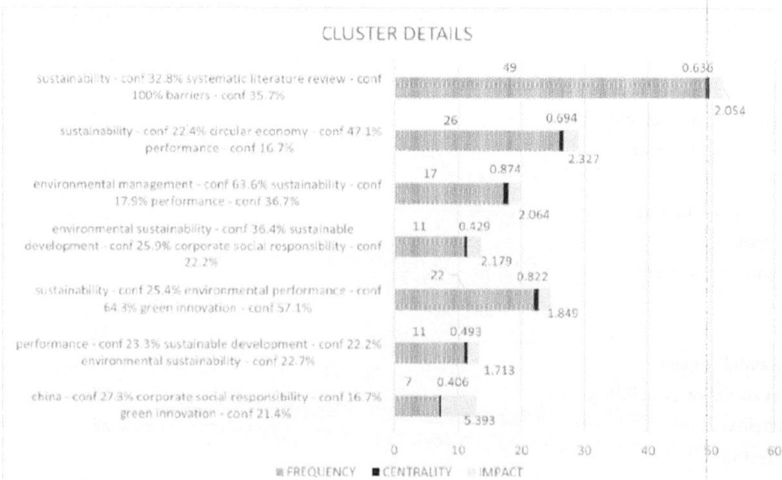

FIGURE 5.14 Cluster map of sources (2005–2024).

four distinct major groups. Figure 5.14 shows the size of the clusters in circles, which represents the contribution to the sustainability and green management literature. Cluster 1 has 44 authors, cluster 2 has 53 authors, and cluster 3 has 42 authors groups. Cluster 4 and cluster 5 have 17 authors each, and cluster 7, the largest, has 83 authors in the network. Predictive analytics play a key role in predicting environmental impacts and resource needs, enabling proactive measures for sustainable development. The term "sustainable development" appears in almost all clusters—here AI and ML play a critical role as large datasets can enable predictions. The score of 2.327 for supply chain management significantly highlights the high potential of digital transformations with blockchain, AI, ML, and automation to enhance SCM activities by improving transparency and efficiency. The high impact score of 1.849 for sustainable development and environmental management shows the potential of AI and ML applications to drive improvements in these areas with data-driven insights. Likewise, the 1.713 score for performance assessments suggests real-time monitoring and data-driven decision-making using digital tools to provide performance metrics aligned with the sustainability goals of the organization, which are monitored and controlled with actionable insights. In the area of sustainable operations, the impact score of 2.064 highlights the role of digitalization tools in innovations and efficiency—by tracking energy consumption, minimizing waste, and ensuring compliance with continuous improvements. These findings emphasize the achievement of key business objectives and operational efficiencies, unlocking partnerships through technological innovations.

The metric of centrality is an effective tool for the identification of influential authors within the network. Figure 5.15 was developed using betweenness centrality as the centrality algorithm to highlight authors who are gatekeepers of information and resources. The circle at the center right is the most central node, having the shortest distance between the other nodes of the network, while the size of the cluster circles represents the frequencies of keyword occurrences.

The factorial analysis using multiple correspondence analysis (MCA) (Figure 5.16) depicts the clusters of similar research dimensions that evolved over the study period. Using the author's keywords for the factor analysis, the MCA depicted five

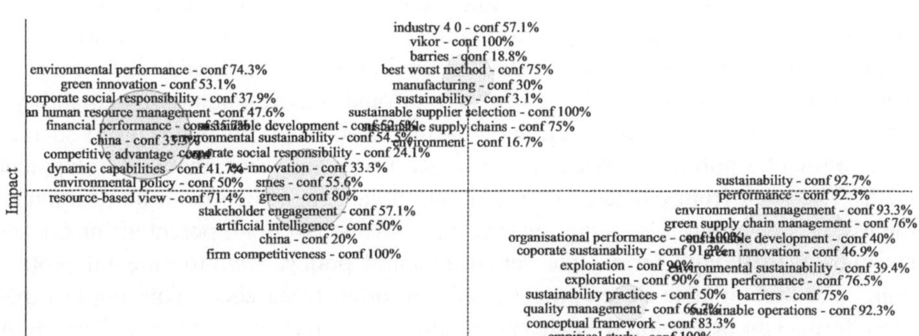

FIGURE 5.15 Major clusters (2005–2024).

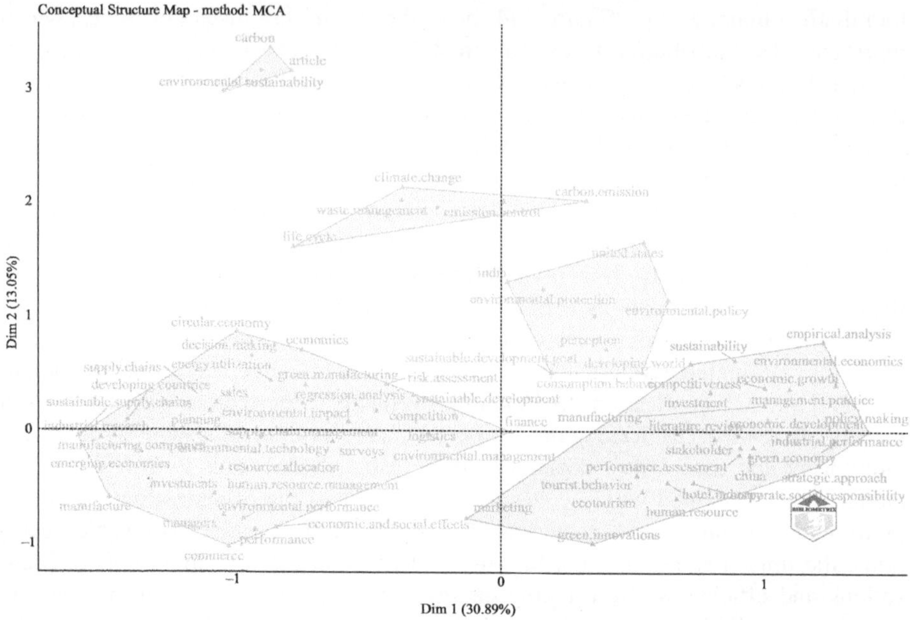

FIGURE 5.16 Conceptual factorial map and keyword clusters (2005–2024).

clusters. The largest cluster contains the highest number of keywords and is scattered broadly with terms that clustered toward the center of the cluster map. These terms have garnered the most attention among the research community on the subject of "sustainability." This cluster has 182 articles of which the terms "environmental technologies" and "manufacturing" are at the center. This large cluster implies linkages of processes and systems to optimize cost and produce sustainable products, as well as the minimization of waste, key areas under research. Cluster 2 has 176 articles, in which notice that "management practices," "economic development," "performance assessment," and "strategic approach" are centered, indicating critical decision areas where technologies of AI, ML, and automation can bring in innovations that can aid ecological and economic growth. The rest of the clusters have fewer articles in comparison to cluster 1 and cluster 2. Clusters 3 and 4 are the less discussed topics. The upper left cluster shows the terms "carbon" and "environmental sustainability," which are areas of potential research opportunities. Cluster 3 includes terms such as "environmental policy," "environmental protection," and "sustainable development goal." The term from cluster 4, "carbon emission," when seen from its position in the MCA conceptual map, suggests that these terms have high potential for further research even though terms like "environmental policy," "environmental protection," and "sustainable development goal" are often researched. This implies that these terms can be explored further in conjunction with newer concepts. The MCA thus reveals four dimensions of research conducted with its future potentials.

A classification tree dendrogram graphically depicts the cluster groups and specific levels and the association between the groups. Figure 5.17 depicts the resultant

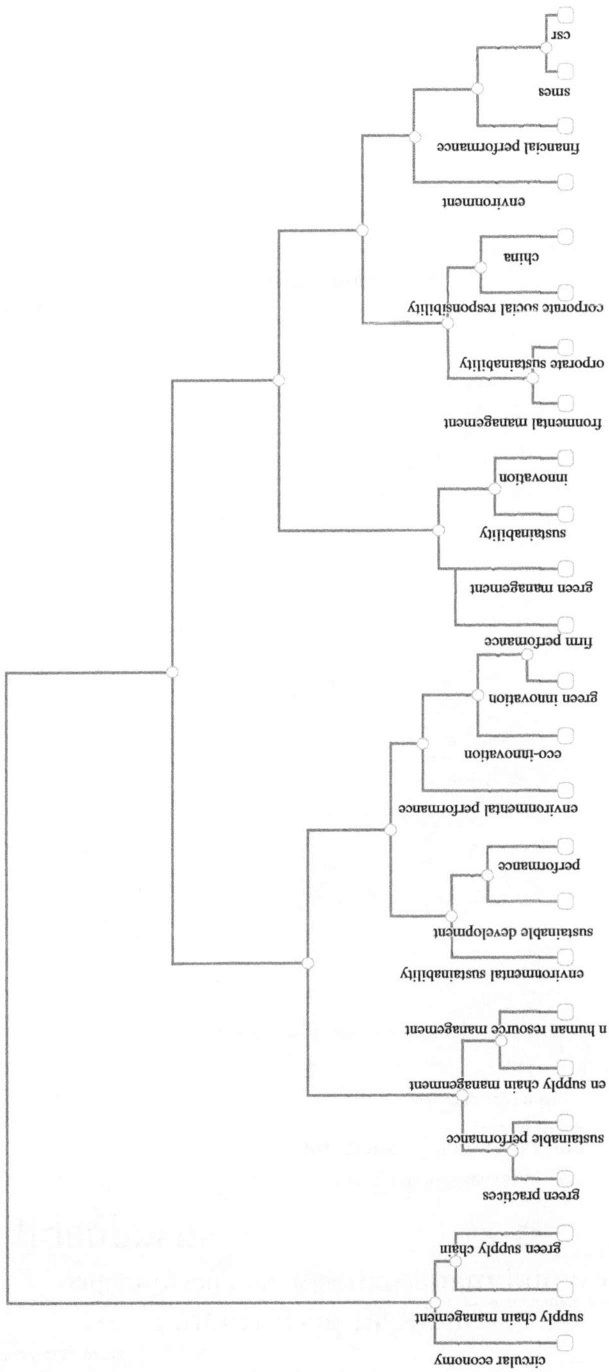

FIGURE 5.17 Dendrogram of hierarchical analysis of author keywords (2005–2024).

dendrogram of the author's keywords used by the method of multidimensional scaling using the top 25 keywords. It details the correlated relationship between keywords. At the cutoff line, notice the clusters identified are two major closely related clusters with similar branch heights, signifying a significant difference between the two distinct clusters. The outlier cluster in the dendrogram has a higher branch height, indicating a significant difference from the rest of the clusters.

The co-occurrence density and network based on the author's keywords are shown in Figure 5.18 and Figure 5.19. Note the co-occurrence network has three main clusters, groups closely connected. Notice that the clusters "environmental management" and "financial performance" have a small distance between them, indicating the influence of each theme and their connections. The theme "financial performance" in cluster 1 showed the highest betweenness score of 9, indicating its connections

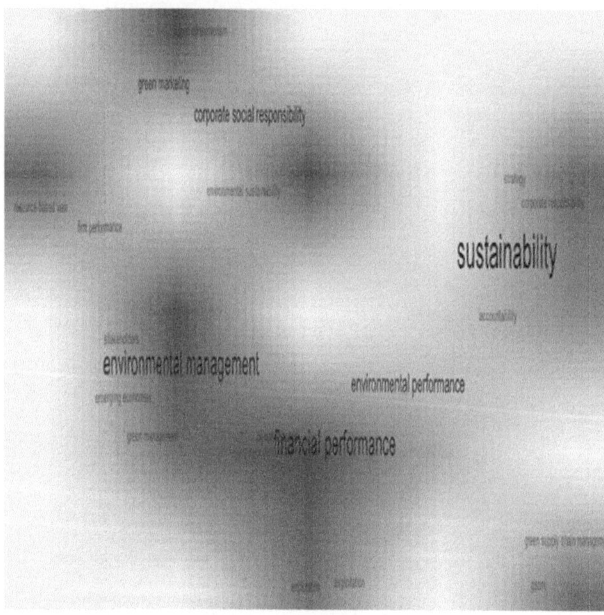

FIGURE 5.18 Co-occurrence density and network diagram.

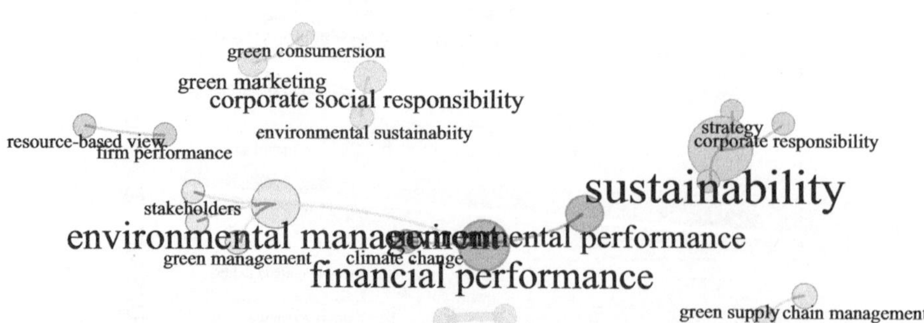

FIGURE 5.19 Co-occurrence network.

with other clusters, even though it had moderate importance of 0.080. Likewise, in cluster 2, the term "environmental management" showed the highest betweenness of 12 with a high importance of 0.105. Similarly, the term "sustainability" in cluster 4 showed a betweenness score of 3.0 and an importance page rank of 0.091.

5.6 FUTURE RESEARCH AND IMPLICATIONS

This research review has identified the earliest significant studies on sustainability with a focus on its development over the past 20 years. The keyword synthesis revealed groups of research literature on the study area. The pertinent authors, journals, key terms, themes, and their evolution over the years reveal that the existing research community's focus on sustainability research is converging into two main clusters—concepts of sustainability and impacts on areas of an organization, for example, financial performance, human resource, and SCM.

It transpired that a few terms associated with the least important clusters, such as "environmental policy" and "carbon emission," have significant potential for further research. Furthermore, the study has highlighted high-potential areas for future research in the areas of "carbon," "environmental sustainability," "environmental policy," "environmental protection," and "sustainable development goal." During the early stage of this research, our results indicated that the sustainability theme is poised to undergo major transformation in the upcoming years, given the technological developments, artificial intelligence, and other digital transformation techniques firms undertake. With digital transformation in firms patchy, incomplete, and undergoing a major shift, management actions and managerial practices should be geared toward the creation of organizational sustainability as a consciously managed, structured process used to leverage the digital potential of the entire organization. Organizations that develop a new business model in light of digitalization may fail in the long run (Geissdoerfer et al., 2018). Unless boards and top management adopt a mindset of sustainability and embrace change to create value for their clients, customers, and society (Bocken et al., 2014; Biloslavo et al., 2018), they will would spiral out of business. Operationally, firms should focus on sustainable initiatives in their supply chain partners and providers, and develop internal capabilities in data intelligence resulting in more stable and adaptive firm performance (Moghrabi, 2023).

Pertinently, this study has identified key themes of research on green management and its key areas. The relevant areas for future research are shown in Table 5.9.

5.7 CONCLUSION

Valuable insights about the evolution and present state of scholarly work in the area of green management and its role in driving sustainable organizational performance have been offered in this chapter. With sustainability and environmental management emerging as central themes, there has been a significant rise in publications related to this area since 2015. Sustainable development, supply chain management, and environmental performance have emerged as key research clusters. The analysis

TABLE 5.9
Key Literature on Technology Impacting Green Management, Findings, and Future Research Agenda

Reference	Main Theme	Future Research Agenda
Riggs et al. (2024)	Information system capabilities linkages with sustainable activities enable circular practices and enhance business performance.	Longitudinal studies of firms belonging to different industries were suggested, including how information technology interacts with other firm capabilities and how adoption of technologies in different industries is considered. Further research was suggested to investigate the antecedents of information service systems in helping sustainability.
Lee and Kim (2024)	Investigated the elements influencing Korean business-to-business (B2B) sustainable corporate performance.	Given the incredible technical developments in recent years, a company's sustainable performance mostly relies on the continuous application of new technologies, which results in a new focus on the function of the technological capabilities of the company in the sustainability of its business.
Nagadi (2022)	Implementation of green, lean, and six sigma operations for sustainable manufacturing.	Investigate the integration of AI and machine learning in optimizing green, lean, and six sigma operations for enhanced sustainability outcomes.
Brinken et al. (2023)	Comparing decarbonization potential of digital and green technologies.	Explore the long-term environmental and social impacts of digital vs. green technologies across different industries and geographical contexts.
Udokporo et al. (2020)	Impact of lean, agile, and green (LAG) on business competitiveness.	Examine the synergies between LAG practices and circular economy principles for improved sustainability performance.
Biçakcioğlu-Peynirci and Tanyeri (2022)	Stakeholder and resource-based antecedents of green export performance.	Investigate the role of institutional pressures and cultural differences in shaping green export strategies across different markets.
Cherrafi et al. (2018)	Lean, green practices, and process innovation in green supply chain performance.	Study the impact of blockchain technology on enhancing transparency and efficiency in green supply chain management.
Astuty, Sudirman, and Aryanto (2024)	To model predictable approaches that enable midsized firms to meet the challenges of the "dual transition era"—both digitalization and sustainability challenges. Small businesses create a long-term plan to be resilient and support the SDGs by using digital tools.	Microbusinesses need to be very aware of the problems that the SDGs pose. Microbusiness actors need green DTA (digital technology adoption) to promote a sustainable mission to support the success of SEPRAC (sustainable entrepreneurship practices).

(*Continued*)

TABLE 5.9 (CONTINUED)
Key Literature on Technology Impacting Green Management, Findings, and Future Research Agenda

Reference	Main Theme	Future Research Agenda
Ali Alshehail et al. (2022)	Total quality management and sustainability in the public service sector.	Explore the role of citizen engagement and digital platforms in enhancing TQM and sustainability in public services.
Khonje et al. (2021)	Sustainable environmental practices in Malawian hotels.	Conduct comparative studies on sustainable practices in hospitality sectors across different developing countries, focusing on barriers and enablers.
Tamayo-Torres et al. (2019)	Supply chain controversies' role in sustainability and financial performance.	Analyze the long-term effects of supply chain controversies on firm reputation, stakeholder trust, and financial performance using longitudinal studies.
Martínez-del Río et al. (2012)	High-involvement work practices and environmental capabilities.	Investigate the role of remote work and digital collaboration tools in fostering environmental capabilities and employee engagement.
Hermelingmeier and von Wirth (2021)	Nexus of business sustainability and organizational learning.	Explore how AI and big data analytics can enhance organizational learning for improved sustainability performance.
Biçakcioğlu et al. (2020)	Green business strategy and export performance.	Examine the impact of global sustainability regulations on green business strategies and export performance across different industries.
Ariana et al. (2020)	ICT human resources capacity in e-government implementation.	Study the role of continuous learning and upskilling programs in improving ICT capacity for e-government initiatives in developing countries.
Tingbani et al. (2023)	Environmental tax, SME financing constraint, and innovation.	Investigate the long-term effects of environmental taxes on SME innovation, growth, and sustainability performance across different economic contexts.
Boulay et al. (2020)	Multi-unit franchising from franchisor and franchisee perspectives.	Explore the role of sustainability practices in franchise performance, expansion, and brand value in the post-pandemic era.
Pies and Schultz (2023)	Governance of sustainable business model innovation.	Study the impact of different governance structures on the success and scalability of sustainable business models in various industries.
Ogunlela (2018)	Green supply chain management in fast-moving consumer goods.	Investigate consumer perceptions and willingness to pay for products with transparent, sustainable supply chains in the FMCG sector.
Alkaraan et al. (2024)	Reinforcing green business strategies with Industry 4.0 and governance.	Explore the challenges and opportunities of integrating AI, IoT, and blockchain in green business strategies across different sectors.

(*Continued*)

TABLE 5.9 (CONTINUED)
Key Literature on Technology Impacting Green Management, Findings, and Future Research Agenda

Reference	Main Theme	Future Research Agenda
Sá et al. (2023)	Science-based targets and sustainable development of the aviation industry.	Study the effectiveness and challenges of implementing science-based targets in other carbon-intensive industries like shipping and manufacturing.
Fernández-Alles et al. (2022)	Impact of human, social, and psychological capital on academic spin-off internationalization.	Investigate the role of university ecosystems and support systems in enhancing spin-off success, sustainability, and international growth.
Martinez (2021)	Ecological responsiveness and business performance compatibility.	Explore the role of stakeholder engagement and pressure in enhancing ecological responsiveness and its impact on long-term business performance.
Sancha et al. (2022)	Corporate governance to sustainability outcomes through operations management.	Investigate the role of board diversity and sustainability expertise in driving sustainable operations management practices and outcomes.
Lee and Min (2015)	Green R&D for eco-innovation and its impact on emissions and firm performance.	Study the long-term effects of green R&D investments on firm competitiveness, market share, and environmental performance across different industries.
Singh et al. (2019)	Environmental ethics, performance, and competitive advantage.	Explore the role of leadership styles and organizational culture in fostering environmental ethics and driving competitive advantage through sustainability.
Leonidou et al. (2017)	Internal drivers and performance consequences of small firms' green business strategies.	Investigate the role of digital technologies and platforms in enabling small firms to implement and benefit from green business strategies.
Puppim de Oliveira and Jabbour (2017)	Environmental management, CSR, and governance in small firm clusters.	Study the effectiveness of collaborative approaches to sustainability in small firm clusters, focusing on knowledge sharing and resource pooling.
Aslam et al. (2021)	Environmental management practices and financial performance.	Explore the impact of environmental management practices on non-financial performance metrics and long-term organizational resilience.
Güner (2018)	Evolution of green management with a Kuhnian perspective.	Investigate the potential for a paradigm shift in management theory toward sustainability-centered approaches, considering the impact of global environmental challenges.

brings to the fore authors who are influential, and journals and countries contributing to this field. The findings underscore the increasing importance of merging green practices with organizational strategies to further both the environmental and financial performance of firms.

Albeit its contribution, the work has a few limitations. As the analysis was restricted to articles and research papers from the Scopus database, relevant research from other sources may have missed being analyzed. There might have been foundational works in this field before 2005 that were not been considered for this study. The keyword search might have hampered the selection of articles, inadvertently excluding a few relevant studies. The study is biased toward publications in the English language only, resulting in non-English research work being underrepresented.

Notwithstanding the aforementioned limitations, further studies could expand this analysis by including other databases and non-English publications, resulting in a comprehensive view. In-depth qualitative studies may be carried out using the key publications highlighted in this work. Practical implementation and outcomes of green management strategies across different industries and geographical contexts may be investigated. Emerging business trends like the circular economy and stakeholder capitalism and their intersection with green management may be examined.

REFERENCES

Ali Alshehail, O., M. Khan, and M. Ajmal. "Total Quality Management and Sustainability in the Public Service Sector: The Mediating Role of Corporate Governance." *Benchmarking: An International Journal* 29, no. 10 (2022): 3162–3184.

Alkaraan, F., M. Elmarzouky, K. Hussainey, V. G. Venkatesh, Y. Shi, and N. Gulko. "Reinforcing Green Business Strategies with Industry 4.0 and Governance towards Sustainability: Natural-Resource-Based View and Dynamic Capability." *Business Strategy and the Environment* 33 no. 4 (2024): 3588–3606.

Ansari, Z. N., and R. Kant. "Exploring the Framework Development Status for Sustainability in Supply Chain Management: A Systematic Literature Synthesis and Future Research Directions." *Business Strategy and the Environment* 26, no. 7 (2017): 873–892. https://doi.org/10.1002/bse.1945.

Anwar, N., N. H. N. Mahmood, M. Y. Yusliza, T. Ramayah, J. N. Faezah, and W. Khalid. "Green Human Resource Management for Organisational Citizenship Behaviour towards the Environment and Environmental Performance on a University Campus." *Journal of Cleaner Production* 256 (2020): 120401.

Aria, M., and C. Cuccurullo. "A Brief Introduction to Bibliometrix." *Journal of Informetrics* 11, no. 4 (2017): 959–975.

Ariana, S., C. Azim, and D. Antoni. "Clustering of ICT Human Resources Capacity in the Implementation of E-Government in Expansion Area: A Case Study from Pali Regency." *Cogent Business & Management* 7 no. 1 (2020): 1754103.

Aslam, S., M. H. Elmagrhi, R. U. Rehman, and C. G. Ntim. "Environmental Management Practices and Financial Performance Using Data Envelopment Analysis in Japan: The Mediating Role of Environmental Performance." *Business Strategy and the Environment* 30, no. 4 (2021): 1655–1673.

Astuty, E., I. D. Sudirman, and R. Aryanto. "Sustainable Resilience Strategy: Unleash the Micro-Businesses's Potential in the Digitalization and Sustainability Era." *Cogent Business & Management* 11 no. 1 (2024): 2313672.

Atif, S. (2023). "The Role of Industry 4.0-Enabled Data-driven Shared Platform as an Enabler of Product-Service System in the Context of Circular Economy: A Systematic Literature Review snd Future Research Directions." *Business Strategy & Development* 6, no. 3 (2023): 275–295.

Beckmann, M., and S. Schaltegger. "Sustainability in Business: Integrated Management of Value Creation and Disvalue Mitigation." In *Oxford Research Encyclopedia of Business and Management*. Oxford University Press, 2021.

Biçakcioğlu, N., V. Theoharakis, and M. Tanyeri. "Green Business Strategy and Export Performance: An Examination of Boundary Conditions from an Emerging Economy." *International Marketing Review* 37, no. 1 (2020): 56–75.

Bıçakcıoğlu-Peynirci, N., and M. Tanyeri. "Stakeholder and Resource-Based Antecedents and Performance Outcomes of Green Export Business Strategy: Insights from an Emerging Economy." *International Journal of Emerging Markets* 17 no. 1 (2022): 1–46.

Biloslavo, Roberto, Carlo Bagnoli, and David Edgar. "An Eco-Critical Perspective on Business Models: The Value Triangle as an Approach to Closing the Sustainability Gap." *Journal of Cleaner Production* 174 (2018): 746–762.

Bocken, Nancy M. P., Samuel W. Short, Padmakshi Rana, and Steve Evans. "A Literature and Practice Review to Develop Sustainable Business Model Archetypes." *Journal of Cleaner Production* 65 (2014): 42–56.

Boulay, J., B. Caemmerer, H. Evanschitzky, and K. Duniach. "Multi-Unit Franchising from Franchisor and Franchisee Perspectives: Antecedents, Performance Outcomes, and the Optimal Mini-Chain Size." *Journal of Business Research* 113 (2020): 49–58.

Brinken, J., F. Behrendt, and S. Trojahn. "Comparing Decarbonization Potential of Digital and Green Technologies." *Sustainable Futures* 6 (2023): 100125.

Cherrafi, A., J.A. Garza-Reyes, V. Kumar, N. Mishra, A. Ghobadian, and S. Elfezazi. "Lean, Green Practices and Process Innovation: A Model for Green Supply Chain Performance." *International Journal of Production Economics* 206 (2018): 79–92.

Cioffi, R., M. Travaglioni, G. Piscitelli, A. Petrillo, and A. Parmentola. Smart Manufacturing Systems and Applied Industrial Technologies for a Sustainable Industry: A Systematic Literature Review. *Applied Sciences* 10, no. 8 (2020): 2897.

Equator Principles. " Davivienda Bank Becomes a Signatory to the Equator Principles." 2024. https://equator-principles.com/davivienda-bank-becomes-a-signatory-to-the-equator-principles/.

Fernández-Alles, M., D. Hernández-Roque, M. Villanueva-Flores, and M. Díaz-Fernández. "The Impact of Human, Social, and Psychological Capital on Academic Spin-Off Internationalization." *Journal of International Entrepreneurship* 20 no. 3 (2022): 433–473.

Galbreath, J. "Drivers of Green Innovations: The Impact of Export Intensity, Women Leaders, and Absorptive Capacity." *Journal of Business Ethics* 158 (2019): 47–61.

Geissdoerfer, M., S. N. Morioka, M. M. de Carvalho, and S. Evans. "Business Models and Supply Chains for the Circular Economy." *Journal of Cleaner Production* 190 (2018): 712–721.

Ghobakhloo, M., M. Iranmanesh, A. Grybauskas, M. Vilkas, and M. Petraitė. Industry 4.0, Innovation, and Sustainable Development: A Systematic Review and a Roadmap to Sustainable Innovation. *Business Strategy and the Environment* 30, no. 8 (2021), 4237–4257.

Gill, S. S., M. Xu, C. Ottaviani, P. Patros, R. Bahsoon, A. Shaghaghi, and S. Uhlig. AI for Next Generation Computing: Emerging Trends and Future Directions. *Internet of Things* 19 (2022): 100514.

Güner, S. "Evaluation of the Evolution of Green Management with a Kuhnian Perspective." *Business Research* 11, no. 2 (2018): 207–229.

Harkawat, P. "An Integrated Quality Management System (QMS) Combined with AI/ML Guidelines for the Design and Development of Sustainable Products and Services for IT and Engineering R&D Companies Providing Solutions to Global Mining Organizations." *Journal of Mines, Metals and Fuels* 72, no. 4, (2024): 323–332.

Hermelingmeier, V., and T. von Wirth. "The Nexus of Business Sustainability and Organizational Learning: A Systematic Literature Review to Identify Key Learning Principles for Business Transformation." *Business Strategy and the Environment* 30, no. 4 (2021): 1839–1856.

Hsu, L. C., S. L. Ou, and Y. C. Ou. "A Comprehensive Performance Evaluation and Ranking Methodology under a Sustainable Development Perspective." *Journal of Business Economics and Management* 16, no. 1 (2015): 74–92.

Jamwal, A., R. Agrawal, M. Sharma, and A. Giallanza. "Industry 4.0 Technologies for Manufacturing Sustainability: A Systematic Review and Future Research Directions." *Applied Sciences* 11, no. 12 (2021): 5725.

Khonje, L. Z., F. G. Bello, and M. B. Sepula. Sustainable Environmental Practices in Malawian Hotels, 2021.

Lee, S., and Y. J. Kim. "Analyzing Factors That Affect Korean b2b Companies' Sustainable Performance." *Sustainability* 16, no. 5 (2024): 1719. https://doi.org/10.3390/su16051719.

Lee, K.-H., and B. Min. "Green R&D for Eco-Innovation and Its Impact on Carbon Emissions and Firm Performance." *Journal of Cleaner Production* 108 (2015): 534–542.

Leonidou, L. C., P. Christodoulides, L. P. Kyrgidou, and D. Palihawadana. "Internal Drivers and Performance Consequences of Small Firm Green Business Strategy: The Moderating Role of External Forces." *Journal of Business Ethics* 140, no. 3 (2017): 585–606.

Loiseau, E., L. Saikku, R. Antikainen, N. Droste, B. Hansjürgens, K. Pitkänen, P. Leskinen, P. Kuikman, and M. Thomsen. "Green Economy and Related Concepts: An Overview." *Journal of Cleaner Production* 139 (2016): 361–371.

Lusthaus, C. *Enhancing Organizational Performance: A Toolbox for Self-Assessment.* IDRC, 1999.

Martinez, F. "Ecological Responsiveness and Business Performance: Four Scenarios of Compatibility and a Multi-Faceted Strategies Framework." *Business Strategy and the Environment* 30 no. 8 (2021): 3606–3625.

Martínez-del Río, J., J. Céspedes-Lorente, and E. Carmona-Moreno. "High-Involvement Work Practices and Environmental Capabilities: How HIWPS Create Environmentally Based Sustainable Competitive Advantages." *Human Resource Management* 51 no. 6 (2012): 827–850.

Moghrabi, Issam A. R., Sameer Ahmad Bhat, Piotr Szczuko, Rawan A. AlKhaled, and Muneer Ahmad Dar. "Digital Transformation and Its Influence on Sustainable Manufacturing and Business Practices." *Sustainability* 15, no. 4 (2023): 3010.

Nagadi, K. "Implementation of Green, Lean and Six Sigma Operations for Sustainable Manufacturing." *International Journal of Production Management and Engineering* 10, no. 1 (2022): 1–10.

Ogunlela, G. O. "Green Supply Chain Management as a Competitive Tool in the Fast-Moving Consumer Goods Manufacturing Industry." *Journal of Business and Retail Management Research* 12, no. 4 (2018): 167–176.

Pedersen, C. S. "The UN Sustainable Development Goals (SDGs) Are a Great Gift to Business!" *Procedia CIRP* 69 (2018): 21–24.

Pies, I., and F. C. Schultz. "The Governance of Sustainable Business Model Innovation—An Ordonomic Approach." *Scandinavian Journal of Management* 39 no. 1 (2023): 101246.

Puppim de Oliveira, J. A., and C. J. C. Jabbour. "Environmental Management, Climate Change, CSR, and Governance in Clusters of Small Firms in Developing Countries: Toward an Integrated Analytical Framework." *Business & Society* 56 no. 1 (2017): 130–151.

Raut, R. D., S. Luthra, B. E. Narkhede, S. K. Mangla, B. B. Gardas, and P. Priyadarshinee. "Examining the Performance Oriented Indicators for Implementing Green Management Practices in the Indian Agro Sector." *Journal of Cleaner Production* 215 (2019): 926–943.

Riggs, R., C. M. Felipe, J. L. Roldán, and J. C. Real. "Deepening Big Data Sustainable Value Creation: Insights Using IPMA, NCA, and cIPMA." *Journal of Marketing Analytics* (2024): 1–20.

Sá, J. C., A. Carvalho, L. Fonseca, G. Santos, and J. Dinis-Carvalho. "Science Based Targets and the Factors Contributing to the Sustainable Development of an Organisation from a Literature Review to a Conceptual Model." *Production Engineering Archives* (2023): 241–253.

Sancha, C., L. Gutierrez-Gutierrez, I. Tamayo-Torres, and C. Gimenez Thomsen. "From Corporate Governance to Sustainability Outcomes: The Key Role of Operations Management." *International Journal of Operations & Production Management* 42, no. 13 (2022): 307–337.

Singh, P. K. "Digital Transformation in Supply Chain Management: Artificial Intelligence (AI) and Machine Learning (ML) as Catalysts for Value Creation." *International Journal of Supply Chain Management* 17 (2023): 57–63.

Singh, S. K., J. Chen, M. Del Giudice, and A.-N. El-Kassar. "Environmental Ethics, Environmental Performance, and Competitive Advantage: Role of Environmental Training." *Technological Forecasting and Social Change* 146 (2019): 203–211.

Smith, J., and K. Lee. "The Role of Industry 4.0 in Enhancing Corporate Sustainability." *Journal of Business Strategy* 42, no. 3 (2021): 45–58. https://doi.org/10.1108/JBS-03-2021-0045

Smith, J., and L. Brown. "The Role of AI and ML in Sustainable Manufacturing." *Journal of Cleaner Production* 250 (2022): 119–130. https://doi.org/10.1016/j.jclepro.2022.119130

Sroufe, R. *Integrated Management*. Emerald Publishing Limited, 2018. https://www.perlego.com/book/740697/integrated-management-pdf.

Tamayo-Torres, I., L. Gutierrez-Gutierrez, and A. Ruiz-Moreno. "Boosting Sustainability and Financial Performance: The Role of Supply Chain Controversies." *International Journal of Production Research* 57 no. 11 (2019): 3719–3734.

Thanki, Shashank, Kalyan Govindan, and Jitesh Thakkar. "An Investigation on Lean-Green Implementation Practices in Indian SMEs Using Analytical Hierarchy Process (AHP) Approach." *Journal of Cleaner Production* 135 (2016): 284–298.

Thiele, L. P. *Sustainability*. John Wiley & Sons, 2024.

Tingbani, I., S. Salia, J. G. Hussain, and Y. Alhassan. "Environmental Tax, SME Financing Constraint, and Innovation: Evidence from OECD Countries." *IEEE Transactions on Engineering Management* 70, no. 4 (2023): 1750–1763.

Udokporo, C. K., A. Anosike, M. Lim, S. P. Nadeem, J. A. Garza-Reyes, and C. P. Ogbuka. "Impact of Lean, Agile and Green (LAG) on Business Competitiveness: An Empirical Study of Fast Moving Consumer Goods Businesses." *Resources, Conservation and Recycling* 156 (2020): 104714.

United Nations. "Sustainable Development Goals." Sustainable Development Knowledge Platform, 2015.

Wang, C. H., and W. J. Juo. "An Environmental Policy of Green Intellectual Capital: Green Innovation Strategy for Performance Sustainability." *Business Strategy and the Environment* 30, no. 7 (2021): 3241–3254.

Wernerfelt, B. "A Resource-Based View of the Firm." *Strategic Management Journal* 5, no. 2 (1984): 171–180.

Yu, Y., M. Zhang, and B. Huo. "The Impact of Relational Capital on Green Supply Chain Management and Financial Performance." *Production Planning & Control* 32, no. 10 (2021): 861–874.

6 Impact of Green Management, AI, ML, and Automation on Firm Performance

A Review of Empirical Evidence

Sanjay Kumar Mangla, Rajiv Divekar, Pravin Kumar, Nandita Mishra, Shehnaz Lockhat, and Ritika Paul

6.1 INTRODUCTION

Global warming and climate change have emerged as two crucial issues and have been brought to the limelight by international media and many international organizations. The Sustainable Development Goals (SDGs) as prescribed by the United Nations touch on many economic, social, and environmental issues, of which many concern environmental betterment (Fallah Shayan et al. 2022). Sustainability in terms of the environment gave birth to the concept of green management, leading to the adoption of environmentally friendly practices in companies. An equilibrium between sustainability and profitability can be brought about if companies try to mitigate their environmental effects while operating.

Companies can maximize profitability by reducing waste generation and by maximum utilization of resources. For instance, using more renewable energy products is beneficial because it reduces overall effective expenditure and has no negative impact on the environment. According to Dewi and Sari (2023), green management can help improve customer satisfaction, increase loyalty, better the brand image, and give a firm a competitive edge. Green activities may lead to innovation, which can provide new goods and manufacturing methods that satisfy customers' increasing needs for sustainable products (Nurhidayat et al. 2023).

Global climate change has also prompted governments to impose stricter compliance. Green management aligns with the corporate social responsibility (CSR) initiative of firms. Green management incorporates a wide range of practices that help mitigate the adverse impact businesses have on the environment (Kusdi 2018).

DOI: 10.1201/9781003626527-6

These practices can include minimizing energy consumption, effective utilization of resources, waste minimization, and adopting sustainable supply chain management. The main objective is not only to promote sustainability but to use it as an asset to maximize profitability.

The efficiency of green management techniques has been greatly increased by the incorporation of artificial intelligence (AI), data analytics, machine learning (ML), and automation. Huge datasets on energy use, waste creation, and pollutants have been analyzed by AI and ML algorithms on a massive scale, helping businesses improve resource use and lessen their environmental effect (Balcıoğlu et al. 2024). AI-driven systems can assess data in real time to predict energy usage patterns and make changes that reduce waste and emissions. According to Turner et al. (2023), autonomous production systems facilitate life cycle management and predictive maintenance, thereby mitigating environmental impacts and advancing the circular economy. Implementing these cutting-edge technologies has helped businesses meet their sustainability targets and improve their overall performance and competitive posture in the market (Jankovic and Curovic 2023).

The extensive literature on the relationship between green management and business performance has demonstrated a mixed outcome. Some studies demonstrate a positive influence of green management on a firm's performance, while others show a negative or minimal impact. These diverse findings can be due to various factors like firm size, industry characteristics, and market conditions.

In light of the importance of sustainable business practices, which not only protect the environment but also affect a firm's performance, this chapter aims to present a systematic literature review that delves deep into the study of the current body of literature on the relationship between green management and firm performance, identifying the different patterns and explanations for it. It also aims at finding the gaps in the current literature for future studies. The findings of this chapter aim to enhance comprehension of the strategic importance of environmental sustainability in the corporate setting.

6.1.1 Research Questions

The study aims to look into the influence green management has on the firm's performance, and the following three research questions help dig deep into the subject:

- How do green management techniques affect companies' financial and non-financial performance in different sectors? Further, do AI, ML, and automation affect firm performance?
- Which processes and pathways connect green management practices with business performance, and how do they differ based on organizational size and geography? This question recognizes that the effectiveness and outcomes of green management practices can be greatly affected by organizational size (e.g., small and medium-sized enterprises vs. large corporations) and geography (developing vs. developed countries).
- How do regulatory contexts and policies influence the relationship between green management practices and corporate performance?

6.1.2 Frameworks for Selecting Studies for Review

A system implementation requires systematic planning, uniform execution, and proper documentation to ensure that the process can be replicated, which is essential for conducting systematic literature reviews. Anticipating challenges, mitigating subjective choices, promoting accountability, and upholding the credibility of the research can be ensured by developing a methodological approach. Several frameworks like PRISMA, SPAR-4, and TCCM can be employed to ensure transparent, reproducible, and comprehensive reviews and thus enhance the validity, reliability, and quality of the research.

6.2 PRISMA FRAMEWORK

PRISMA (Preferred Reporting Items for Systematic Reviews and Meta-Analyses) is a streamlined set of criteria designed for reporting evidence-based systematic reviews and meta-analyses. PRISMA can serve as a foundational framework for systematic reviews across a wide array of objectives, and it also aims to detail studies that assess the efficiency of various interventions. PRISMA helps study the papers concerning the effect that green management has on firm performance. This investigation not only helps in understanding what each study entails, but it also helps in finding gaps in the literature for future contributions. PRISMA was thus used to categorize research papers that were commonly referenced to address specific questions and propose future research directions (Table 6.1).

Using PRISMA modeling, as demonstrated by Figure 6.1, the relevant papers were identified for the review of literature. The registered papers (constituting a total of n = 1241) were downloaded from the Scopus (n = 393) and Web of Science (n = 848) databases. Of the total, 627 papers were excluded in the identification stage: 201 papers were excluded for being duplicates in both databases; and 426 non-academic papers, books, book chapters, and conference proceedings were excluded (only journal articles were included). The screening stage follows the identification stage. After a thorough reading of the abstracts of all the papers, 441 reports were excluded, followed by the removal of 88 papers as they were not ABDC listed, leaving 85 papers that were relevant to the topic. Finally, 5 papers could not be retrieved from the databases, leaving a total of 80 papers. Hence, 80 papers were screened for our systematic literature review. Table 6.2 provides the bibliometric profile of the 80 papers that were screened.

The 80 screened papers are derived from 36 journals, of which 60 are A, 7 are A*, 6 are B, and 7 are C from the ABDC-listed journals. The geographic distribution of the study reveals that the majority of the papers are from China (n = 10), Spain (n = 8), and India (n = 6). Also, the yearwise distribution indicates that about 12 papers were published in the year 2016, 10 in 2018, and 8 in 2019.

6.2.2 SPAR-4 Framework

PRISMA helps to record articles in an organized, clear, and systematic manner, but it does not provide justifications and reasons for decisions taken during the review process. Thus, it is best suited for systematic literature reviews. To overcome the

TABLE 6.1
Journals Included in the Review

Journal	Number of Papers
Journal of Cleaner Production	20
Business Strategy and the Environment	7
International Journal of Production Economics	7
Journal of Environmental Management	5
IEEE Transactions on Engineering Management	2
Industrial Management & Data Systems	2
International Journal of Contemporary Hospitality Management	2
International Journal of Hospitality Management	2
International Journal of Production Research	2
Journal of Business Ethics	2
Management Decision	2
Technological Forecasting and Social Change	2
Technology in Society	2
Corporate Social Responsibility and Environmental Management	1
Ecological Economics	1
Economic Analysis and Policy	1
Economies	1
International Journal of Energy Economics and Policy	1
International Journal of Operations & Production Management	1
International Journal of Productivity and Performance Management	1
International Journal of Technology Management	1
International Journal of Tourism Research	1
Journal of Business & Industrial Marketing	1
Journal of Engineering and Technology Management	1
Journal of Engineering, Design and Technology	1
Journal of Environmental Planning and Management	1
Journal of Global Responsibility	1
Journal of Korea Trade	1
Journal of Manufacturing Technology Management	1
Journal of Operations Management	1
Journal of Supply Chain Management	1
Long Range Planning	1
Management Accounting Research	1
Management Science	1
Production and Operations Management	1
Tourism Management	1
Total	**80**

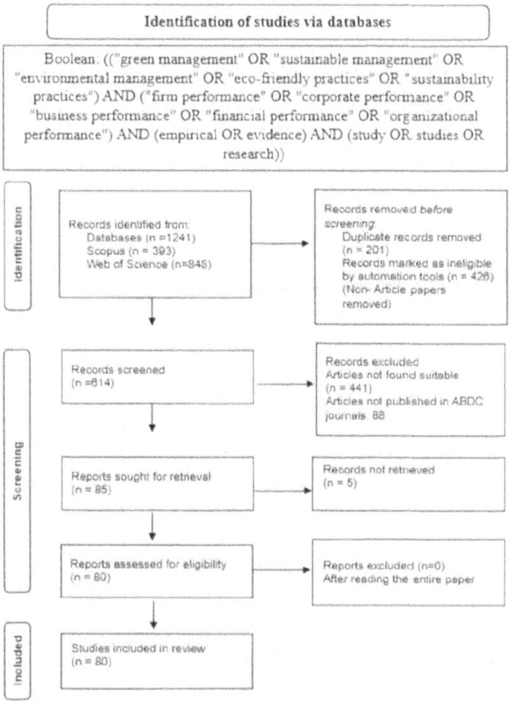

FIGURE 6.1 PRISMA flowchart for selecting studies that are included in the review.

aforementioned limitations, an alternate method of review called the SPAR-4 SLR (Scientific Procedures and Rationales for Systematic Literature Review) protocol can be used. The process comprises three main stages and six substages that follow each other in a sequential manner. The assembly stage involves exploring and finding non-synthesized literature, the arranging stage involves organizing and cleaning the semi-synthesized literature, and assessing the synthesized literature. The SPAR-4 SLR procedure made it possible to complete tasks that were supported by useful and rational arguments, enabling comprehension of the review domain. Figure 6.2 and Table 6.3 methodically explain the SPAR-4 process.

6.3 INSIGHTS FROM REVIEW USING TCCM FRAMEWORK

The TCCM (Theory, Context, Characteristic, and Method) approach conceptualized by Paul and Rosado-Serrano (2019) is recognized as an organizing framework for delving into the theories, context, and methodology present in the literature. The information is gathered from various research papers and presented in the form of an Excel spreadsheet with the headings Authors, Title name, Year, Source, In-text citations, Theory, Variables used, Data, Geography, Methodology, and Results. TCCM is used to answer the research question systematically. Figure 6.3 is a demonstration of the TCCM framework and is explained in the following sections.

TABLE 6.2
Year, Journal ABDC Rank, and Geography of Papers Included in the Review

Year	Number of Papers	Geography	Number of Papers
1996	1	China	10
2003	1	Spain	8
2006	1	India	6
2007	1	Italy	3
2010	3	Pakistan	3
2011	5	Korea	2
2012	9	United Kingdom	2
2013	2	United States	2
2014	1	Australia	1
2015	11	Brazil	1
2016	12	Egypt	1
2017	7	Europe	1
2018	10	Finland	1
2019	8	Germany	1
2020	4	Ghana	1
2021	1	Indonesia	1
2022	2	Kenya	1
2023	1	Malaysia	1
Total	80	Mexico	1
		Netherlands	1
Journal Rank	**Number of Papers**	Oman	1
A*	7	Palestine	1
A	60	Slovakia	1
B	6	Taiwan	1
C	7	Tunisia	1
Total	80	USA	1
		Multiple countries	7
		Others	19
		Total	80

6.3.1 Theories

In the context of our present study looking into the impact green management has on the firm's performance, the TCCM framework plays a crucial role in arranging and analyzing the research output via numerous theoretical perspectives. Some of the theories are listed next:

- Contingency theory states that the success of an organization depends on the link between the organization's external environment and its internal capacities. In green management, the efficiency of various environmental strategies could depend on several regulatory climates or the industry's background.

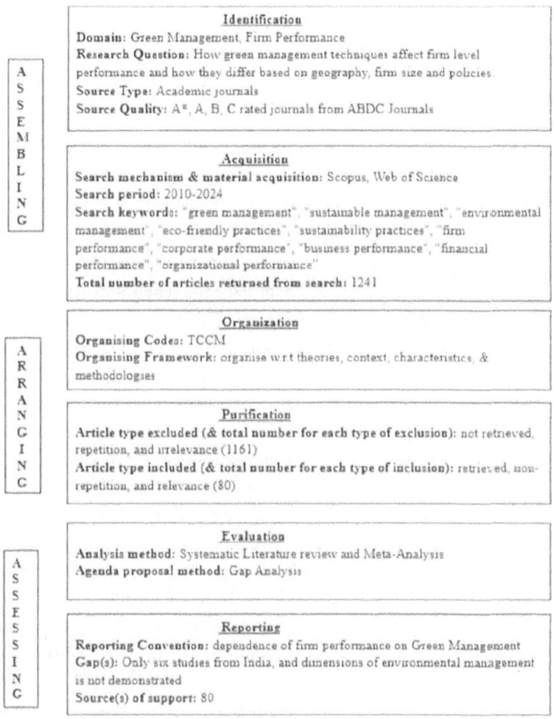

FIGURE 6.2 SPAR-4 framework for selecting studies to be included in the review.

- Resource-based view theory suggests that a firm's control over resources that are unique, sparse, valuable, and non-substitutable allows it to gain a competitive advantage. With its direct influence on company performance, green management should be seen as a positive factor that can improve firm development.
- Super modularity theory demonstrates how systems in organizations can adopt changes effectively. For green management, this could refer to technological changes or changes in rules and regulations concerning environmental practices.
- Practice-based view aims at studying the practices and routines that are followed in an organization. This could thus help understand how green management impacts the performance of a firm.
- Dynamic capabilities theory helps an organization keep up with the rapidly ever-changing environment by focusing on the firm's ability to reconfigure, integrate, and build. It paves the way for the idea of better economic and environmental performance.
- Stakeholders theory highlights the importance of all stakeholders, not just shareholders. Green management improves relationships with all stakeholders in matters concerning environmental regulators and communities.

TABLE 6.3
Rationales for SPAR-4 Framework

Stage	Substage	Criterion	Rationale
Assembling	Identification	Domain	The study will focus on research conducted in the domain of sustainability, particularly in relation to green management adoption in firms.
		Research question	How green management techniques affect firm-level performance, and how they differ based on geography, firm size, and policies.
		Source type	For the source type, only academic journals were considered. Research papers from other journals, including non-academic journals, books and book chapters, and conference proceedings, are being excluded.
		Source quality	The scholarly journals selected for review in this study are mentioned in Table 6.1. All the journals are ABDC-listed journals that rank as A*, A, B, C, etc. The inclusion of ABDC journals was based on their recognition as the most popular and reputable publications in the field of business, as determined by the Australian Business Deans Council Journal Quality List (ABDC JQL). By including more journals, the ABDC JQL expands its collection of papers in developing fields from reputable journals, resulting in a wider pool of resources. Research publications with low rankings (D and O) in the ABDC Journal are excluded from the evaluation process.
	Acquisition	Search mechanism and material acquisition	Researchers used search engines such as Scopus or Web of Science (WOS) databases. Downloadable bibliometric data is available from WOS and Scopus, which is helpful for reviews that primarily rely on bibliometric analysis. When faced with a significant number of articles in the domain of energy economics, the WOS database was utilized to select a smaller, more manageable collection of articles for review. In contrast, when dealing with a smaller number of articles in the field of energy economics related to the adoption of renewable energy (RE), the Scopus database has a wider collection of papers from developing fields that are published in prestigious journals.
		Search period	The review consists of research papers from 1996 to 2023.
		Search keyword(s)	Boolean words that are used for the selection of the papers are "green management," "sustainable management," "environment management," "corporate performance," "firm performance," "financial performance," and "organizational performance."
Arranging	Organization	Organizing codes	Our framework-based review relies on the TCCM framework that can code articles in terms of theories, context, characteristics, and methodologies.

(Continued)

TABLE 6.3 (CONTINUED)
Rationales for SPAR-4 Framework

Stage	Substage	Criterion	Rationale
		Organizing framework	The articles were organized in terms of different theories (stakeholder theory, resource-based theory, practice-based theory) used in their research literature, different contexts (green management parameters, firm performance parameters, mediating variables that help green management improve firm performance, policy factors) with which research is approached, different characteristics (return on sales, firm size, eco-design, gender diversity) that are the parameters of the context in our framework-based reviews, and the methodology (hierarchical regression, stochastic frontier analysis, partial least squares) with which the study is approached.
	Purification	Article type	Certain papers were excluded that are irrelevant to the context of the proposed TCCM framework, with irrelevance in the theories, context, characteristics, and methodologies.
Assessing	Evaluation	Analysis method	A deliberate and comprehensive technique to examine the body of research already conducted on a particular subject or question is the systematic literature review. It entails setting inclusion and exclusion criteria, clearly stating the goals of the study, and methodically looking up pertinent literature. Once the studies that fit the criteria have been chosen, the review assesses their quality and combines their results to give a thorough summary of the state of the field today.
		Agenda proposal method	A review of the body of existing literature is used to determine which factors of green management affect which factors of firm performance. These factors and the research gap inform our future research strategies, ensuring that the recommendations for curating new literature and reviewing the body of existing material are not mutually exclusive. The research question and proposition are examined, and recommendations for further investigation are made.
	Reporting	Reporting convention	A systematic literature review of 80 papers provided an overview of green management adoption.

- Social exchange theory involves resource exchange during social interactions. Improved social capital and corporate reputation are how green management can help improve firm performance.
- Resource dependence theory centers around the strategy formation by companies, which is driven by their requirements. In connection with the theory, application of green management practices helps corporations decrease their reliance on natural resources, thereby lowering the risk arising due to their restricted supply.

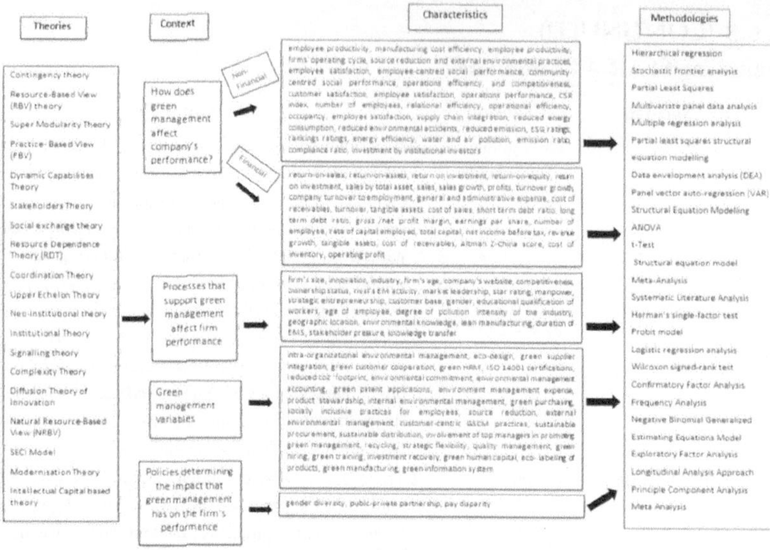

FIGURE 6.3 TCCM framework.

- Coordination theory focuses on the integration of activities and information in organizations. This theory explains the enhancement of corporate performance through applying green management practices.
- Upper echelon theory states that firm performance is affected by the cognitive biases of top management, whose commitment toward sustainability also drives the application of effective green management practices.
- Neo-institutional theory studies how organizational behavior is affected by the institutional structure. Pressures of an establishment can lead to the adoption of green management.
- Institutional theory studies how coercive, normative, and mimetic pressures can lead a firm to adopt green practices, like legitimacy-seeking behavior.
- Signaling theory states organizations can implicate their intentions and quality using their actions. Social responsibility and ethical governance can signal the sustainable practices of an organization.
- Complexity]theory demonstrates how outcomes can be achieved by communication among complex systems. It shows how the complex interplay of green management can help achieve better outcomes.
- Diffusion of innovation theory looks into how technologies and ideas can be dispersed in a society. It examines the adoption of a new idea like green management across sectors.

6.3.2 Context

In this study, four important aspects are looked into, the primary one being what the main factors of green management are. There is a variation in the perception of green management practices in different sectors or industries. For example, service industries will have a different set of challenges compared to heavy industries like those engaged in manufacturing. The following part concerns which firm performance indicators are affected by green management. The firm performance is broadly classified into financial and non-financial parameters.

Next are the mediating factors that help green management improve firm performance. For example, different geographic locations may have varying environmental conditions, rules, and regulations. The final aspect is the policies that enable sustainable performance to improve the firm's performance. This includes specific laws and policies concerning sustainability that may not be legally binding but are adopted by the government to promote sustainable practices.

- The impact of green management on the financial performance of the company: Research has shown that environmentally conscious actions generally lead to better economic outcomes, though the tenure and impact of these advantages can differ. This suggests a relationship between sustainable management and firm financial performance. Research by Zhang et al., (2019) and Chen et al. (2015) shows a clear correlation between improved financial indicators and green supply chain management methods as well as senior management's environmental commitment. The environmental orientation of a company and the level of sophistication of its environmental management systems (EMS) have been found to favorably affect financial performance (Lo et al. 2012; Martin-de Castro et al. 2016). According to Esfahbodi et al. (2016), there are challenges in adopting sustainable supply chain management (SSCM) in emerging economies. The advantages in terms of the economy are real, but they are context-specific and do not necessarily materialize right away. Wahba (2010) highlights the importance that certifications like ISO 14001 and EMAS play in improving both environmental and financial performance, with a particularly strong impact when linked to environmental innovation. Longitudinal research by Song et al. (2017) and Lannelongue et al. (2015) indicates that environmental responsibility and long-term economic success are aligned, implying that while improvements may not always be apparent right away, they can eventually materialize. Martí-Ballester and Simon (2017) study supports this viewpoint by showing how management system integration improves business financial performance.
- Context of the influence of green management on the company's non-financial performance: Recent studies have found the impact that green management has on companies. Wong et al. (2015), Yu and Ramanathan (2016), and Treacy et al. (2019) discovered that customer satisfaction, operational efficiency, and financial performance are improved by green management. Lee et al. (2012) and Sheyadi et al. (2019) stress the importance of environmental management as a means to ensure the social, environmental, and overall success of a firm. Moreover, Gonzalez-Benito et al. (2011) found that strategic flexibility helps strengthen competitive advantage and organizational legitimacy.

To improve performance while also being environmentally sound, and to ensure corporate transparency, Ikram et al. (2019) and Mungai et al. (2020) believe that environmental certifications such as ISO 14001 are not just a requirement but also a necessity. In contrast, Zaid et al. (2018) and Xu and Gursoy (2015) demonstrate the need for green human resources and hospitality management to improve various performance-related parameters. Also, the ISO 14001 certification brings about standardization in the domain of green management, which is crucial to ensure its efficiency (Lundgren and Zhou 2017; Curkovic and Sroufe 2011). Finally, Zobel (2016), and Prajogo et al. (2012) studied the complexities that come in the way of adopting ISO 14001 and its importance in improving the economic well-being of a firm.

- Mediating factors: Erauskin-Tolosa as al. (2020) and Afum et al. (2023) noted export orientation to be one of the primary factors that encourage enterprises to use EMS. Integration of internal and external GSCM gravely depends on IT infrastructure. The works of Mazzi et al. (2016) and Yu and Huo (2019) show that internal corporate factors like job satisfaction and lean management practices are key mediators in helping green management improve firm performance. An environmental management plan doesn't directly cater to improving competitiveness but improves operational and environmental performance, which is backed by social and human capital. Studies by Al-Sheyadi et al. (2019), Shahab et al. (2018), Petrova et al. (2021), Yang et al. (2015), and Martin-de Castro et al. (2016), among others, stress the importance of firm size and firm age as mediating factors, whereby larger and older firms show a stronger relationship with stakeholders' interest in implementing environmental management programs.

Stakeholder involvement and corporate incentives also affect the impact green management has on firm performance, according to the works of Kasim (2015), Paulraj et al. (2017), and Papagiannakis et al. (2019). In contrast, Hamdoun et al. (2018) and Singh et al. (2015) discovered that stakeholder engagement and the practice of a voluntary environmental system improve environmental innovation and performance. The composition of top management is a key factor that determines what energy efficiency policies will be adopted and their ultimate impact on financial stability and productivity (Buffa et al. 2018; Vijayvargy et al. 2017). Customer satisfaction is ensured by green management, which in turn improves the environmental and economic performance of a firm, as noted by Xu and Gursoy (2015) and Feng and Wang (2016). Further, customer satisfaction boosts customer loyalty and willingness to pay a premium.

- Policies that influence how green management affects a company's performance: Shahab et al. (2018) look at how policy mechanisms that facilitate the use of the SECI (socialization, externalization, combination, and internalization) model can improve environmental knowledge, which is essential for the adoption of green management practices later on. Policies that promote a greater commitment to combining quality management and environmental management processes have the potential to significantly increase performance, according to Grekova et al. (2014). Leonidou and Leonidou (2011) and Ozusaglam et al. (2018) address how state regulation and the implementation of internally or externally certified EMS are dependent on one another. They contend that although externally certified EMS can occasionally take the place of state regulations, direct regulation encourages businesses to build internal EMS.

6.3.3 Characteristics

A review of academic literature indicates that ISO 14001 is a critical variable to consider when analyzing corporate performance in the context of environmental management systems. A rich tapestry is shown by research examining the impact of ISO 14001 on corporate performance. While the incorporation of ISO 14001 improves the performance and standardization of environmental management, Curkovic and Sroufe (2011) found that this does not always translate into better commercial success. Although there may be subtle advantages to certification for waste and energy management, Link and Naveh (2006) found no appreciable differences in performance between ISO 14001-certified and non-certified businesses. According to Zobel (2016), companies with large financial reserves and limited investment opportunities—which may be attributed to institutional ownership—were more likely to adopt ISO 14001, especially in Egypt. According to Wahba (2010), ISO 14001 and EMAS have a generally good impact on sustainability, with more advantages going to established and creative businesses. Erauskin-Tolosa et al. (2020) conducted an adjustment for publication bias and concluded that exporting enterprises have a real motivation to achieve ISO 14001 certification. This conclusion is corroborated by higher-quality research.

According to recent research, green recruiting and training strategies used in green human resource management (GHRM) are strategically important and may significantly improve the performance of sustainable businesses. Zaid et al. (2018) demonstrate that while external GSCM appears to have less of an influence on social and economic outcomes, internal GSCM practices enhance the beneficial effects of GHRM on sustainability. Green human and relational capital, which is supported by knowledge and connections, has a greater influence on the successful implementation of GHRM within enterprises than does green structural capital, according to Yong et al. (2019).

ISO 14001 and EMAS certifications are associated with green management and higher return on assets (ROA) and return on equity (ROE). Lopez-Gamero et al. (2010) state that the improvement in financial performance resulting from ISO 14001 and EMAS certifications implies a positive impact on ROA and ROE. Corporate performance is not considerably impacted by the straightforward implementation of EMS without certification, nevertheless. Treacy et al. (2019) claim that consumer cooperation and green purchasing might boost financial performance by raising ROA and ROE. However, Feng and Wang (2016) point out that the benefits of eco-friendly management strategies—like product stewardship—depend on the environmental management capability of suppliers, suggesting a variable impact on company performance. Normative and mimetic pressure-driven eco-innovation is a sign of enhanced market competitiveness, which might lead to increased ROA and ROE, according to Wong et al. (2012). Finally, Martí-Ballester and Simon (2017) demonstrates that integrating management systems, including environmental standards, may enhance a company's financial performance and positively impact financial ratios like ROA and ROE, even in the face of specific challenges.

Studies indicate that the implementation of green management practices has a noteworthy impact on sales revenue. Client-centric GSCM methods, especially those that enhance product quality and delivery, boost customer satisfaction and sales profitability (Yu and Ramanathan 2016). But Feng and Wang (2016) caution that the success of environmental programs, like product stewardship, is dependent on suppliers' ability to manage the environment, which has an impact on sales. Chen et al. (2015) claim that senior management's environmental commitment, which also has the ability to boost organizational effectiveness and sales, is the reason why green practices are embraced. Martí-Ballester and Simon (2017) note that integrating management systems can boost sales income despite challenges. Green management strategies are strongly associated with corporate performance, even though the outcomes differ. Heras-Saizarbitoria et al. (2011) state that industry, company size, and R&D expenditure have a positive impact on the adoption of EMS. On the other hand, the results might differ in terms of profitability. Studies by Frondel et al. (2018) and Lannelongue et al. (2015) indicate that improved environmental performance and EMS adoption benefit bottom lines for enterprises. Hofer et al. (2012) warn that although successful environmental outputs might raise profitability, early environmental management inputs may decrease it. According to Paulraj et al. (2017) and Afum et al. (2023), business goals and IT resources promote e-GSCM integration and SSCM practices, which both significantly increase profitability.

Green management may lower CO_2 emissions and improve business success. According to research by Lannelongue et al. (2017), there may be trade-offs between strict environmental requirements and short-term market success. The study indicated a negative link between environmental requirements (in regard to CO_2 emissions) for providers and sales growth. According to Lundgren and Zhou (2017), CO_2 emissions may be decreased by implementing GSCM techniques, particularly in eco-design and green buying.

The age and size of the company have a big impact on how well environmental practices work for the company. Heras-Saizarbitoria et al. (2011) note that larger companies with higher R&D spending are more likely to use EMS, taking advantage of their greater resources and influence. Yang et al. (2015) discovered that the association between stakeholders' influence and environmental management practices is moderated by the size and age of the hotel, with older and larger hotels exhibiting a greater correlation.

Green management requires business innovation, which has a big influence on the company's performance. According to Al-Sheyadi et al. (2019), a company's commitment to sustainability and green innovation significantly boosts performance. In a similar way, market competitiveness is enhanced by eco-innovation, which is driven by normative and mimetic factors (Wong et al. 2012). However, Lannelongue et al. (2017) discovered that while most environmental management practices do not directly increase financial performance, they do have an association with creativity, which is an essential component in the creation of new goods and better processes. Hamdoun et al. (2018) demonstrate that EMS alone is insufficient for environmental product innovation in the absence of stakeholder interaction.

When a corporation has previously used quality management systems, this effect is increased. According to Tari et al. (2010), EMS may increase the link between environmentally conscious product creation and commercial success, underlining the challenges associated with both compliance and innovation.

Competitiveness is substantially increased by green management techniques. Green management practices (GMPs), when supported by institutional frameworks, enhance competitive advantages through greater organizational legitimacy and resource relocation efficacy. High strategic flexibility enterprises are more likely to utilize GMPs (Gonzalez-Benito et al. 2011). Businesses that adopt ecologically friendly methods may stand out in the marketplace, according to Wong et al. (2012). Eco-innovation success is positively correlated with market competitiveness and is influenced by normative and mimetic forces.

6.3.4 Methodologies

A wide variety of methodologies have been adopted to systematically analyze the impact green management has on firm performance. ANOVA (analysis of variance) was employed by Kasim (2015) and Grekova et al. (2014) to find differences in their datasets. Both Vijayvargy et al. (2017) and Wong et al. (2015) used DEA (data envelopment analysis), a non-parametric technique used in economics and operational research, to analyze the effectiveness of green supply chain initiatives.

Tan and Zhu (2022), Zhu et al. (2012a), and Kasim (2015) adopted exploratory factor analysis (EFA) to understand the primary relationship between observable factors and latent components. Treacy et al. (2019), Lundgren and Zhou (2017), and many more used multiple regression analyses to examine the determinants of green management and how it impacts the performance of a firm. Regression analysis helps to demonstrate the incremental effect of an independent variable. Studies by Wahba (2010), Buffa et al. (2018), Prajogo et al. (2012), and many more use meta-analyses that help synthesize information from many studies to offer a comprehensive understanding of green management practices. Gonzalez-Benito et al. (2011) and Laari et al. (2016) used systematic equation modeling (SEM), a statistical technique that reflects the complex connections between observed and latent variables. Treacy et al. (2019), Lo et al. (2012), Wong et al. (2012), Kim et al. (2012), Hofer et al. (2012), Tari et al. (2010), and Lundgren and Zhou (2017) used hierarchical multiple regression analyses, which helps understand the additional effect of adding a set of variables in a regression model. Heras-Saizarbitoria et al. (2011), Al-Sheyadi et al. (2019), Shahab et al. (2018), Zaid et al. (2018), and many more used partial least squares equation analysis, which becomes an important tool when dealing with a complex model and smaller sample sizes. Exploratory factor analysis was adopted by Tan and

Zhu (2022), Kasim (2015), and Zhu et al. (2012b), which helps identify the structure of a large set of variables.

In conclusion, these methodical applications demonstrate that, while most academics concur that green management improves business performance, the precise nature of this impact varies based on the industry, size of the organization, and particular management techniques employed. The use of advanced statistical techniques indicates that the field is expanding as research builds upon increasingly robust and sophisticated analytical foundations.

6.4 IMPACT OF AI, ML, AND AUTOMATION ON FIRM PERFORMANCE

Automation, machine learning, and artificial intelligence have become key technologies impacting business performance in a variety of fields. According to Mikalef et al. (2023), AI competencies boost business-to-business (B2B) marketing capabilities, which in turn improve organizational performance by simplifying operations and providing insights into client behavior. In a similar vein, Singh et al. (2024) demonstrate that generative AI (GenAI) encourages both exploitative and exploratory innovation, which enhances organizational performance, especially when environmental dynamism and ethical quandaries are taken into account. Ye et al. (2024), concentrating on the Chinese market, show that generative AI enhances firm performance via improved managerial connections, which are mediated by technology developments.

Samadhiya et al. (2024) emphasize how combining AI with the Internet of Things (IoT) and big data analytics (BDA) promotes creativity and productivity, resulting in a dynamic environment that improves business performance in the context of developing technologies like the Metaverse. In their investigation of the adoption of generative AI and its ethical ramifications, Rana et al. (2024) point out that institutional pressures influence AI adoption, which improves organizational performance by fostering greater accountability, transparency, and justice. According to Fredström et al. (2022), AI improves corporate performance, especially in marketing, by creating more favorable market emotions, especially when businesses prioritize innovation and teamwork.

Mishra et al. (2022) also draw a connection between the use of AI and enhanced operational effectiveness, emphasizing that the use of AI in financial reporting is linked to higher profitability and lower marketing costs. In their analysis of AI's function in knowledge generation, Bag et al. (2021) demonstrate how AI-powered insights into market and consumer behavior result in more logical B2B marketing decisions, which have a big impact on business performance. Last but not least, Mikalef et al. (2023) show that AI capabilities in public companies improve cognitive insights and process automation, both of which have a favorable impact on performance.

Visani et al. (2024) highlight the importance of machine learning in performance management by showing how machine learning makes it easier to discover business value drivers, which in turn improves firm performance by facilitating more objective decision-making. According to Ranta and Ylinen's (2024) machine learning analysis of employee perks, businesses that provide uncommon advantages like support for family members report increases in employee satisfaction and business performance, particularly in high-growth enterprises.

Alharbi (2024) found that automation increases operational efficiency, sustainability, and profitability in Saudi Arabia's manufacturing sector, which further propels business success. Automation is essential for increasing efficiency and profitability, especially for businesses that are subject to global competition, according to Kromann and Sørensen (2019). However, cognitive engagement may have a more complicated, occasionally detrimental effect on results.

In summary, all of the above-mentioned studies show that automation, AI, and machine learning have improved many facets of organizational operations, greatly increasing business performance. AI has played a key role in improving marketing tactics, offering more in-depth understanding of consumer behavior, and expediting decision-making. Better resource allocation and increased productivity result from machine learning's ability to simplify and improve decision-making, especially in performance management. Automation increases productivity, sustainability, and profitability, particularly in manufacturing and process optimization.

6.5 CONCLUSION

This systematic literature review explores how green management affects firm performance by using SPAR-4, PRISMA, and TCCM frameworks, which ensures comprehensive and rigorous study. From the Scopus and Web of Science databases, 1214 papers were gathered. PRISMA was used to remove duplicates, non-articles, and non-ABDC-listed papers, leaving a total of 80 high-quality articles.

The SPAR-4 framework was also used to justify the selection of papers. Papers published from 2010 to 2024 were reviewed, with the distribution analysis showing the most publications from the *Journal of Cleaner Production* (20), followed by *Business Strategy and Environment* (7) and *Journal of Environmental Planning* (5). The peak publication years are 2015, 2018, and 2016, primarily from China, India, and Spain. Out of the 80 selected papers, 60 were published in A-ranked journals, guaranteeing the impact and credibility of the work considered.

The theoretical and contextual aspects were explored by the TCCM framework, including a diverse selection of theories, including the resource-based view theory, contingency theory, and dynamic capabilities theory. A positive correlation was observed between the impact green management has on the financial performance of a firm, such as ROA and ROE profits as demonstrated by Zhang et al. (2019), Chen et al. (2015), Lo et al. (2012), and Martin-de Castro et al. (2016). Wong et al. (2015), Yu and Ramanathan (2016), and Treacy et al. (2019) have also noted similar results in the non-financial aspects of a company's performance, such as an increase in customer satisfaction, operational efficiency, and employee productivity.

One of the key determinants of these relationships is the size of the organization and the level of competition within the market. Other important factors include future innovations in this area as well as some aspects that are specific to green management such as eco-design or ISO 14001 certification. Additionally, policies on gender diversity or public–private partnerships may also impact these results. This review analysis has found that eco-design, ISO certificates, competitiveness, and innovations are among the important mediating variables that decide the extent of impact of green management practices on firm performance. Structural equation modeling, stochastic frontier analysis, hierarchical regression, and many other analytical techniques were identified as the primary methodological techniques in the 80 research papers.

In conclusion, this review focuses on the numerous benefits that green management impacts firm performance advocating its adoption into corporate strategies. Future research would help explore how evolving technological enhancements and market conditions will extend these findings.

6.6 FUTURE RESEARCH AGENDA

Future research can help improve green management's impact on firm performance. Longitudinal studies are required to assess the long-run implications of green management and its impact on sustainability. Geographical diversification is also essential to explore the impact of green management across borders and in emerging economies with varied regulations. Green management has a different impact on various sectors, ranging from service to manufacturing, and a sector-specific analysis would help reveal these variations. With the recent adoption of digital technologies like AI, IoT, and blockchain, studies concerning how these technologies can improve the adoption of green management to further improve firm performance can be conducted.

Several cultural and behavioral factors of an organization affect the adoption and efficiency of green management. Comparative studies across different firms within a sector can reveal which environmental policy is effective in a given regulatory setup. The development of environmental parameters, apart from financial parameters, and emphasis on their importance would provide a more holistic view of a firm's performance. Additionally, research on the impact on stakeholders like consumers, communities, and suppliers can allow firms to tailor their strategies accordingly. Research on the impact of mediating factors like innovation, firm size, and market conditions would further help green management improve firm performance.

REFERENCES

Afum, E., K. Issau, Y. Agyabeng-Mensah, C. Baah, E. Dacosta, E. Essandoh, and E. Agyenim Boateng. "The missing links of sustainable supply chain management and green radical product innovation between sustainable entrepreneurship orientation and sustainability performance." *Journal of Engineering, Design and Technology* 21, no. 1 (2023): 167–187.

Alharbi, S. S. "Gear up for development: The automation advantage for sustainability in manufacturing in the Kingdom of Saudi Arabia." *Sustainability (Switzerland)* 16, no. 11 (2024). https://doi.org/10.3390/su16114386

Al-Sheyadi, A., L. Muyldermans, and K. Kauppi. "The complementarity of green supply chain management practices and the impact on environmental performance." *Management Accounting Research* 242 (2019): 186–198.

Amores-Salvado, J., G. Martin-de Castro, and J. E. Navas-Lopez. "The importance of the complementarity between environmental management systems and environmental innovation capabilities: A firm level approach to environmental and business performance benefits." *Journal of Cleaner Production* 96 (2015): 288–297.

Balcıoğlu, Y. S., A. A. Çelik, and E. Altındağ. "Artificial intelligence integration in sustainable business practices: A text mining analysis of USA firms." *Sustainability* 16, no. 15 (2024): 6334.

Buffa, F., M. Franch, and D. Rizio. "Environmental management practices for sustainable business models in small and medium-sized hotel enterprises." *Journal of Environmental Management* 194 (2018): 656–664.

Bag, S., S. Gupta, A. Kumar, and U. Sivarajah. "An integrated artificial intelligence framework for knowledge creation and B2B marketing rational decision making for improving firm performance." *Industrial Marketing Management* 92 (2024): 178–189.

Chavez, R., W. Yu, M. Feng, and F. Wiengarten. "The effect of customer-centric green supply chain management on operational performance and customer satisfaction." *IEEE Transactions on Engineering Management* 25 (2016): 205–220.

Chen, L., O. Tang, and A. Feldmann. "Applying GRI reports for the investigation of environmental management practices and company performance in Sweden, China and India." *Journal of Cleaner Production* 98 (2015): 36–46.

Chiappetta Jabbour, C. J., Lopes de Sousa Jabbour, A. B., Govindan, K., Teixeira, A. A, and de Souza Freitas, W. R. "Environmental management and operational performance in automotive companies in Brazil: The role of human resource management and lean manufacturing." *Journal of Cleaner Production* 47 (2013): 129–140.

Curkovic, S., and R. Sroufe. "Using ISO 14001 to promote a sustainable supply chain strategy." *Business Strategy and the Environment* 20, no. 1 (2011): 71–93.

Daddi, T., F. Testa, M. Frey, and F. Iraldo. "Exploring the link between institutional pressures and environmental management systems effectiveness: An empirical study." *Journal of Operations Management* 183 (2016): 647–656.

Das, D. "Sustainable supply chain management in Indian organisations: An empirical investigation." *International Journal of Production Research* 56, no. 16 (2018): 5776–5794.

Das, D. "The impact of sustainable supply chain management practices on firm performance: Lessons from Indian organizations." *Journal of Cleaner Production* 203 (2018): 179–196.

De Giovanni, P. "Do internal and external environmental management contribute to the triple bottom line?" *International Journal of Operations & Production Management* 32, no. 3 (2018): 265–290.

Demirel, P., K. Iatridis, and E. Kesidou. "The impact of regulatory complexity upon self-regulation: Evidence from the adoption and certification of environmental management systems." *Journal of Global Responsibility* 207 (2018): 80–91.

Dewi, C. F., and D. Sari. "Green packaging and branding impact on green purchase intention with green brand image as a mediator: Study at Love Beauty and Planet of Gen Z in West Java." *Journal of Business Studies and Management Review* 7, no. 1 (2023): 1–9.

Erauskin-Tolosa, A., E. Zubeltzu-Jaka, I. Heras-Saizarbitoria, and O. Boiral. "ISO 14001, EMAS and environmental performance: A meta-analysis." *Business Strategy and the Environment* 29, no. 3 (2020): 1145–1159.

Esfahbodi, A., Y. Zhang, and G. Watson. "Sustainable supply chain management in emerging economies: Trade-offs between environmental and cost performance." *Technology in Society* 181 (2016): 350–366.

Fallah Shayan, N., N. Mohabbati-Kalejahi, S. Alavi, and M. A. Zahed. "Sustainable development goals (SDGs) as a framework for corporate social responsibility (CSR)." *Sustainability* 14, no. 3 (2022), 1222.

Feng, T., and D. Wang. "The influence of environmental management systems on financial performance: A moderated-mediation analysis." *Journal of Cleaner Production* 135 (2016): 265–278.

Fredström, A., V. Parida, J. Wincent, D. Sjödin, and P. Oghazi. 2022. "What is the market value of artificial intelligence and machine learning? The role of innovativeness and collaboration for performance." *Technological Forecasting and Social Change* 180. https://doi.org/10.1016/j.techfore.2022.121716

Frondel, M., K. Kraetschell, and L. Zwick. "Environmental management systems: Does certification pay?" *Journal of Cleaner Production* 59 (2018): 14–24.

Geng, R., A. Mansouri, and E. Aktas. "The relationship between green supply chain management and performance: A meta-analysis of empirical evidences in Asian emerging economies." *Journal of Cleaner Production* 183 (2017): 245–258.

Gonzalez-Benito, J., G. Lannelongue, and D. Queiruga. "Stakeholders and environmental management systems: a synergistic influence on environmental imbalance." *Management Decision* 19 (2011): 1622–1630.

Grekova, K., H. J. Bremmers, J. H. Trienekens, R. G. M. Kemp, and S. W. F. Omta. "Extending environmental management beyond the firm boundaries: An empirical study of Dutch food and beverage firms." *Journal of Environmental Management* 152 (2014): 174–187.

Hamdoun, M., C. J. C. Jabbour, and H. Ben Othman. "Knowledge transfer and organizational innovation: Impacts of quality IT and environmental management." *Journal of Cleaner Production* 193 (2018): 759–770.

Heras-Saizarbitoria, I., J. F. Molina-Azorin, and G. P. M. Dick. "ISO 14001 certification and financial performance: Selection-effect versus treatment-effect." *Industrial Management & Data Systems* 19 (2011): 1–12.

Hofer, C., D. E. Cantor, and J. Dai. "The competitive determinants of a firm's environmental management activities: Evidence from US manufacturing industries." *International Journal of Hospitality Management* 30 (2012): 69–84.

Ikram, M., P. Zhou, S. A. A. Shah, and G. Q. Liu. "Do environmental management systems help improve corporate sustainable development? Evidence from manufacturing companies in Pakistan." *Journal of Cleaner Production* 226 (2019): 628–641.

Jankovic, S. D., and D. M. Curovic. "Strategic integration of artificial intelligence for sustainable businesses: Implications for data management and human user engagement in the digital era." *Sustainability* 15, no. 21 (2023): 15208.

Johnstone, L. "The construction of environmental performance in ISO 14001-certified SMEs." *Journal of Cleaner Production* 263 (2020): 121559.

Kasim, A. "Environmental management system (EMS) postulating the value of its adoption to organizational learning in hotels." *International Journal of Contemporary Hospitality Management* 27, no. 6 (2015): 1233–1253.

Khalid, J., M. Chuanmin, F. Altaf, M. M. Shafqat, S. K. Khan, and M. U. Ashraf. 'AI-driven risk management and sustainable decision-making: Role of perceived environmental responsibility." *Sustainability* 16, no. 16 (2024): 6799.

Kim, C.-B., S.-N. Jung, and S. E. Ronto. "The impact of green-oriented supply chain management practices and environmental management systems on the organizational performance of Korean manufacturers." *Journal of Environmental Management* 16 (2012): 27–55.

Klassen, R. D., and C. P. McLaughlin. "The impact of environmental management on firm performance." *Management Science* 42, no. 8 (1996): 1199–1214.

Kromann, L., and A. Sørensen. "Automation, performance and international competition: A firm-level comparison of process innovation." *Economic Policy* 34, no. 100 (2019): 691–722.

Kusdi, R. "The role of green management in creating sustainability performance on the small and medium enterprises." *Management of Environmental Guality: An International Journal* (2018).

Laari, S., J. Toyli, T. Solakivi, and L. Ojala. "Firm performance and customer-driven green supply chain management." *Journal of Cleaner Production* 112 (2016): 1960–1970.

Lannelongue, G., J. Gonzalez-Benito, and O. Gonzalez-Benito. "Input, output, and environmental management productivity: Effects on firm performance." *Journal of Cleaner Production* 24 (2015): 145–158.

Lannelongue, G., J. Gonzalez-Benito, and I. Quiroz. "Environmental management and labour productivity: The moderating role of capital intensity." *Journal of Supply Chain Management* 190 (2017): 158–169.

Lee, E., S. H. Jo, and H. Jeong. "Environmental management in small and medium enterprises: The role of customer orientation and firm performance." *International Journal of Production Economics* 34 (2019): 1779–1790.

Lee, S. M., S. T. Kim, and D. Choi. "Green supply chain management and organizational performance." *International Journal of Production Economics* 112 (2012): 1148–1180.

Leonidou, C. N., and L. C. Leonidou. "Research into environmental marketing/management: A bibliographic analysis." *Economic Analysis and Policy* 45 (2011): 68–103.

Lin, H.-F. "IT resources and quality attributes: The impact on electronic green supply chain management implementation and performance." *Technology in Society* 68 (2022): 101833.

Link, S., and E. Naveh. "Standardization and discretion: Does the environmental standard ISO 14001 lead to performance benefits?" *IEEE Transactions on Engineering Management* 53, no. 3 (2006): 508–519.

Lo, C. K. Y., A. C. L. Yeung, and T. C. E. Cheng. "The impact of environmental management systems on financial performance in fashion and textiles industries." *Journal of Business & Industrial Marketing* 135 (2012): 561–567.

Lopez-Gamero, M. D., J. F. Molina-Azorin, and E. Claver-Cortes. "The potential of environmental regulation to change managerial perception, environmental management, competitiveness and financial performance." *Journal of Business Ethics* 18 (2010): 963–974.

Lundgren, T., and W. Zhou. "Firm performance and the role of environmental management." *Journal of Environmental Management* 203 (2017): 330–341.

Martí-Ballester, Carmen P., and Antonio Simon. "Union is strength: The integration of ISO 9001 and ISO 14001 contributes to improve the firms' financial performance." *Management Decision* 55, no. 1 (2017): 81–101.

Martin-de Castro, Gregorio, Jose Amores-Salvado, and Jose Emilio Navas-Lopez. "Environmental management systems and firm performance: Improving firm environmental policy through stakeholder engagement." *International Journal of Production Economics* 23 (2016): 243–256.

Martínez-Martínez, Antonio, Jose-Gines Cegarra-Navarro, and Antonio García-Pérez. "Environmental knowledge management: A long-term enabler of tourism development." *International Journal of Production Economics* 50 (2015): 281–291.

Mazzi, Anna, Stefano Toniolo, Michele Mason, Francesco Aguiari, and Alberto Scipioni. "What are the benefits and difficulties in adopting an environmental management system? The opinion of Italian organizations." *Journal of Cleaner Production* 139 (2016): 873–885.

Mikalef, P., N. Islam, V. Parida, H. Singh, and N. Altwaijry. "Artificial intelligence (AI) competencies for organizational performance: A B2B marketing capabilities perspective." *Journal of Business Research* 164 (2023): 113998.

Mikalef, P., K. Lemmer, C. Schaefer, M. Ylinen, S. O. Fjørtoft, H. Y. Torvatn, M. Gupta, and B. Niehaves. "Examining how AI capabilities can foster organizational performance in public organizations." *Government Information Quarterly* 40, no. 2 (2023): 101797.

Mishra, S., M. T. Ewing, and H. B. Cooper. "Artificial intelligence focus and firm performance." *Journal of the Academy of Marketing Science* 50, no. 6 (2022): 1176–1197.

Mungai, Elizabeth M., Samuel W. Ndiritu, and Tazeeb Rajwani. "Do voluntary environmental management systems improve environmental performance? Evidence from waste management by Kenyan firms." *Journal of Cleaner Production* 265 (2020): 121636.

Nurhidayat, P., S. M. Indah, A. Vebiola, S. Atmanegara, A. Nurhasanah, and A. Octavia. "Eco-innovative strategies: Enhancing environmental awareness and shaping consumer preferences through green marketing." *Journal of Business Studies and Management Review* 7, no. 1 (2023): 127–131.

O'Donohue, Wayne, and Nattavudh Torugsa. "The moderating effect of 'Green' HRM on the association between proactive environmental management and financial performance in small firms." *Journal of Environmental Management* 27 (2016): 239–261.

Ozusaglam, Serdal, Eleni Kesidou, and Chee Yew Wong. "Performance effects of complementarity between environmental management systems and environmental technologies." *Economies* 197 (2018): 112–122.

Papagiannakis, Giorgos, Ioannis Voudouris, Spyros Lioukas, and George Kassinis. "Environmental management systems and environmental product innovation: The role of stakeholder engagement." *Business Strategy and the Environment* 28 (2019): 939–950.

Paul, J., and A. Rosado-Serrano. "Gradual internationalization vs born-global/international new venture models: A review and research agenda." *International Marketing Review* 36, no. 6 (2019): 830–858.

Paulraj, Antony, Ivan J. Chen, and Christoph Blome. "Motives and performance outcomes of sustainable supply chain management practices: A multi-theoretical perspective." *Journal of Business Ethics* 145, no. 2 (2017): 239–258.

Petrova, Eva, Tomas Stofa, and Marek Soltes. "Exploration of the factors that influence the implementation of environmental management systems—The case of Slovakia." *Business Strategy and the Environment* 9 (2021).

Prajogo, D., A. K. Y. Tang, and K.-H. Lai. "Do firms get what they want from ISO 14001 adoption?: An Australian perspective." *Journal of Cleaner Production* 33 (2012): 117–126.

Qiang, G., S. Tang, J. Hao, L. Di Sarno, G. Wu, and S. Ren. "Building automation systems for energy and comfort management in green buildings: A critical review and future directions." *Renewable and Sustainable Energy Reviews* 179 (2023): 113301.

Rana, N. P., R. Pillai, B. Sivathanu, and N. Malik. "Assessing the nexus of generative AI adoption, ethical considerations, and organizational performance." *Technovation* 135 (2024): 103064.

Ranta, M., and M. Ylinen. "Employee benefits and company performance: Evidence from a high-dimensional machine learning model." *Management Accounting Research* 64 (2024): 100876.

Roy, M., and D. Khastagir. "Exploring role of green management in enhancing organizational efficiency in petro-chemical industry in India." *Production and Operations Management* 121 (2016): 109–115.

Samadhiya, A., R. Agrawal, A. Kumar, and S. Luthra. "Bridging realities into organizations through innovation and productivity: Exploring the intersection of artificial intelligence, internet of things, and big data analytics in the metaverse environment using a multi-method approach." *Decision Support Systems* 185 (2024): 114290.

Sanchez-Medina, P. S., R. Diaz-Pichardo, and M. Cruz-Bautista. "Stakeholder influence on the implementation of environmental management practices in the hotel industry." *International Journal of Tourism Research* 18 (2016): 387–398.

Shahab, Y., C. G. Ntim, Y. Chengang, F. Ullah, and S. Fosu. "Environmental policy, environmental performance, and financial distress in China: Do top management team characteristics matter?" *International Journal of Energy Economics and Policy* 27 (2018): 1635–1652.

Singh, K., S. Chatterjee, and M. Mariani. "Applications of generative AI and future organizational performance: The mediating role of explorative and exploitative innovation and the moderating role of ethical dilemmas and environmental dynamism." *Technovation* 133 (2024). https://doi.org/10.1016/j.technovation.2024.103021

Singh, M., M. Brueckner, and P. K. Padhy. "Environmental management system ISO 14001: Effective waste minimisation in small and medium enterprises in India." *Journal of Cleaner Production* 102 (2015): 285–301.

Singh, N., S. Jain, and P. Sharma. "Motivations for implementing environmental management practices in Indian industries." *Ecological Economics* 109 (2015): 1–8.

Somjai, S., R. Fongtanakit, and K. Laosillapacharoen. "Impact of environmental commitment, environmental management accounting and green innovation on firm performance: An empirical investigation." *Tourism Management* 10 (2020): 204–210.

Song, H., C. Zhao, and J. Zeng. "Can environmental management improve financial performance: An empirical study of A-shares listed companies in China." *Corporate Social Responsibility and Environmental Management* 141 (2017): 1051–1056.

Sroufe, R. "Effects of environmental management systems on environmental management practices and operations." *Technological Forecasting and Social Change* 12 (2003): 416–431.

Tan, Y., and Z. Zhu. "The effect of ESG rating events on corporate green innovation in China: The mediating role of financial constraints and managers' environmental awareness." *Journal of Cleaner Production* 68 (2022).

Tari, J. J., E. Claver-Cortes, J. Pereira-Moliner, and J. F. Molina-Azorin. "Levels of quality and environmental management in the hotel industry: Their joint influence on firm performance." *Journal of Cleaner Production* 29 (2010): 500–510.

Tipu, S. A. A., and K. Fantazy. "Exploring the relationships of strategic entrepreneurship and social capital to sustainable supply chain management and organizational performance." *International Journal of Productivity and Performance Management* 67, no. 10 (2018): 2046–2070.

Treacy, R., P. Humphreys, R. McIvor, and C. Lo. "ISO14001 certification and operating performance: A practice-based view." *Business Strategy and the Environment* 20, no. 8 (2019): 319–328.

Turner, C., and J. Oyekan. "Manufacturing in the age of human-centric and sustainable industry 5.0: Application to holonic, flexible, reconfigurable and smart manufacturing systems." *Sustainability* 15 no. 13 (2023): 10169.

Vijayvargy, L., J. Thakkar, and G. Agarwal. "Green supply chain management practices and performance: The role of firm-size for emerging economies." *Journal of Manufacturing Technology Management* 28, no. 3 (2017): 299–323.

Visani, F., A. Raffoni, and E. Costa. "The quest for business value drivers: Applying machine learning to performance management." *Production Planning and Control* 35, no. 10 (2024): 1127–1147.

Wagner, M. "Integration of environmental management with other managerial functions of the firm – Empirical effects on drivers of economic performance." *Long Range Planning* 40, no. 6 (2007): 611–628.

Wahba, H. "How do institutional shareholders manipulate corporate environmental strategy to protect their equity value? A study of the adoption of ISO 14001 by Egyptian firms." *Business Strategy and the Environment* 19, no. 8 (2010): 495–511.

Wong, C. W. Y. "Leveraging environmental information integration to enable environmental management capability and performance." *Technological Forecasting and Social Change* 49 (2013): 114–136.

Wong, C. W. Y., K.-H. Lai, K.-C. Shang, C.-S. Lu, and T. K. P. Leung. "Green operations and the moderating role of environmental management capability of suppliers on manufacturing firm performance." *Industrial Management & Data Systems* 112, no. 3 (2012): 283–294.

Wong, C. Y., C. W. Y. Wong, and S. Boon-itt. "Integrating environmental management into supply chains: A systematic literature review and theoretical framework." *International Journal of Production Economics* 45 (2015): 43–68.

Xu, X., and D. Gursoy. "Influence of sustainable hospitality supply chain management on customers' attitudes and behaviors." *International Journal of Hospitality Management* 49 (2015): 105–116.

Yang, J., F. Zhang, X. Jiang, and W. Sun. "Strategic flexibility, green management, and firm competitiveness in an emerging economy." *International Journal of Technology Management* 101 (2015): 347–356.

Yang, M. G. (Mark), P. Hong, and S. B. Modi. "Impact of lean manufacturing and environmental management on business performance: An empirical study of manufacturing firms." *Business Strategy and the Environment* 129 (2011): 251–261.

Ye, J., S. Wang, and S.-B. Tsai. "Impact of generative AI on enterprise performance in China: Mediating role of managerial relationships." *Journal of Global Information Management* 32, no. 1 (2024): 1–20.

Yong, J. Y., M.-Y. Yusliza, T. Ramayah, and O. Fawehinmi. "Nexus between green intellectual capital and green human resource management." *Journal of Cleaner Production* 215 (2019): 364–374.

Yu, W., and R. Ramanathan. "Environmental management practices and environmental performance: The roles of operations and marketing capabilities." *International Journal of Production Economics* 116 (2016): 1201–1222.

Yu, Y., and B. Huo. "The impact of environmental orientation on supplier green management and financial performance: The moderating role of relational capital." *Journal of Korea Trade* 211 (2019): 628–639.

Zaid, A. A., A. A. M. Jaaron, and A. T. Bon. "The impact of green human resource management and green supply chain management practices on sustainable performance: An empirical study." *Journal of Cleaner Production* 204 (2018): 965–979.

Zhang, M., Y. K. Tse, J. Dai, and H. K. Chan. "Examining green supply chain management and financial performance: Roles of social control and environmental dynamism." *International Journal of Production Economics* 66 (2019): 20–34.

Zhu, Q., J. Sarkis, and K.-H. Lai. "Examining the effects of green supply chain management practices and their mediations on performance improvements." *International Journal of Production Research* 50, no. 5 (2012a): 1377–1394.

Zhu, Q., J. Sarkis, and K.-H. Lai. "Green supply chain management innovation diffusion and its relationship to organizational improvement: An ecological modernization perspective." *Journal of Engineering and Technology Management* 29, no. 1 (2012b): 168–185.

Zobel, T. "The impact of ISO 14001 on corporate environmental performance: A study of Swedish manufacturing firms." *Journal of Environmental Planning and Management* 59, no. 4 (2016): 587–606.

7 The Role of AI and Automation in Green Management

A Bibliometric Approach to Evaluating Firm Performance

Bhartrihari Pandiya, Ankita Srivastava, and Manisha Raj

7.1 INTRODUCTION

A sustainable economy satisfies the current generation's needs and requirements without compromising the future generation's ability to use them. Sustainable practices provide current and future generations with equitable social justice. This is done through policy frameworks like green taxation, cap-and-trade systems, and sustainable development goals, which are specifically designed to realign economic incentives with the goals of the environment. In doing so, governments can bias economies toward more sustainable paths of development, promote innovation, and increase exporting competitiveness. Meadows et al. (1972) in their report "Limits to Growth," argued that if calamity has to be avoided, economic growth must be controlled fairly. However, sustainability is not the same in the contemporary world as it was in the 1970s. The "limits to growth" concept is no longer acceptable from society's or industry's perspectives. Firms have to be ready to encounter issues related to sustainability to support economic growth alongside protecting the environment.

Green management (sustainable, environmental) is an approach introduced to business processes where environmental consideration combines with other aims to enable ecological sustainability, decrease the unavoidable effects on the ecology, and provide more of a social or economic benefit than the expense the environment bears. It recognizes the relatedness among the environment, society, and economy, and pursues value creation for its shareholders, employees, customers, communities, and the environment. Pascale et al. (2024) suggested that the fusion of green and digital service innovations is now becoming central not only for realizing sustainable development but also for firms to acknowledge the consumer demand for low-carbon solutions.

DOI: 10.1201/9781003626527-7

Sustainable management is a wide concept involving green marketing, finance, human resource management, operations, and global supply chain management (GSCM). According to Baki (2018), GSCM practices include internal and external practices for managing the upstream and downstream supply chains inside and between firms to reduce the overall environmental effect of both forward and reverse flows. Green packaging (GP) uses packaging practices with a low negative environmental footprint. Maziriri (2020) contends that for a package to be considered green, it should adapt to the different needs of logistics and supply chains. It should be designed for the environment (reduce, reuse, and recycle). In the same manner, a green package should have an eco-label. Zailani et al. (2015) agree that eco-design, eco-labeling, and packaging integration with logistics and supply chain management are pivotal green packaging practices (part of green logistics). Similarly, green human resource management practices such as remote working and going paperless in daily operations are believed to enhance an organization's performance (Khuwaja et al., 2024). Many firms are examining how to consider environmental problems as there is a lot of awareness of sustainable development (Bevilacqua et al., 2007). The approach of developing green management practices by firms typically involves several strategies such as:

Environmental compliance: Keeping up with the environmental regulations and standards established by local, national, and international agencies. It includes monitoring and managing the activities that cause emissions, waste, and pollution to avoid damage to our ecosystem and maintain human health.

Resource efficiency: To reduce waste and costs by saving energy, water, and raw materials. This involves incorporating energy-efficient technologies, implementing recycling and waste minimization strategies, and advocating sustainable sourcing principles.

Prevention of pollution: Implementing changes in how things are designed and used so they create less or no pollution or waste, global warming, or other degradation of the environment. This could involve redesigns of processes, cleaner production methods, or substitution of hazardous materials for less dangerous alternatives.

Product stewardship: Ensuring that a product causes less harm and contributes to more good throughout its life cycle.

Impact investing: Investing money not only with a financial return in mind but also to increase social and environmental value.

Containerization: Packing goods into a shipping container, which is then loaded onto ships, trains, or trucks to be transported around the world.

Initial coin offering: An innovative mode to raise capital in which start-ups distribute their cryptocurrency to investors in exchange for capital.

Social lending: Loans with terms that enable social finance. Social lending, also known as peer-to-peer (P2P) lending, allows people to lend and borrow money directly from each other.

7.1.1 Economic and Environmental Performance

Economic performance and environmental performance are two pillars of sustainable development and the well-being of societies all over the globe. This relationship is a reminder of the significance of the harmonious coexistence of economic development and environmental sustainability for record growth and preservation of the environment. In other words, sustainable development is gaining importance because it does not risk the capacity for economic development of people living now or in the future. Nonetheless, unfortunately, economic achievement is also per se a threat to the environment.

At its basic level, the concept of economic performance refers to the efficiency of a country's economic system in generating products and services, leading to economic growth and prosperity. Economic performance may be determined in various ways, using traditional metrics such as gross domestic product (GDP), employment rates, and income distribution. However, indicators like GDP fail to consider many indirect aspects of the health of an economy, such as income inequality, resource depletion, or environmental harm.

On the other hand, environmental performance refers to natural resource and ecosystem management and conservation, as well as protection from adverse environmental impacts on human health. Environmental performance needs to focus on reducing pollution, protecting biodiversity, mitigating climate change mitigation, and ensuring responsible resource use. Environmentally focused performance is assessed through several metrics like carbon emissions, air and water quality, deforestation rates, and waste generation.

There is a need to balance economic and environmental performance for sustainable development in contrast to the archaic idea of building an economy at the cost of the environment. The data increasingly suggest that protecting the environment can boost economic well-being. Investments in renewable energy, sustainable agriculture, and clean technologies promote economic growth, generate employment, and drive innovation, in addition to reducing environmental harm. Industries are now making a trade-off between economic performance and environmental performance. Economic performance measures profitability, whereas environmental performance measures how successful a firm is in managing its negative impact on the environment. However, managers believe environmental performance management is a compliance requirement with environmental regulations. As a part of the trade-off, the cost of environmental protection needs to be considered, which has significantly increased since the 1970s and is likely to grow further (Christmann and Taylor, 2002). It infers that gainful green management practice is the main factor in deciding a firm's performance in the market. Therefore, research on environmental issues has shifted to green management practices rather than pollution, etc. (Klassen and McLaughlin, 1993). The literature on environmental management suggests that companies can improve their competitive advantage by implementing strategies that give them an edge over their rivals while simultaneously reducing the adverse effect of their actions on nature by employing green management practices. A study on cross-functional green management and industrial output has suggested that green strategy alignment can enhance market and environmental outcomes (Yue et al., 2023).

Yet, many companies appear, understandably, to resist this change—seeing it as too proactive—since "the associated benefits may not be proved adequate to offset the expenses of indulging in GMP [green management practice] initiatives" (Montabon et al., 2007). Much more work needs to be done to study the potential opportunities—both in the perception and reality—of performance improvement through the implementation of green management practices. Initiatives regarding environmental management can either be a good or a bad thing for the financial performance of the firm. For instance, some argue that profitability is reduced by the cost of being environmentally friendlier in production, while others refer to rare cases where the financial performance is increased by evidence from friends or relatives. A positive correlation has been established between energy efficiency and environmental performance in different studies (Lundgren et al., 2017).

Overall, economic and environmental performance are interdependent, and the two go hand in hand for the goal of sustainable development. Thus, through a combination of integrated approaches, nations can develop stronger, fairer, and more prosperous futures for mankind in the years ahead, future-proofing their economies and protecting their environments. In light of this discussion, this study aims to conduct a detailed literature review via bibliometric analysis to understand the trends followed in this area. The study will use literature extracted from the Scopus database. The study will focus on understanding the association between green or sustainable management and firms' performance. It will also attempt to understand and interpret trends and recent developments in the field of green management and firms' performance.

7.1.2 The Role of AI, ML, Data Analytics, and Automation in Green Management

In recent years, the increasing need to combat climate change and environmental degradation has brought green management to the top of business strategies in most parts of the world (Dwivedi et al.,2022). Green management is the process whereby environmental concerns are integrated into the functions of organizations to mitigate impacts on the environment while creating economic value. As organizations grapple with this paradigm shift, the adoption of advanced technologies, including AI and automation, has become a disruptive force in pursuing sustainability goals (Goralski and Tan, 2020). Not only are these technologies transforming most business models, but they are also changing how sustainability works and opening doors for usage that were not there in the past. The emergence of AI and automation is transforming how organizations think about and execute green initiatives. From minimizing resource utilization to curbing emissions, these technologies enable resource-intensive businesses to achieve enhanced accuracy and efficiency in their sustainable efforts. This may include AI-enabled insights, for example, to predict the energy consumption of devices (Ahmad et al., 2021), leading to minimized waste and reduced reliance on fossil fuel energy. Likewise, automation technologies such as the Internet of Things (IoT) and robotics have allowed organizations to track and mitigate environmental impacts in real time, helping ensure compliance and reducing operational inefficiencies.

There are challenges associated with the adoption of AI and automation in green management. The concerns related to high resources for implementation, data security, and adapting the workforce are significant barriers faced by organizations eager to introduce these technologies in their path toward sustainable practices (Campion et al., 2022). However, the application of such tools works well when sufficiently robust data is available and organizational goals align with long-term environmental goals. The potential benefits outweigh the challenges by a factor of hundreds, especially considering the profound impacts AI and automation have on the sustainable development of the global economy. AI in green management not only impacts the efficiency of operations but also affects strategic decision-making and innovation. The volume of data that machine learning algorithms can analyze to identify trends and derive insights for evidence-based decision-making is substantial. For example, organizations can use AI to evaluate the environmental performance of their supply chains, find solutions to enhance performance, and allow more precise interventions to be placed. Furthermore, AI-driven tools like natural language processing can explore the market sentiment on eco-friendliness, providing firms with precise data to pursue green initiatives without overstepping, thereby minimizing waste driven by less-than-ideal customer engagement data. In contrast, automation performs repetitive activities and aids in resource optimization for strategic endeavors while promoting a platform of innovation within organizations (Koppisetti, 2024).

Additionally, one of the most important uses of AI and automation in green management is the facilitation of compliance with environmental regulations. With more stringent sustainability regulations imposed by governments globally, organizations will inevitably adopt measures to comply while ensuring operational excellence. Systems powered by AI can now help analyze regulatory requirements and provide actionable insights to enable firms to meet standards, which in turn eases the risk of incurring penalties and more destructive harm to reputational standing. Moreover, because automation technologies can observe emissions along with resource usage as they occur, this allows organizations to mitigate issues of noncompliance before problems arise (Solanke et al., 2024).

7.2 CONCEPTUAL OVERVIEW

Green management refers to practices that reduce or eliminate environmental harm. These efforts consist of energy conservation, waste minimization, pollution control, sustainable utilization of resources, and utilization of power produced by clean technologies (Nižetić et al., 2019; Li et al., 2023). Business practices are associated with environmental sustainability with intentions to reduce the ecological footprints and to perform the business (Amjad et al., 2022; Wong et al., 2012).

Firm performance can be measured from various dimensions including financial, operational, and environmental dimensions, but not limited to profitability, return on investment, market share, efficiency, productivity, as well as emission reductions and resource conservation (Yang et al., 2011; Le and Ikram, 2022). The combination of green-oriented management practices is expected to exert a positive impact on these dimensions. For example, energy-efficient processes can reduce operational costs, and sustainable practices may improve brand reputation, increasing market share and profitability. One of the key figures in the field is Michael Porter, who developed the Porter hypothesis, which generally states and argues that the strictness of environmental regulations could provoke competitiveness and innovation among firms (Porter and van der Linde, 1995). Hart and Ahuja (1996) examined financial returns of environmental performance, and Blanco et al. (2009) studied environment and economic performance. One major issue that came up from both reviews was the strategic role of green management in competitiveness. Christmann and Taylor (2002) agree with these studies by showing that environmental strategies can be a source of competitive differentiation for firms, attract ecological consumers, and generate market barriers. The resource-based view (RBV) is commonly used to examine how sustainability practices can increase firm capabilities and resources (Lubis, 2022). Both the impact of regulatory frameworks and the success of voluntary initiatives in terms of firm performance have been ascertained over the years, and these should be included when identifying barriers as well. Similarly, Ambec et al. (2013) point out that adherence to environmental standards can lead to innovation and productivity. Studies have shown that voluntary initiatives like ISO 14001 and participation in the Global Reporting Initiative enhance corporate reputation, causing corporate stakeholders to trust these companies, in turn expecting an improvement in the financial performance of these companies (Nazari et al., 2015). The need for innovation and technology in green management has been focused on by various researchers (Singh et al., 2020; Dangelico, 2016). One work linking green technologies (such as renewable energy and sustainable supply chain practices) to the bottom line is that of Eltayeb et al. (2011). To have economic progress along with sustainability, innovation is key. The involvement of stakeholders is important for the successful implementation of green management practices. Some studies support the belief that firms involving stakeholders as part of their environmental efforts not only show greater support but also have improved performance outcomes (Wassmerelsj et al., 2014; Baah et al., 2021; Makhdoom et al., 2023). Trust and cooperation allow for positive stakeholder engagement, indicating that it can be made sustainable for the long run (Deviney et al., 2023; Beck and Ferasso, 2023).

The variables under study, which have been previously discussed, concerning the impact of green management on firm performance have been summarized. Green management and financial performance is a relationship where the cause and effect are not always definite (Elshaer et al., 2023). A few research works have reported a positive correlation, which means industrial eco-initiatives can contribute to profit and share price (Taha et al., 2023; Doni and Fiameni, 2024). Still, others report either neutral or negative effects, indicating that financial benefits are contingent on the industry context and implementation quality (Antony et al., 2023; Maganga and Taifa, 2023). When viewed from the perspective of operational performance, green management practices are believed to lead to better operational performance, decreased costs, and increased productivity (Aslam and Jawaid, 2023). Energy efficiency and waste reduction measures result in substantial cost savings, which in turn positively affects financial performance (Bosu et al., 2023; Gao and Wan, 2023). Additionally, the application of green management practices positively influences environmental performance, with firms reporting decreases in emissions, waste, and resource consumption. Positive environmental performance also leads to better regulatory compliance and decreases environmental liabilities (Kalyar et al., 2020; Lutfi et al., 2023).

Thus, the literature on green management and firm performance is highly fragmented and complex. Existing studies provide evidence for the positive impact of green management practices on firm performance, but the nature and extent of the impact are likely context-specific and conditional on the type of firm performance. Although a large amount of attention is paid in the literature to green management and its implications for firm performance, several avenues deserve to be further explored. These gaps may guide future research to further knowledge and enhance the practical applicability of green management practices. The most important gap in the literature is that findings about the effect of green management on financial performance are inconsistent. These inconsistencies may be due to differences in the measurement process, differences in industry contexts, or differences in geography (regulatory environments and market conditions vary globally, and this will impact the financial outcomes of greening initiatives). Longitudinal studies are also an area lacking in research, as the most quality research in the area is cross-sectional. Cross-sectional studies provide a snapshot of what is happening at that time, with no consideration of what is happening over the years. Consequently, longitudinal studies are needed to reveal how much green management influences firm performance in the long run. Combined with a focus on career length that will bridge different eras, longitudinal studies can help uncover the impact of sustainable practices on long-term performance through changes in market conditions, regulatory changes, and technological advancements. In addition, this research tends to examine only large corporations, and the experiences and problems of small and medium-sized enterprises (SMEs) are often not considered. SMEs have an important role in the worldwide economy and can affect local, national, and global-scale environmental management. Further research should be done to examine the SME-specific barriers and drivers to practice sustainability and how it impacts their performance. Most of the published studies have been performed in developed countries, and only a few of them have focused on green management practices in developing regions. Identification of commonalities and differences in literature that is based on regional and cultural content is crucial. Given the significant differences between cultures, economies, and regulations, the adoption and efficacy of green management practices could be influenced. Research is required to understand these regional variances better and establish region-specific plans to promote sustainability globally. Otherwise, knowledge of green management leading to firm performance could not be fully grasped, so they need to be included in research to fill these literature gaps. Overall, these analyses demonstrate that future research may benefit from examining sustainability practices in detail, targeting a specific area of sustainability, and investigating how a specific type of sustainability practice contributes to both the environmental and economic performance of firms. Therefore, bibliometric analysis will identify trends, significant authors, and principal themes that drive the research landscape.

Thus, the objectives of the research work are:

1. To identify and analyze the thematic clusters within the research domain of green management and firm performance
2. To highlight the emerging challenges and other variables related to green management and firm performance, providing theoretical implications along with the scope for future research

7.3 METHODOLOGY

This study aims to provide a systematic review of the available literature-based evidence. The use of bibliometric analysis allows researchers to explore the existing literature, which additionally provides an organized and unambiguous understanding of the field. Bibliometric analysis, initially presented as a scientific methodology by Pritchard (1969), has served to comprehend the change of research regions' insight variants. These types of analyses help in the assessment of impact by examining author data, research area data, journal data, and emerging topics. The software tool VOSviewer from Van and Waltman (2011) allows for conducting bibliometric analysis.

Consequently, bibliometric analysis was employed for the identification of the various research themes. Additionally, recent trends in this research area will be explored using bibliometric analysis, highlighting various emerging research areas and leading to new research directions. The Scopus database was considered for the bibliometric analysis. The literature search period was from 2020 to 2024. The last search query was as follows: Abstract-Title-Keyword (Green Management) AND Abstract-Title-Keyword (Firm Performance). The total number of research papers finalized was 201. A comprehensive record containing the bibliographical information, citation details, abstract, and related data was downloaded in CSV format for further analysis.

7.4 DISCUSSION

Co-authorship analysis serves as a valuable tool for understanding research collaboration networks and mapping collaborations within and across institutions, disciplines, and geographic regions. Visualizing these networks makes it easier to identify key researchers and influential groups within a particular field. Co-authorship networks can highlight central figures with numerous connections, indicating their influential role. In Figure 7.1, the connected authors can be seen, suggesting that collaboration is limited and that there is a huge opportunity for collaboration in this area with many others in the network. The seven clusters show groups of authors who are more closely connected, indicating strong collaborative groups. This network can help identify key researchers, potential collaborators, and the structure of research

FIGURE 7.1 Co-citation analysis.

collaboration within a specific field or institution. Understanding these relationships can aid in strategic planning for research projects, funding allocations, and academic networking.

The co-occurrence network analysis facilitated the identification of clusters, revealing the research themes that previous researchers explored (refer to Figure 7.2). These identified themes can guide future research efforts. Researchers have examined different aspects of this field and shared their findings. Through comprehensive screening and evaluation of these works, certain variables related to the research topic were understood. The emerging research themes, identified through clusters in green management and firm performance, can be further investigated using empirical datasets. Figure 7.1 illustrates the keywords co-occurrence network analysis, showing macro clusters identified through bibliometric analysis. The different colors in the co-occurrence diagram represent the connectivity between related keywords within each cluster. Keywords such as "environmental management"

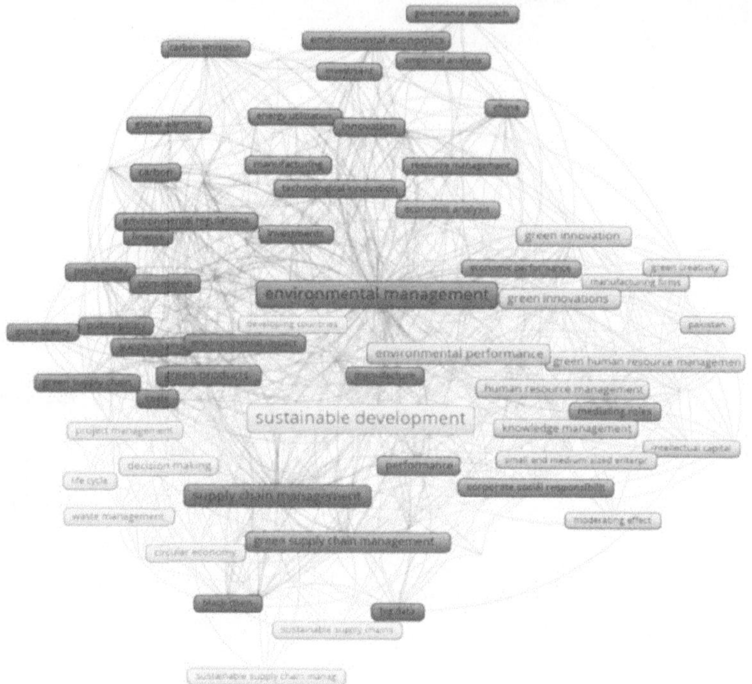

FIGURE 7.2 Co-occurrence keyword analysis.

and "sustainable management" appear most prominently, indicating their central role, with other related keywords. The optimization algorithm detects communities within large networks by combining nodes using heuristics (Kumar et al., 2020). The keywords co-occurrence analysis provides unique insights into various areas and their applications across multiple research domains.

A thorough reading of the research papers provided insights and identified variables in the research area, paving the way for emerging research domains. These findings are summarized next.

Cluster on right-hand side: This cluster depicts the connection between manufacturing firms/small and medium-sized firms and environmental performance, as keywords like green innovation, green creativity, and intellectual capital are present in it. The presence of such keywords depicts the importance and trend of these emerging concepts.

Cluster at top: Keywords like environmental, carbon emission, and global warming underline the importance of these concepts in today's time. These keywords connect environmental management and economics. Environmental regulations and governance approaches are the need of the hour.

Cluster at bottom: The keyword "sustainable development" is the major keyword in this cluster. Sustainable development and decision-making through project management are about long-term strategies leading to a circular economy.

Cluster on right-hand side: The blue cluster discusses green products, and public policy is associated with our keyword "green supply chain." Usage of big data and blockchain are also mentioned. Overall, the keywords in this cluster are associated with regulatory aspects, profitability, and public policy.

Cluster in centre: This cluster mentions corporate social responsibility and economic analysis, along with economic performance and other variables that might have a mediating role on the rest of the factors.

Global firms are realizing the significance of sustainable development at a rapid pace and are adopting this approach to benefit social and business contexts. Although the ultimate goal of sustainable development is to save the earth in an environmental and social context, an additional functionality of the approach can be observed. With the literature review and detailed bibliometric analysis, a few sustainable/green initiatives by firms enhancing their performance are:

Generation of renewable energy: A popular way of sustainable development for a firm is to invest in renewable resources, such as solar, wind, or hydro energy; in such a way, a firm cannot only reduce their harmful emissions but also potentially the costs of operation. In addition, a firm can ensure its steady energy supply in the future, as well as willingly supply energy to the public grid to benefit the environment.

Green supply chain management: Running a sustainable supply chain involves much less waste and is much more cost-efficient due to the reduced need for resources. This implies responsible resource sourcing, emission reduction via shortening the routing between firms and destinations, and involvement of suppliers in the process.

Resource efficiency and reduction of waste: This step allows the cutting of the tremendous costs involved in waste processing and allows recycling of the resources used, in case they are limited goods. This implies the usage of technologies that reduce the usage of water and so on. The concept of lean management processes can be applied in this case.

Eco-product and service design: A product or service developed by a firm today is expected to be evaluated from an environmental context as well, mainly attributable to the lower costs. Of considerable import are the concept of eco-design, the design of environmentally friendly goods, and the concept of creating products that would cause the smallest possible damage to the environment.

CSR activities: Considering the involvement of firms today in several projects to contribute to the social and business life of the population or the needs of the community, the waste generated by a firm's activity can be reduced via engagement in CSR activities. The latter can be, for example, the construction of wells in poor regions.

Stakeholder engagement and transparency: Another significant reason for dissolution between a firm and stakeholders is also the reduced trust in companies by the public publicizing the information about the firm's current state. The waste in such cases includes not only the natural waste emitted in the process of the application of the technologies but also publicity and overhead rope.

7.5 THEORETICAL IMPLICATIONS

Green management, also known as environmental and sustainable management, focuses on improving a firm's environmental performance. It addresses an issue that has sparked a considerable amount of discussion, given the growing global awareness of the environment and the rise of sustainable development. The theoretical insights gained from the bibliometric analysis of green management on firm performance are important in providing an overview of the emerging literature in this thread. One common theoretical framework used in the green management literature is the firm's RBV (Barney, 1991). It defines how firms achieve these resources and capability-based competitive advantages as valuable, rare, inimitable, and non-substitutable. Despite the attempts made so far to appraise strategic management practices that increase the organizational capacities and sustainable competitive advantages of firms, limited previous literature addressed the question of whether green management practices can be envisaged among such strategic resources. Also, green management practices, such as waste reduction, energy efficiency, and sustainable supply chain management, can be seen as a natural area within which particular sorts of capabilities of firms can be developed. These two measures may lead to lower costs. This is also reinforced by the RBV focus on exploiting unique capabilities. RBV can be generalized to encompass green innovations as well. Companies that use green technologies and practices successfully can gain innovation capabilities that are likely to be hard for their rivals to duplicate, thereby finding a competitive advantage. This corresponds to the notion of dynamic capabilities, where companies adapt and renew their resources to remain competitive. Institutional theory states that the behavior of organizations is primarily driven by the shared norms, values, and rules of the institutional environment. Regulatory pressures, societal objectives, and stakeholder demands drive green management practices. Several institutional theory implications are offered based on the bibliometric analysis. Firms implement green management practices not only to comply with regulations but also to achieve legitimacy and trust with stakeholders. This aligns with an argument from institutional theory, suggesting that as normative social pressure changes, organizations over time adapt their behavior to accommodate such change. One is institutional theory and its concept of mimetic isomorphism, where firms imitate the successful practices of peers, helping explain why so many firms adopt green management practices. As more and more leading corporations prove the economic and reputational advantages associated with sustainability, we witness a green wave ripple across most sectors of the economy, with lagging firms feeling the pressure to green their product offerings to stay competitive. Stakeholder theory states that organizations must meet the needs of customers, employees, investors, and the community. The inclusion of green management practices fits well with the precepts of stakeholder theory and merits several theoretical implications, such as improved stakeholder relationships by promoting closer linkage with stakeholders and recognizing their environmental interests. This alignment can enhance loyalty, increase worker morale, and improve public brand perception. Additionally, the issue of conflicting economic, environmental, and social targets that we are expected to pursue adds to the argument for stakeholder management. Stakeholder theory is complex and firms face conflicting interests, for example, short-term profitability versus long-term sustainability. Corporate social responsibility (CSR) relates to voluntary actions undertaken by firms to improve social goals and the natural environment. Given CSR is integrated with green management philosophy, the bibliometric analysis provides some theoretical insights into green management practices within the literature of CSR. According to contingency theory, the extent of the fit (interaction) between the characteristics of an organization and external environmental conditions determines the organization's effectiveness. In particular, the features of green management practices driven by external pressures like regulatory changes and market requirements may be interpreted as consistent with contingency theory. The success of green management strategies is influenced by factors such as the sector, the regulatory environment, and market characteristics. This emphasizes a context-dependent strategy rather than a general strategy, hence the contingency perspective. These changes also frequent further changes and uncertainties, requiring businesses to develop adaptive strategies. Green management is a moving target due to its need for evolving and dynamic practices; however, the concept of green management is bolstered by contingency theory. The bibliometric analysis of green management and firm performance yields many theoretical implications extending to multiple frameworks. In terms of grounding, the model extends the RBV by focusing on the strategic significance of green management, contributes to institutional theory with underlying regulatory and normative pressures, connects to stakeholder theory with an emphasis on key relationships with stakeholders, adds to the wider CSR literature, which contains strategic sustainability, while aligning with contingency theory regarding the need for adaptive strategies. These conceptualizations can significantly explain the results of empirical research that investigated the relationship between green management and firm performance, to better inform research and practice in what is a nascent content domain.

Barring applications in practice, the intersection of AI and automation with green management practices has important theoretical implications for both business management and sustainability literature. These technologies disrupt the conventional paradigms of integration by shaping data-oriented decision-making and reconfigurable optimization in roadmaps to sustainability. They facilitate predictive modeling and real-time analytics that are consistent with new theories of adaptive management and systems thinking in green practices. The use of automation also endorses theories of operational efficiency and process innovations as it reduces redundant work and enables process flow. This integration also reshapes stakeholder theory by emphasizing technological collaboration as a key driver of value creation for stakeholders, including employees, consumers, and regulators. The role that AI and automation play in greening facilities presents an excellent opportunity to understand how organizations are successfully balancing the need to be economically viable with their inherent environmental responsibility, which adds to the limited literature available on corporate sustainability performance and technology adoption. This topic serves to advance interdisciplinary theories that align environmental and business priorities needed in an accelerated and rapidly deteriorating world by reducing the distance between technology and sustainability.

7.6 IMPEDIMENTS TO CURRENT RESEARCH

The sustainability of the environment is increasingly becoming the focus of both businesses and policymakers (Bakos et al., 2020; Arvidsson, 2023). The relationship between green management and firm performance is of particular interest due to the increasing need for businesses and policymakers to strike a balance between environmental sustainability and economic objectives. Although extensive research addresses this issue, a number of challenges remain in the development and operation of successful green management strategies. In this analysis, we summarize a systematic review and content analysis of publications in the literature and identify the major challenges that both researchers and practitioners face in this area.

Lack of systematic theoretical frameworks: Although there have been many empirical studies on green management and firm performance, there is a lack of comprehensive theoretical frameworks. Segmented studies are found in the literature that concentrate on different elements of green management, such as energy efficiency, reduction of waste, and sustainability in the supply chain. Nevertheless, there is still a need for a more overarching theory that integrates the multifaceted character of green management and its effect on firm performance. The lack of well-established theoretical models is an obstacle to such findings to make general recommendations and to generate coherent strategies. Future work needs to build theoretically grounded frameworks that can meaningfully inform empirical research and policy.

Lack of data availability and quality: Data availability and quality still remain the most important constraints in the way of advancing the field of green management and firm performance. Most of them are based on self-reported data, which is potentially biased or false. On the other hand, very few standard metrics for green management practices and their outcomes also contribute to the difficulty of comparing results from different studies and contexts. In developing nations, where green management practices are typically much less mature, the data itself is even more scarce. Resolution can be obtained through the establishment of standardized data collection, followed by collective consolidation of large databases that capture environmental and performance metrics.

Limited cross-disciplinary research: The research area of green management and firm performance is inherently interdisciplinary and intersects with the fields of environmental science, economics, organizational behavior, and strategic management. However, scientists often do not get involved in collaborative efforts, so the research efforts are often siloed. Greater collaboration between scholars in various disciplines will hopefully lead to a more comprehensive understanding of the relationship between green management and firm performance. An interdisciplinary team of researchers can use different, complementary, and/or enriching methodologies and perspectives to tackle complex research questions and thereby develop an improved understanding.

Different areas of the brain (regional biases): Prior research on green management and sustainable performance is largely focused on developed countries, which have propitious external and internal pressures for sustainability practices through their regulatory frameworks and market conditions. Consequently, very few studies have been conducted in relation to these questions in developing and emerging economies where the obstacles and enablers of green management may be very different. These regions generally have unique environmental, economic, and social considerations that affect the efficacy of green management strategies. Increased research in areas outside of sub-Saharan Africa would give more in-depth perspectives that can contribute to localized responses.

Performance measurement as a near-term focus: Most studies on green management and firm performance focus on short-term financial measures, such as profitability and stock returns. Although these metrics are relevant, they may be insufficient to capture the payoff of green management practices over time. Investing in the environment often involves a lot of capital, and the dividend it pays is not returned for several years. As a result, the report highlights the need for research into long-term performance measures incorporating sustainable growth, resource efficiency, and resilience to environmental risks. Taking a longer-term perspective should result in a more realistic valuation of the benefits derived from the practices of green management.

Inappropriate research in the industry: The effect of green management practices on firm performance could differ considerably across industries. Environmental initiatives are influenced by regulations, the market, and the limits of technology. Yet in practice, very few studies do this, and when they do, they tend to use multi-industry datasets that are better for finding effects that are generally true. This creates a risk that industry-specific dynamics are masked and overly broad conclusions are drawn. Future research should address this gap by examining differences between industries, taking into account the specificity and challenges of the various sectors. Similarly, industry comparisons or benchmarking studies can also enlighten us on best practices and industry-based strategies.

Insufficient attention to behavioral and organizational factors: This boils down to not focusing on the behavioral and organizational aspects of getting people to play along. Much of the previous research on green management and firm performance has focused on the technical and financial aspects of the implementation of sustainable technologies and their economic benefits. Nonetheless, to the extent that green management initiatives are successful, it is in part because of behavioral and organizational elements like leadership commitment, the engagement of employees, and corporate culture. These are important factors that determine the take-up and impact of environmental practices. Further research investigating the interaction of organizational behavior and green management outcomes is necessary. It has certainly highlighted that such research can help in informing strategies to build an organizational landscape that facilitates multiple initiatives to achieve sustainability.

Fragmented policy and regulatory analysis: Although this domain is vital to the study of green management practices, the literature often presents such policy and regulation as fragmented and detached. The adoption of environmentally friendly practices and the consequent improvement in the performance of the firm depend mainly on the efficiency of environmental policies. Studies on policy impacts are largely fragmented and lack thorough analyses of the interactions between different regulatory frameworks and market conditions. Whether the same is true of other contexts is unclear, and future research should seek more comprehensive evaluations of policy effects, taking into account the specific mechanisms in place (e.g., policy design or enforcement strategies, industry responses).

Thus, these obstacles are critical for the development of green management and firm performance literature. When we develop more inclusive theoretical frameworks, data sources, and cross-disciplinary collaboration, we can study stronger geographical and industrial differences. Moreover, emphasizing long-term measures of performance, behavioral and organizational dimensions along with policy interplay analysis would provide us with more insights into how green management practices drive sustainable and resilient business performance.

7.7 SCOPE FOR FUTURE RESEARCH

In the era of unstoppable climate change, green management is essential to any firm's business strategy, as businesses need to manage their profitability and sustainability in parallel. Existing research on the relationship between green management practices and firm performance offers valuable insights. However, several unanswered questions and areas remain underexplored. In this section, the extent of future research is explored based on the insights derived from the bibliometric analysis. Most studies are cross-sectional, meaning they capture the relationship between green management practices and firm performance at the moment of the particular analysis. First, there is a need for longitudinal studies capturing how green management practices provide value for the firm over time. It is crucial to investigate whether the benefits arising from greener practices last in the long term and whether any potential disadvantages would remain unrevealed in short-term analyses. Second, there is a need for industry-specific analyses. Future research should explore how green management practices impact firm performance in different sectors, such as manufacturing and services, technology, and agriculture. This will provide more specific insights, leading to more detailed and feasible recommendations. Last, there is a need for more research on developing countries. Most existing studies focus on developed or high-income countries, meaning there is a gap in understanding what factors and frameworks apply to developing countries, which often have unique economic, regulatory, and even cultural environments. It is important to investigate the role of culture in the adoption and success of green management practices. Empirical studies containing this information remain scarce, meaning that this area requires more attention. The evaluation of how these factors apply to firms in the developing world can also be an area of future research. In addition, most studies focus on large and/or public firms, meaning there is a gap in understanding how SMEs adopt and implement greener practices. There is a need to investigate the context-specific barriers to greener practices' adoption and strategies for their implementation in the SME context. Also, the contribution of innovation and technology should be evaluated. It is also an underexplored area and can provide valuable insights for green technology companies endeavoring to build more sustainable solutions.

Future research should develop a more nuanced understanding of this relationship and best practices for integrating green management into a business strategy. There remains a wide variety of opportunities for future research on green management and firm performance. The questions asked by these pieces would add to a better understanding of these concepts, providing researchers with deeper insights and firms with more salient recommendations on how to address the challenge of combining environmental sustainability with good economic performance. Undoubtedly, the theoretical understanding and applied implementation of green management practices will be greatly enhanced as work in this area continues.

REFERENCES

Ahmad, T., D. Zhang, C. Huang, H. Zhang, N. Dai, Y. Song, and H. Chen. Artificial Intelligence in Sustainable Energy Industry: Status Quo, Challenges and Opportunities. *Journal of Cleaner Production* 289 (2021): 125834.

Ambec, Stefan, Mark A. Cohen, Stewart Elgie, and Paul Lanoie. "The Porter Hypothesis at 20: Can Environmental Regulation Enhance Innovation and Competitiveness?" *Review of Environmental Economics and Policy* 7, no. 1 (2013): 2–22.

Amjad, Adeel, Khurram Abbass, Yasir Hussain, Fahad Khan, and Sadiq Sadiq. "Effects of the Green Supply Chain Management Practices on Firm Performance and Sustainable Development." *Environmental Science and Pollution Research* 29, no. 44 (2022): 66622–66639.

Antony, Jiju, Manoj Sony, Orla McDermott, Sandra Furterer, and Michael Pepper. "How Does Performance Vary Between Early and Late Adopters of Industry 4.0? A Qualitative Viewpoint." *International Journal of Quality & Reliability Management* 40, no. 1 (2023): 1–24.

Aria, Massimo, and Corrado Cuccurullo. "bibliometrix: An R-Tool for Comprehensive Science Mapping Analysis." *Journal of Informetrics* 11, no. 4 (2017): 959–975.

Arvidsson, Susanne. "CEO Talk of Sustainability in CEO Letters: Towards the Inclusion of a Sustainability Embeddedness and Value-Creation Perspective." *Sustainability Accounting, Management and Policy Journal* 14, no. 7 (2023): 26–61.

Aslam, Waqar, and Shamsuddin Tariq Jawaid. "Green Banking Adoption Practices: Improving Environmental, Financial, and Operational Performance." *International Journal of Ethics and Systems* 39, no. 4 (2023): 820–840.

Baah, Charles, Dennis Opoku-Agyeman, Isaac Samuel Kwame Acquah, Yaw Agyabeng-Mensah, Emmanuel Afum, Daniel Faibil, and Abdoulaye F. A. M. Abubakar. "Examining the Correlations Between Stakeholder Pressures, Green Production Practices, Firm Reputation, Environmental and Financial Performance: Evidence from Manufacturing SMEs." *Sustainable Production and Consumption* 27 (2021): 100–114.

Baki, R. "Literature Review on Green Supply Chain Management Concept and Problems During Its Implementation." *The Journal of International Scientific Researches* 3, no. 2 (2018): 114–122.

Bakos, Jason, Meaghan Siu, Arturo Orengo, and Narges Kasiri. "An Analysis of Environmental Sustainability in Small & Medium-Sized Enterprises: Patterns and Trends." *Business Strategy and the Environment* 29, no. 3 (2020): 1285–1296.

Barney, Jay. "Firm Resources and Sustained Competitive Advantage." *Journal of Management* 17, no. 1 (1991): 99–120.

Beck, David, and Marcos Ferasso. "Bridging 'Stakeholder Value Creation and 'Urban Sustainability': The Need for Better Integrating the Environmental Dimension." *Sustainable Cities and Society* 89 (2023): 104316.

Bevilacqua, Maurizio, Filippo Emanuele Ciarapica, and Giacomo Giacchetta. "Development of a Sustainable Product Lifecycle in Manufacturing Firms: A Case Study." *International Journal of Production Research* 45, no. 18–19 (2007): 4073–4098.

Blanco, Esther, Javier Rey-Maquieira, and José Lozano. "The Economic Impacts of Voluntary Environmental Performance of Firms: A Critical Review." *Journal of Economic Surveys* 23, no. 3 (2009): 462–502.

Bosu, Isaac, Hoda Mahmoud, and Hossam Hassan. "Energy Audit and Management of an Industrial Site Based on Energy Efficiency, Economic, and Environmental Analysis." *Applied Energy* 333 (2023): 120619.

Campion, A., Gasco-Hernandez, M., Jankin Mikhaylov, S., & Esteve, M. (2022). Overcoming the challenges of collaboratively adopting artificial intelligence in the public sector. *Social Science Computer Review, 40*(2), 462–477.

Christmann, Petra, and Glen Taylor. "Globalization and the Environment: Strategies for International Voluntary Environmental Initiatives." *Academy of Management Perspectives* 16, no. 3 (2002): 121–135.

Dangelico, Rosa Maria. "Green Product Innovation: Where We Are and Where We Are Going." *Business Strategy and the Environment* 25, no. 8 (2016): 560–576.

Deviney, Amanda, Khara Grieger, Andrea Merck, Jennifer Classen, and Amanda M. Marshall. "Phosphorus Sustainability through Coordinated Stakeholder Engagement: A Perspective." *Environment Systems and Decisions* 43, no. 3 (2023): 371–378.

Doni, Francesca, and Margherita Fiameni. "Can Innovation Affect the Relationship Between Environmental, Social, and Governance Issues and Financial Performance? Empirical Evidence from the STOXX200 Index." *Business Strategy and the Environment* 33, no. 2 (2024): 546–574.

Dwivedi, Y. K., L. Hughes, A. K. Kar, A. M. Baabdullah, P. Grover, R. Abbas, and M. Wade. "Climate Change and COP26: Are Digital Technologies and Information Management Part of the Problem or the Solution? An Editorial Reflection and Call to Action. *International Journal of Information Management* 63 (2022): 102456.

Elshaer, Ibrahim Ahmed, Aladdin M. Azazz, and Samir Fayyad. "Green Management and Sustainable Performance of Small-and Medium-Sized Hospitality Businesses: Moderating the Role of an Employee's Pro-Environmental Behaviour." *International Journal of Environmental Research and Public Health* 20, no. 3 (2023): 2244.

Eltayeb, Tamer Khalifa, Suhaiza Zailani, and T. Ramayah. "Green Supply Chain Initiatives among Certified Companies in Malaysia and Environmental Sustainability: Investigating the Outcomes." *Resources, Conservation and Recycling* 55, no. 5 (2011): 495–506.

Gao, Lijun, and Lingling Wan. "Does Corporate Environmental Responsibility Contribute to Financial Performance? A Dual Path Analysis Through Operational Efficiency and the Cost of Debt." *Corporate Social Responsibility and Environmental Management* 30, no. 1 (2023): 308–323.

Giddings, Bob, Bill Hopwood, and Geoff O'Brien. "Environment, Economy and Society: Fitting Them Together into Sustainable Development." *Sustainable Development* 10, no. 4 (2002): 187–196.

Goralski, M. A., and T. K. Tan. "Artificial Intelligence and Sustainable Development." *The International Journal of Management Education* 18, no. 1 (2020): 100330.

Hart, Stuart L., and Gautam Ahuja. "Does It Pay to Be Green? An Empirical Examination of the Relationship Between Emission Reduction and Firm Performance." *Business Strategy and the Environment* 5, no. 1 (1996): 30–37.

Kalyar, Muhammad Nawaz, Afshan Shoukat, and Imran Shafique. "Enhancing Firms' Environmental Performance and Financial Performance through Green Supply Chain Management Practices and Institutional Pressures." *Sustainability Accounting, Management and Policy Journal* 11, no. 2 (2020): 451–476.

Khuwaja, Bushra, Samiuddin Shaikh, Syeda Dania Azhar, and Abdul Jaleel Mahesar. "Green Human Resource Management Practices and Firm Performance in Private Banks in Pakistan." *Remittances Review* 9, no. 1 (2024).

Klassen, Robert D., and Curtis P. McLaughlin. "TQM and Environmental Excellence in Manufacturing." *Industrial Management and Data Systems* 93, no. 6 (1993): 14–22.

Koppisetti, V. S. K. "Robotic Process Automation: Streamlining Operations in the Digital Era. *ESP International Journal of Advancements in Computational Technology (ESP-IJACT)* 2, no. 2 (2024): 74–81.

Kumar, Tarun, Shankar Vaidyanathan, Harish Ananthapadmanabhan, Subramanian Parthasarathy, and Balaraman Ravindran. "Hypergraph Clustering by Iteratively Reweighted Modularity Maximization." *Applied Network Science* 5, no. 1 (2020): 52.

Le, Thu Thao, and Muhammad Ikram. "Do Sustainability Innovation and Firm Competitiveness Help Improve Firm Performance? Evidence from the SME Sector in Vietnam." *Sustainable Production and Consumption* 29 (2022): 588–599.

Li, Chuang, Shah Faisal Ahmad, Ahmad Y. B. A. Ayassrah, Muhammad Irshad, Abdullah A. Telba, Eyad M. Awwad, and Mohammad Irfan Majid. "Green Production and Green Technology for Sustainability: The Mediating Role of Waste Reduction and Energy Use." *Heliyon* 9, no. 12 (2023). https://doi.org/10.1016/j.heliyon.2023.e22496

Liu, Zongran, Yixue Yin, Wei Liu, and Michael Dunford. "Visualizing the Intellectual Structure and Evolution of Innovation Systems Research: A Bibliometric Analysis." *Scientometrics* 103 (2015): 135–158.

Lubis, Nurul Widya. "Resource-Based View (RBV) in Improving Company Strategic Capacity." *Research Horizon* 2, no. 6 (2022): 587–596.

Lundgren, Tommy, and Wenchao Zhou. "Firm Performance and the Role of Environmental Management." *Journal of Environmental Management* 203 (2017): 330–341.

Lutfi, Abdulaziz, Hisham Alqudah, Musab Alrawad, Ahmad Fawzi Alshira'h, Mohammad Hazza Alshirah, Mohammad Anas Almaiah, and Mohammad Faisal Hassan. "Green Environmental Management System to Support Environmental Performance: What Factors Influence SMEs to Adopt Green Innovations?" *Sustainability* 15, no. 13 (2023): 10645.

Maganga, D. P., and I. W. Taifa. "The Readiness of Manufacturing Industries to Transit to Quality 4.0." *International Journal of Quality & Reliability Management* 40, no. 7 (2023): 1729–1752.

Makhdoom, Z. H., Y. Gao, X. Song, W. M. Khoso, and Z. A. Baloch. "Linking Environmental Corporate Social Responsibility to Firm Performance: The Role of Partnership Restructures." *Environmental Science and Pollution Research* 30, no. 16 (2023): 48323–48338.

Maziriri, E. T. "Green Packaging and Green Advertising as Precursors of Competitive Advantage and Business Performance among Manufacturing Small and Medium Enterprises in South Africa." *Cogent Business & Management* 7, no. 1 (2020): 1719586.

Meadows, D. H., D. L. Meadows, J. Randers, and W. W. Behrens III. *The Limits to Growth: A Report for the Club of Rome's Project on the Predicament of Mankind*. New York: Universe Books, 1972.

Montabon, Frank, Robert Sroufe, and Ram Narasimhan. "An Examination of Corporate Reporting, Environmental Management Practices and Firm Performance." *Journal of Operations Management* 25 (2007): 998–1014.

Nazari, J. A., I. M. Herremans, and H. A. Warsame. "Sustainability Reporting: External Motivators and Internal Facilitators." *Corporate Governance* 15, no. 3 (2015): 375–390.

Nižetić, S., N. Djilali, A. Papadopoulos, and J. J. Rodrigues. "Smart Technologies for Promotion of Energy Efficiency, Utilization of Sustainable Resources and Waste Management." *Journal of Cleaner Production* 231 (2019): 565–591.

Pascale, De. G., N. Faccilongo, M. Riefolo, A. Romagno, and R. Silvestri. *Digital Innovation and Sustainable Development: Two Sides of the Same Coin*. IntechOpen, 2024. https://doi.org/10.5772/intechopen.112294

Porter, M. E., and C. V. D. Linde. "Toward a New Conception of the Environment-Competitiveness Relationship." *Journal of Economic Perspectives* 9, no. 4 (1995): 97–118.

Pritchard, A. "Statistical Bibliography or Bibliometrics." *Journal of Documentation* 25 (1969): 348.

Sarin, S., C. Haon, M. Belkhouja, A. Mas-Tur, N. Roig-Tierno, T. Sego, and S. Carley. "Uncovering the Knowledge Flows and Intellectual Structures of Research in Technological Forecasting and Social Change: A Journey through History." *Technological Forecasting and Social Change* 160 (2020): 120210.

Singh, S. K., M. Del Giudice, R. Chierici, and D. Graziano. "Green Innovation and Environmental Performance: The Role of Green Transformational Leadership and Green Human Resource Management." *Technological Forecasting and Social Change* 150 (2020): 119762.

Solanke, B., F. B. Onita, O. J. Ochulor, and H. O. Iriogbe. "The Impact of Artificial Intelligence on Regulatory Compliance in the Oil and Gas Industry." *International Journal of Science and Technology Research Archive* 7, no. 1 (2024): 61–72.

Sun, Y., and S. Grimes. "The Emerging Dynamic Structure of National Innovation Studies: A Bibliometric Analysis." *Scientometrics* 106 (2016): 17–40.

Taha, R., A. Al-Omush, and M. Al-Nimer. "Corporate Sustainability Performance and Profitability: The Moderating Role of Liquidity and Stock Price Volatility-Evidence from Jordan." *Cogent Business & Management* 10, no. 1 (2023): 2162685.

Van Eck, N. J., and L. Waltman. "Text Mining and Visualization Using VOSviewer." *arXiv preprint arXiv:1109.2058* (2011).

Wang, M. "How Much Do Green and Digital Service Innovations Matter for Firm Performance? Understanding the Mediating Role of Product Creativity." *International Journal of Technology Management* 94, no. 1 (2023): 31–55.

Wassmer, U., R. Paquin, and S. Sharma. "The Engagement of Firms in Environmental Collaborations: Existing Contributions and Future Directions." *Business & Society* 53, no. 6 (2014): 754–786.

Wong, C. W., K. H. Lai, K. C. Shang, C. S. Lu, and T. K. P. Leung. "Green Operations and the Moderating Role of Environmental Management Capability of Suppliers on Manufacturing Firm Performance." *International Journal of Production Economics* 140, no. 1 (2012): 283–294.

Yang, M. G. M., P. Hong, and S. B. Modi. "Impact of Lean Manufacturing and Environmental Management on Business Performance: An Empirical Study of Manufacturing Firms." *International Journal of Production Economics* 129, no. 2 (2011): 251–261.

Yue, X., B. Huo, and Y. Ye. "The Impact of Coercive Pressure and Ethical Responsibility on Cross-Functional Green Management and Firm Performance." *Journal of Business & Industrial Marketing* 38, no. 5 (2023): 1015–1028.

Zailani, S., M. R. Shaharudin, V. Govindasamy, M. Ismail, and S. F. A. S. Mahdzar. "The Eco-Efficiency Practices of the Sustainable Packaging and Its Effect Towards Sustainable Supply Chain Performance." In *2015 International Symposium on Technology Management and Emerging Technologies (ISTMET)*, 448–453. IEEE, 2015.

8 Impact of Green Management on Firm Performance in the Purview of AI and Data Analytics

A Comprehensive Review

Anand Patil and Swathi Shekar

8.1 INTRODUCTION

There is no business on a dead planet.

—David Brower

Social responsibility initially emerged as a mechanism opted by businesses to differentiate themselves from others; it later transitioned into a yardstick that certain businesses were obligated to meet. Many organizations initially viewed social responsibility as separate from business activities (Greening & Turban, 2000) or pertained to organizational levels only (Frederiksen, 2019; Jamali & Karam, 2018). Though conceptually social responsibility has been in practice for over three decades, it is in the last decade that a crucial emphasis on sustainability has been evident. Sustainability refers to a range of methods, procedures, and approaches used to protect the environment. It is a requirement for today and a necessity for the future (Linton et al., 2007; Van Hoek & Johnson, 2010). Companies are integrating conservation strategies and environmental conservation into their daily operations. This development ushers in a new era in which companies consider their stakeholders and the environment positively, looking beyond their corporate goals.

Corporate sustainability aims to uphold both corporate objectives and moral standards (Linnenluecke et al., 2010). It is not a single concept but an insightful holistic approach that integrates different aspects of business (Bansal, 2005; Lloret, 2016). Organizations must use cutting-edge technology to thrive in the fiercely competitive and dynamic corporate world. It is imperative to reconcile economic objectives with present and future societal demands (Ayuso & Navarrete-Báez, 2018; Günerergin et al., 2012).

 DOI: 10.1201/9781003626527-8

Green management is often viewed as a culmination of the ethical practices of a business along with its corporate social responsibility (Cruz & Pedrozo, 2009; Luchs et al., 2010). These practices refer to strategically implementing environmentally sustainable actions within a firm's operations (Zhu et al., 2013; Aguilera-Caracuel et al., 2012). The goal of green management is to make all business decisions based on sustainable practices. It entails adopting dependable practices, procedures, and initiatives that prioritize conservation and environmental protection in addition to preventing environmental deterioration. It includes the architectural design of buildings and project sites necessary for business operations (Zhang, 2012; C. Wu et al., 2016).

Artificial intelligence (AI) and data analytics reshape management by offering powerful tools that aid in tackling diverse business challenges and enhance decision-making processes. As industries face more complex issues, AI's rapid advancements and ability to analyze large datasets make it essential for optimizing and automating tasks, personalizing customer experiences, and improving overall efficiency (Trzaska et al., 2021). Through AI, it is possible to absorb repetitive tasks through automation, which allows employees to focus on higher-value activities. This enables better resource management and helps incorporate customer insights into business decisions. Management applications for AI go beyond advanced analytics and big data, extending to simpler processes that streamline operations (Nowak et al., 2022).

Through the lens of sustainable development, AI drives the progress of organizations by enhancing speed, efficiency, productivity, and innovation (Pinzaru et al., 2022). By integrating AI-driven solutions, organizations can address sustainability challenges more effectively—the data-driven insights of AI help to make impactful, eco-friendly decisions. AI's potential to foster competitive advantages and support sustainable initiatives becomes evident, making it a valuable asset in the evolving management landscape.

Green practices promote the effective and efficient use of resources and waste reduction. It is emphasized in many studies that when a firm incorporates environmental or green practices, operational efficiency is positively impacted (M. Yang et al., 2017; Melnyk et al., 2003; Gobeli, 2009). Through green practices, both the organization and the environment benefit in the long run (Shrivastava, 1995). It is said to improve operational efficiency and increase savings in businesses (Porter & van der Linde, 1995). By aligning sustainability with core business strategies, companies can achieve better financial performance through cost savings, innovation, and improved market positioning (Ghadimi et al., 2019).

A company's position in the market, operational efficacy, and financial performance are typically used to assess its performance. These tactics aid businesses in projecting a positive image by strengthening stakeholder connections and gaining the support of consumers (Luo & Bhattacharya, 2006). Better risk management, reduction in cost, and innovation boost the success of the business; this is achieved through green management practices (Hart & Ahuja, 1996). Also, as consumers are increasingly interested in eco-friendly items, businesses that adopt sustainable practices generally gain a competitive advantage (Leonidou et al., 2013).

Although sustainability is propagated and emphasized, its adoption at the grassroots level is questionable. Large organizations and multinational corporations are obligated by law and their reporting requirements to absorb corporate social responsibility and ESG (environmental, social, and governance) initiatives (Torres et al., 2023; Lu et al., 2021). The same logic applies to developed and industry-oriented nations that invariably adapt to dynamic environmental changes (Holt & Ghobadian, 2009; Zhu & Sarkis, 2006, 2008; Jansson et al., 2017).

Meanwhile, small and medium-sized enterprise (SMEs) and sole proprietorship concerns are not under the scrutiny and compulsion of green practices. But they contribute vastly toward pollution as they are great in number (Revell et al., 2010; OECD, 2018; De et al., 2020). Though governments introduce regulatory mechanisms to push small businesses to adopt sustainable practices, they are still in the infant stage (Holt & Ghobadian, 2009). Firms are yet reluctant to view the bigger picture of the adoption of green management practices (Montabon et al., 2007). For environmental conservation and improvement, green practices have to be the way of life for all entities, irrespective of how big or small. Through this alone, worldwide positive changes can be witnessed (Olawumi et al., 2018).

Green management techniques ensure that businesses focus on a holistic approach. They need to carefully balance business, societal, and environmental needs. This is necessary to build stakeholder trust and loyalty, which in turn improves firm performance in the long run (J. Yang et al., 2015; Mustapha et al., 2017). Hence, green management is extremely beneficial to organizations to sustain in a rapidly growing and demanding corporate environment. This makes green management a necessity rather than an option. AI and data analytics help in meeting this need innovatively and effectively. The futuristic perspective and flexibility of AI help organizations to mold it to suit their requirements. In summary, sustainable business practices are key essentials. They are critical for business survival and growth. The present study provides a comprehensive understanding of a large corpus of data. It analyzes the scope and development of the area of study. This, in turn, helps to identify the nuances and niche areas, which help in further research.

8.2 METHODS

Bibliometrics offers insightful information about a subject of study by analyzing academic literature through a variety of lenses (Benckendorff & Zehrer, 2013; De Bellis, 2009). It is an essential tool to reveal the structure and dynamics of scientific research (Waltman et al., 2010). It offers an overview of the information gathered from a database about a specific subject, including the primary writers, the nations that publish the most, and the evolution of literature throughout time (Van Raan, 2005).

The approaches are quantitative in nature and include but are not limited to co-word analysis for keywords, co-author and co-citation analysis, bibliographic linking, and quotation. Bibliometric analysis helps identify the nuances and niche areas by analyzing the related studies to date, providing a broader perspective to researchers.

8.2.1 Bibliometric Maps

VOSviewer (version 1.6.20), a potent program made by the Centre for Science and Technology Studies at Leiden University (Netherlands), was used to construct the bibliometric maps. This program, which is exceptional in its capacity to deconstruct and visualize intricate bibliometric data, examined author keywords, bibliographies, and citations from 1,555 publications in order to provide insightful information on the current study.

Each map showcases a definitive focal object, such as authors, keywords, countries associated with authors, and/or citations. Links are established for each focal object. The strength of these links, signifying a connection in the selected studies, is represented by a positive numerical value. These connections aid in determining the co-authorship study's strength of the link, the citation analysis's strength of the relationship between important writers, and the frequency of co-occurrence in publications (keywords analysis).

8.2.2 Selection of Studies

Scopus search query: TITLE-ABS-KEY ((Green AND Management) AND (Firm AND performance) AND (Sustainability))

Data extraction date: June 30, 2024

These keywords helped in the extraction of 1,555 documents from 1994 to 2024.

For the present study, the Scopus database was chosen alone. This decision was made for some definite reasons. The database is very well received by the research community with an extensive set of publications (Pranckutė, 2021; Pham-Duc et al., 2023). It has a highly convenient outlook and aids researchers in data segregation and literature analysis (Mongeon & Paul-Hus, 2016; Lima & Bonetti, 2020). It is also regarded as the biggest "peer-reviewed" database (Karagiannopoulou et al., 2023). Considering the aforementioned reasons, the selection of the database and studies was made by the authors.

Inclusion criteria are essential for bibliometric analysis to ensure that the selected research papers and the objectives of the study align. It also conforms with the relevance of the study. Additionally, proper inclusion criteria help to maintain consistency, which is essential to improve the study's validity and reliability. Hence, the following inclusion criteria were considered for this research:

1. Research articles, conference papers, and book chapters are the only forms of publication taken into account.
2. Publications between 1994 and 2024 are included.
3. Only research articles from related disciplines are taken into account.
4. Only research papers written in English will be included in this study.

After applying the criteria mentioned above, 1,404 results were obtained.

Further, a screening was employed, where document titles, keywords, and abstracts were examined to decide their inclusion for the present examination. In this step, 126 articles were omitted. Finally, 1,278 research studies published between 1994 and 2024 were selected for bibliometric analysis.

The present analysis is conducted in various facets encompassing numerous aspects related to research on green management and firm performance. Conclusions and recommendations will be made based on the following analyses:

1. The publishing trend of the topic
2. Citation analysis
3. Productive journals in the discipline
4. Top contributing countries
5. Major supporting institutions
6. Keyword analysis
7. Prominent authors
8. The co-authorship analysis
9. Current active publication sources
10. Relevant independent studies

8.3 DATA ANALYSIS

8.3.1 Publishing Trend

The concept of sustainability started to enter the mainstream in the late 2000s. From Figure 8.1, an increasing trend in the papers published in the Scopus database can be identified. The highest peak was in 2023, with 256 articles. By mid-2024, the number of articles was 158. The keenness showcased by the researchers is reflected in the constant rise in the number of publications, which depicts the topic's relevance to date and the future.

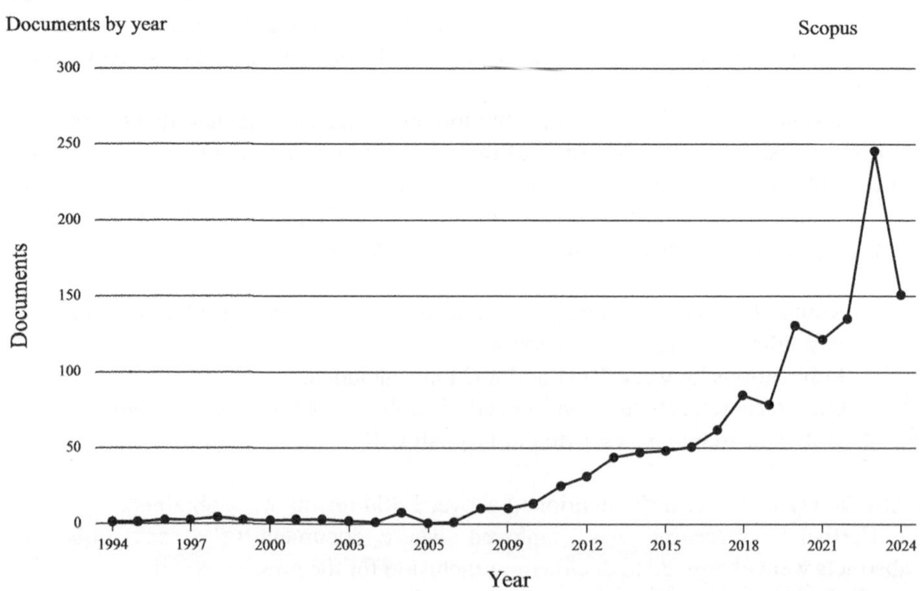

FIGURE 8.1 Documents published on the current topic in the Scopus database between 1994 and 2024.

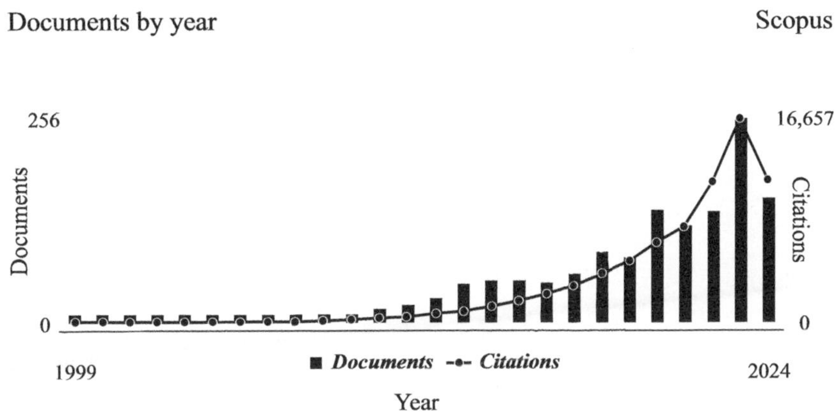

FIGURE 8.2 Documents published and citations on the current topic in the Scopus database between 1999 and 2024.

8.3.2 Citation Analysis

To conduct the citation analysis, Figure 8.2 and Table 8.1 show the yearly citation pattern for the current topic. Articles published in 2023 garnered the most citations (n = 16,652). Data for 2024 is examined up to June 30, 2024, with 70% of the citations 2023 already received. This upward trend signifies that the impact of the current topic is gradually expanding, and it is also increasingly gathering the interest of researchers.

TABLE 8.1
Total Citations on the Current Topic in the Scopus Database between 1999 and 2024

Year	Number of Citations	Year	Number of Citations
Up to 1999	20	2012	666
1999	13	2013	889
2000	34	2014	1,281
2001	45	2015	1,768
2002	36	2016	2,311
2003	63	2017	2,996
2004	64	2018	3,981
2005	82	2019	5,068
2006	117	2020	6,512
2007	122	2021	7,766
2008	187	2022	11,424
2009	218	2023	16,652
2010	317	2024	11,577
2011	418	**TOTAL (<1999–2024)**	**74,627**

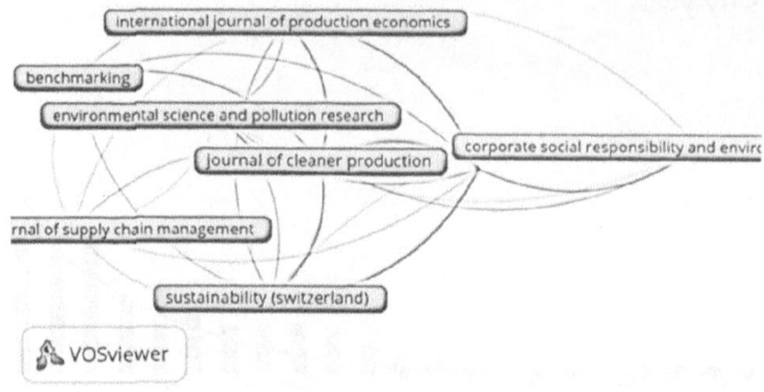

FIGURE 8.3 Top publishing journals.

TABLE 8.2
Top Publishing Journals with the Number of Documents Published and Citations Rceived

Source	Documents	Citations	Total Link Strength
Journal of Cleaner Production	122	12,849	227
Business Strategy and the Environment	67	4,984	190
International Journal of Production Economics	32	3,482	110
Sustainability (Switzerland)	81	2,060	121
Environmental Science and Pollution Research	37	847	73
Benchmarking	23	786	58
Corporate Social Responsibility and Environmental Management	20	593	39
International Journal of Supply Chain Management	20	564	22

8.3.3 Top Publishing Journals

Figure 8.3 and Table 8.2 depict the top journals with the maximum number of citations. The data was derived from the VOSviewer application, with the minimum citation criterion being 20.

8.3.4 Top Contributing Countries

The data retrieved for the present study has contributions from 98 countries on the subject of green management and sustainability aiding firm performance. The large number of countries contributing to the topic reveals the global importance and relevance of sustainability. To arrive at Figure 8.4, the criteria selected were at least 50 documents from a particular country incorporated. This has resulted in a dataset of the top 13 contributing countries. The details on the number of documents produced

FIGURE 8.4 Top contributing countries to the current field of study.

and the total citations for the same, coupled with the link strength, are shown in Table 8.3.

The list of nations mentioned in Table 8.3 has contributed mainly toward global sustainability. For instance, the United States, the United Kingdom, Italy, Spain, France, and Australia, through advancements in renewable energy and environmental policies (Chau et al., 2022), have all made notable contributions to sustainability. China, India, Malaysia, Pakistan, Indonesia, South Korea, and Taiwan have made substantial progress in sustainability, driven by governmental policies, industrial innovation, and international cooperation (Jahanger et al., 2023; Hansen et al., 2018).

TABLE 8.3
Top Contributing Countries with Documents Produced and Citations Received

Country ID	Country	Documents	Citations	Total Link Strength
14	China	433	16,785	328,088
94	United States	186	11,025	187,508
47	Malaysia	156	6,273	171,913
93	United Kingdom	146	12,384	192,402
61	Pakistan	132	4,842	162,074
32	India	114	3,646	120,940
37	Italy	86	5,948	82,483
78	Spain	82	6,196	114,864
84	Taiwan	82	7,829	92,382
3	Australia	69	5,165	79,243
25	France	68	5,958	78,161
77	South Korea	57	2,483	76,919
33	Indonesia	55	1,541	48,545

The countries of the world, regardless of their geographical origin, have a strong commitment to sustainability. This is demonstrated through a variety of measures, including the use of renewable energy and strict environmental policies, as well as breakthrough technologies and international collaborations. To prove this, a lot of literature has been published on the necessity and impact of sustainability on various geographical locations. As sustainability equals survival, no country can function without adopting and adapting to green practices (Phillis et al., 2011).

8.3.5 Top Contributing Institutions

Institutional support is an important criterion for research development. Collaboration among organizations allows researchers to share their expertise. It also supports resource and knowledge sharing and cooperation. Figure 8.5 and Table 8.4, generated from VOSviewer, showcases clusters of affiliated institutions actively engaged in research on green management and firm performance.

Strong associations among the various universities located in different provinces of China can be observed. There exist evident collaborations between Regent University College of Science and Technology (Ghana) and Dalian Maritime University (China), which have resulted in prominent research work. There is a wide range of affiliated work between universities in Thailand, Vietnam, and Pakistan and the universities in China.

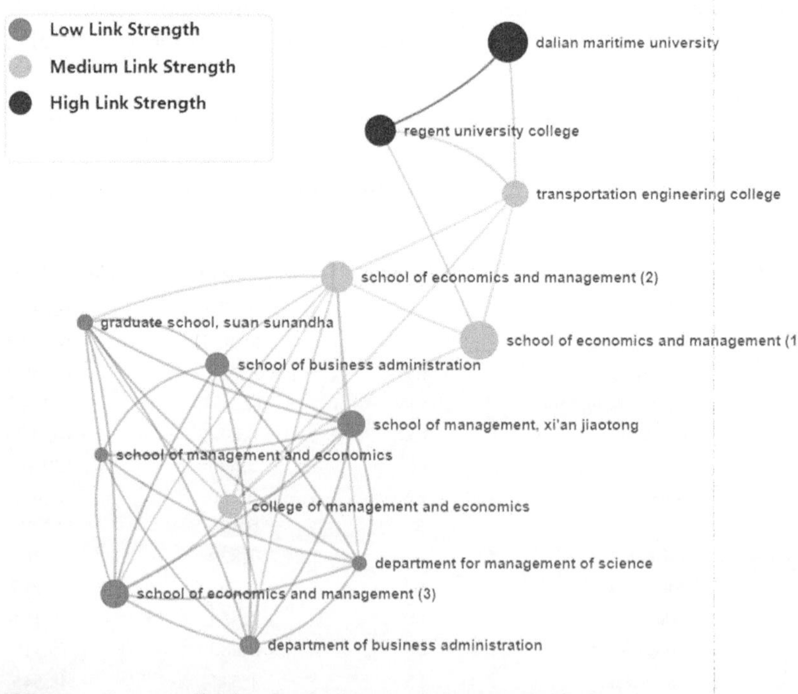

FIGURE 8.5 Top contributing institutions to the current field of study.

TABLE 8.4
Top Contributing Institutions with Documents Produced and Citations Received

Institution	Documents	Citations
Dalian Maritime University, China	7	762
Regent University College of Science and Technology, Ghana	5	450
Chang'an University, China	6	340
Tianjin University, China	8	695
Tsinghua University, Beijing, China	8	277
Ilma University, Pakistan	5	467
Suan Sunandha Rajabhat University, Thailand	5	173
Xi'an Jiaotong University, China	5	129
Harbin Institute of Technology (Weihai), China	7	353
South China University of Technology, China	5	381
Ton Duc Thang University, Vietnam	5	273
Beijing Institute of Technology, China	5	109

The articles considered for the study are associated with 3,422 institutions. A criterion of a minimum of five articles is considered to generate and find the most collaborative institutions in the present field of study.

8.3.6 Keywords Analysis

A word cloud represents the number of times a particular word appears in an article. It also shows the link between the keywords. The size of the word on the map corresponds with the number of publications in which the term appears. Each color represents a collection of terms integrated into clusters; the length of curved lines specifies the repetition of the term. The thickness of the words indicates the strength of pairs of topic areas or keywords. The clusters indicate how one topic relates to another.

The documents selected for the study showcase a total of 5,928 keywords. Figure 8.6 depicts 37 keywords that have occurred more than 50 times. These keywords help us understand the prime focal areas of the field. Three prominent clusters can be identified from Figure 8.6: Cluster 1, Cluster 2, and Cluster 3 (also see Table 8.5).

Cluster 1 represents the aspects related to manufacturing, supply chain, and decision-making, infusing green practices and economic performance. It encompasses all the factors that firms need to incorporate to maintain environmental stability along with industry requirements.

Cluster 2 focuses on the green economy; here, prominence is given to innovation. As China has published the highest number of articles in the area, it has also appeared as a keyword. As research articles are the major component

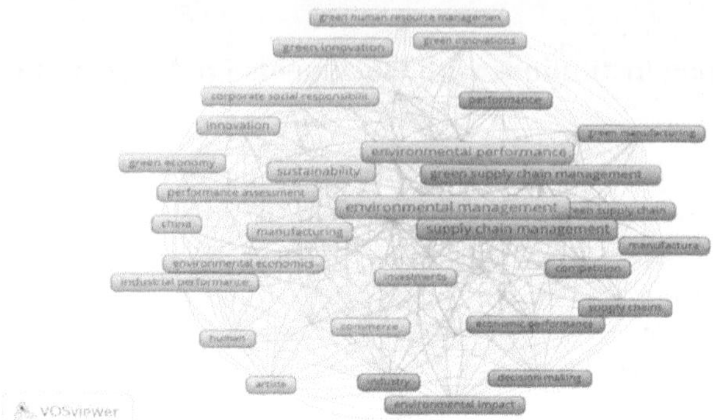

FIGURE 8.6 Keywords mentioned in the selected studies.

based on the inclusion criteria and the prime domain is commerce for the present study, they are keywords in this cluster.

Cluster 3 is more inclusive in nature. It covers the impact of the environment on different aspects like innovation, technology, human resources, and investments.

In a nutshell, these three clusters together give a picture of the vast coverage and importance of the present area of study. It also helps researchers ascertain the key areas and identify niche areas for future studies.

8.3.7 Most Productive Authors

Table 8.6 and Figure 8.7 contain details of the most prominent authors in the present field of study. The names of the authors, their affiliated institutions, and the total citations received are reflected. The table is arranged in the order of the decreasing number of documents published in the subject area of green management. Ebenezer Afum has authored 13 articles and is the first author on 4. Yaw Agyabeng-Mensah has contributed extensively, collaborating with other significant researchers in the field, and also produced 13 articles in the Scopus database, of which he is the first author in 6.

It is noteworthy to observe the affiliations of these top authors. Foreign collaborations are absolutely essential for holistic research development. This provides different perspectives and expertise and increases the impact of the study by making it richer. China is the pioneer in the present study; notable authors from Australia, Indonesia, Malaysia, France, Taiwan, and Hong Kong have made rich contributions, proving the global scope and relevance of the subject.

Table 8.6 also provides information regarding the total publications made by these authors and their h-indexes. This is included to uphold the diverse contributions of the authors and their overall research acumen.

TABLE 8.5
Keywords Mentioned in the Selected Studies

Cluster 1		Cluster 2		Cluster 3	
Keyword	**Occurrences**	**Keyword**	**Occurrences**	**Keyword**	**Occurrences**
Sustainable Development	326	Sustainability	255	Environmental Management	460
Supply Chain Management	279	Innovation	145	Environmental Performance	345
Green Supply Chain Management	253	Manufacturing	121	Green Innovation	126
Firm Performance	122	Environmental Economics	116	Financial Performance	74
Competition	89	China	110	Investments	61
Green Supply Chain	83	Commerce	110	Green Human Resource Management	58
Performance	83	Corporate Social Responsibility	96	Environmental Technology	55
Environmental Impact	80	Performance Assessment	85	Green Innovations	52
Supply Chains	74	Green Economy	77		
Manufacture	72	Industrial Performance	71		
Green Manufacturing	66	Environmental Protection	67		
Environmental Sustainability	65	Human	58		
Industry	58	Article	50		
Economic Performance	56				
Manufacturing Firms	56				
Decision Making	53				

TABLE 8.6
Prominent Authors in the Present Field of Study with Their Affiliation and Publishing Details

Author Name	Affiliation	Country	TC	PD	TPD	H
Ebenezer Afum	Dalian Maritime University	China	1,612	13	48	19
Yaw Agyabeng-Mensah	Curtin University	Australia	1,649	13	46	20
Syed Abdul Rehman Khan	Xuzhou University of Technology	China	9,196	13	160	57
Hengky Latan	Universitas Diponegoro	Indonesia	3,294	12	56	30
T Ramayah	Universiti Sains Malaysia	Malaysia	18,970	12	597	70
Charbel Jose Chiappetta Jabbour	NEOMA Business School	France	19,321	10	310	74
Taiwen Feng	Harbin Institute of Technology	China	3,658	9	114	34
Minglang Tseng	Asia University	Taiwan	18,258	9	428	74
Christina W. Y. Wong	The Hong Kong Polytechnic University	Hong Kong	6,421	9	119	43
Surajit Bag	Leonardo da Vinci University Center	France	5,435	8	127	39

TC = total citations, PD = published documents, TPD = total published documents, H = h-index

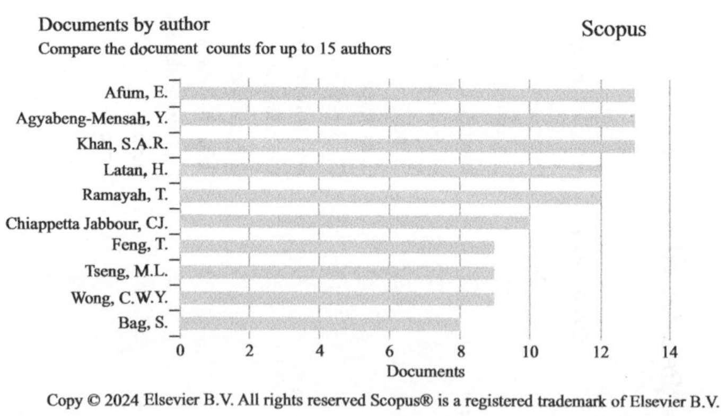

FIGURE 8.7 Prominent authors in the present field of study.

8.3.8 Co-Citation Analysis

Co-citation analysis is essential to identify and associate the most important papers in a field. It also aids researchers in locating the group of papers or clusters that cover similar themes or ideas. It is necessary to help understand how the research topics are connected and to develop a structure of knowledge about a particular field (Borini & Gattaz, 2018). Papers discussing similar topics are often cited together (Donthu et al., 2021). This establishes a link between the articles. The frequency or the number of times they are listed in references showcases the strength of the link.

The clusters in co-citation analysis reveal the patterns of how research works are interconnected with each other. They show the organization of research within a discipline. From Figure 8.8, five clusters are noticeable (also see Table 8.7). These clusters reflect the themes and key topics of the present study. Significant contributions can be identified, and research communities can be observed through this. It is necessary for researchers to understand the structure of the research landscape. They can also understand the areas that are prominently studied and influential work from prominent researchers.

8.3.9 Active Sources of Publication

It is essential to know the active publication sources for a field of study. They provide current information about recent research and advancement and legitimate sources to publish the same. Active publication sources highlight the current trends; this helps researchers identify gaps and niche areas for further research growth. This gives room for budding authors to know where they can get their research published (see Table 8.8). It also aids in creating a network and collaboration among authors of similar interests. These are vital for the growth of the discipline.

8.3.10 Findings of Relevant Independent Studies

The discipline of sustainability is studied extensively by researchers under different dimensions. The researchers are keen on exploring the various facets since the Sustainable Development Goals (SDGs) were incorporated in 2012 at the UN Conference. The coverage of this study is from 1994 to 2024. Table 8.9 consists of selected articles from the past decade conforming to the research.

FIGURE 8.8 Co-citation analysis in the present field of study.

TABLE 8.7
Co-Citation Cluster Analysis in the Present Field of Study

Cluster 1

Authors	Article	Publisher	Year
Zhu Q., Sarkis J.	"Relationships between operational practices and performance among early adopters of green supply chain management practices in Chinese manufacturing enterprises"	*Journal of Operations Management*	2004
Green K.W., Zelbst P.J., Meacham J., Bhadauria S.	"Green supply chain management practices: Impact on performance"	*Supply Chain Management: An International Journal*	2012
Rao P., Holt D.	"Do green supply chains lead to competitiveness and economic performance?"	*International Journal of Operations & Production Management*	2005
Zhu Q., Sarkis J.	"The moderating effects of institutional pressures on emergent green supply chain practices and performance"	*International Journal of Production Research*	2007
Seuring S., Muller M.	"From a literature review to a conceptual framework for sustainable supply chain management"	*Journal of Cleaner Production*	2008

Cluster 2

Authors	Article	Publisher	Year
Barney J.	"Firm resources and sustained competitive advantage"	*Journal of Management*	1984
Fornell C., Larcker D.F.	"Evaluating structural equation models with unobservable variables and measurement error"	*Journal of Marketing Research*	1981
Banerjee S.B., Iyer, E.S., Kashyap R.K.	"Corporate environmentalism: Antecedents and influence of industry type"	*Journal of Marketing*	2003
Chen Y.S., Lai S.B., Wen C.T.	"The influence of green innovation performance on corporate advantage in Taiwan"	*Journal of Business Ethics*	2006
Longoni A., Luzzini D., Guerci M.	"Deploying environmental management across functions: The relationship between green human resource management and green supply chain management"	*Journal of Business Ethics*	2008

(*Continued*)

TABLE 8.7 (CONTINUED)
Rationales for SPAR-4 Framework

Cluster 3

Authors	Article	Publisher	Year
Hart S.L.	"A natural-resource-based view of the firm"	*Academy of Management Review*	1995
Russo M.V., Fouts P.A.	"A resource-based perspective on corporate environmental performance and profitability"	*Academy of Management Journal*	1997
Klassen R.D., Whybark D.C.	"The impact of environmental technologies on manufacturing performance"	*Academy of Management Journal*	1997
Hart S.L., Ahuja G.	"Does it pay to be green? An empirical examination of the relationship between emission reduction and firm performance"	*Business Strategy and the Environment*	1996
Buysse K., Verbeke A.	"Proactive environmental strategies: A stakeholder management perspective"	*Strategic Management Journal*	2003

Cluster 4

Authors	Article	Publisher	Year
Vachon S., Klassen R.D.	"Environmental management and manufacturing performance: The role of collaboration in the supply chain"	*International Journal of Production Economics*	2008
Klassen R.D., Mclaughlin C.P.	"The impact of environmental management on firm performance"	*Management Science*	1996
Rao P., Holt D.	"Do green supply chains lead to competitiveness and economic performance?"	*International Journal of Operations and Production Management*	2005
Carter C.R., Rogers D.S.	"A framework of sustainable supply chain management: Moving toward new theory"	*International Journal of Physical Distribution & Logistics Management*	2008
Bowen F.E., Cousins P.D., Lamming R.C., Faruk A.C.	"The role of supply management capabilities in green supply"	*Production and Operations Management*	2001

Cluster 5

Authors	Article	Publisher	Year
Freeman R.E.	"Strategic management: A stakeholder approach"	Cambridge University Press (book)	1984

(*Continued*)

TABLE 8.7 (CONTINUED)
Rationales for SPAR-4 Framework

Cluster 4			
Authors	**Article**	**Publisher**	**Year**
Nunnally J.C., Bernstein I.H.	"Psychometric theory"	McGraw-Hill Companies (book)	1994
Onzalez-Benito J., Gonzalez-Benito O.	"Environmental proactivity and business performance: An empirical analysis"	Omega, Elsevier	2005
Barney J.	"Firm resources and sustained competitive advantage"	*Journal of Management*	1991
Hart S.L.	"A natural-resource-based view of the firm"	*Academy of Management Review*	1995

TABLE 8.8
Active Publication Sources in the Present Field of Study

Journal	Publisher	Years	Cite Score	Total Documents	Total Citations
Journal of Cleaner Production	Elsevier	1993 to 2024	20.4	39,961	1,755,348
Sustainability (Switzerland)	MDPI*	2009 to 2024	6.8	78,985	977,312
Business Strategy and the Environment	John Wiley & Sons	1992 to 2024	22.5	2,422	130,420
Environmental Science and Pollution Research	Springer Nature	1994 to 2024	8.7	44,130	808,092
International Journal of Production Economics	Elsevier	1991 to 2024	21.4	7,283	435,517
Benchmarking	Emerald Publishing	1999 to 2024	10.4	2,008	46,050
Corporate Social Responsibility & Environmental Management	John Wiley & Sons	2003 to 2024	17.2	1,688	67,922
*International Journal of Construction Supply Chain Management***	ExcelingTech	2012 to 2020	—	88	334
International Journal of Manufacturing Technology and Management	Emerald Publishing	2004 to 2024	16.3	1,004	5,871
Technological Forecasting and Social Change	Elsevier	1970 to 2024	21.3	6,533	293,123

*MDPI = Multidisciplinary Digital Publishing Institute.

**The *International Journal of Construction Supply Chain Management*, by ExcelingTech, was discontinued from the Scopus database in 2020.

TABLE 8.9
Relevant Studies Connected to the Present Field of Study Selected from the Scopus Database

Authors	Study Summary: "Green Management" and "Firm Performance"
Khare et al., 2023	Extensive training and awareness help in the proper implementation of green management, which in turn offers increased organizational performance.
Nugent et al., 2023	Considering labor productivity as a measure, environmental practices highly influence firm performance.
Ansari et al., 2022	Green management and innovative practices cast an influence on organizational performance in general and financial performance in particular.
Farza et al., 2021	A linear relationship is witnessed between financial performance and environmental innovation practices.
Bartolacci et al., 2020	Sustainable practices and firm performance are positively correlated, as green practices enhance competitiveness and help in reducing expenditures.
Chang et al., 2020	When organizations incorporate green practices and link them to their CSR practices, they gain a competitive edge.
Raharjo, 2019	With adequate industry support and policy changes, SMEs greatly benefited from green practices by making them efficient.
Huang et al., 2017	Strong organizational skills, together with an inclusive culture, are needed for leveraging green innovation to improve organizational performance.
Verrier et al., 2014	When lean management principles and green practices come together, it enhances organizational performance and also addresses sustainable development goals.
S. Wu et al., 2014	Enterprises with high knowledge of green management principles are more concerned about the environment; this concern, in the long run, increases firm performance.

8.4 CONCLUSION

Under the umbrella of sustainability, the present study focuses on green management practices and their impact on organizational performance. This research was exclusively conducted on the publications made in the Scopus database. The period for which the data is analyzed is between 1994 and 2024. This period of three decades has seen remarkable changes and advancements related to sustainability. With the infusion of AI and data analytics, green management techniques can be robust and innovative. The method adopted for the study is bibliometric analysis. VOSviewer, software, version 1.6.20, is utilized for generating bibliometric maps and analysis. Further, the Scopus database is used for collecting current information on authors, journals, and publications.

The discipline considered for the study is rich and broad. This is reflected through the number of publications made and the vast number of researchers actively engaged. Also, as the subject area is not specific but is applicable globally, involvement by a huge number of countries is seen. Similarly, the institutions sponsoring studies and the great number of quality journals publishing the studies in this area reflect this.

This bibliometric review considers the implications of green management linked to firm performance. The findings highlight that incorporating green management practices to promote sustainable development has a positive influence on firm performance.

Each study has its own set of limitations, such as this. The first to note is the selection of a single database. Though Scopus is a very comprehensive database, others do not fall far behind. A comparative analysis or an inclusive study comprising major databases can be done. An extensive period of three decades is considered; this era can be subdivided, and the study can reveal the development of the discipline. This study links green management with firm performance; other mediating factors can be taken into account.

REFERENCES

Aguilera-Caracuel, Javier, Nuria Esther Hurtado-Torres, and Juan Alberto Aragón-Correa. "Does International Experience Help Firms to Be Green? A Knowledge-Based View of How International Experience and Organizational Learning Influence Proactive Environmental Strategies." *International Business Review* 21, no. 5 (2012): 847–861.

Ansari, Nabeel, Zill-E-Huma, Raza Ali, Shakira Huma, and Adnan Baig. "The Role of Green Human Resource Management Practices and Eco-Innovation in Enhancing the Organizational Performance." *Vision* (2022): 09722629221092133.

Ayuso, S., and F. E. Navarrete-Báez. "How Does Entrepreneurial and International Orientation Influence SMEs' Commitment to Sustainable Development? Empirical evidence from Spain and Mexico." *Corporate Social Responsibility and Environmental Management* 25, no. 1 (2018): 80–94.

Bansal, Pratima. "Evolving Sustainably: A Longitudinal Study of Corporate Sustainable Development." *Strategic Management Journal* 26, no. 3 (2005): 197–218.

Barin Cruz, Luciano, and Eugênio Avila Pedrozo. "Corporate Social Responsibility and Green Management: Relation Between Headquarters and Subsidiary in Multinational Corporations." *Management Decision* 47, no. 7 (2009): 1174–1199.

Bartolacci, Francesca, Andrea Caputo, and Michela Soverchia. "Sustainability and Financial Performance of Small and Medium-Sized Enterprises: A Bibliometric and Systematic Literature Review." *Business Strategy and the Environment* 29, no. 3 (2020): 1297–1309.

Benckendorff, Pierre, and Anita Zehrer. "A Network Analysis of Tourism Research." *Annals of Tourism Research* 43 (2013): 121–149.

Chang, Tai-Wei, Yen-Li Yeh, and Hung-Xin Li. "How to Shape an Organization's Sustainable Green Management Performance: The Mediation Effect of Environmental Corporate Social Responsibility." *Sustainability* 12, no. 21 (2020): 9198.

Chau, Ka Yin, Massoud Moslehpour, Yu-Te Tu, Nguyen Tan Tai, Nguyen Hoang Tien, and Pham Quang Huy. "Exploring the Impact of Green Energy and Consumption on the Sustainability of Natural Resources: Empirical Evidence from G7 Countries." *Renewable Energy* 196 (2022): 1241–1249.

De Bellis, Nicola. *Bibliometrics and Citation Analysis: From the Science Citation Index to Cybermetrics*. Scarecrow Press, 2009.

De, Debashree, Soumyadeb Chowdhury, Prasanta Kumar Dey, and Sadhan Kumar Ghosh. "Impact of Lean and Sustainability-Oriented Innovation on the Sustainability Performance of Small and Medium-Sized Enterprises: A Data Envelopment Analysis-Based Framework." *International Journal of Production Economics* 219 (2020): 416–430.

Donthu, Naveen, Satish Kumar, Debmalya Mukherjee, Nitesh Pandey, and Weng Marc Lim. "How to Conduct a Bibliometric Analysis: An Overview and Guidelines." *Journal of Business Research* 133 (2021): 285–296.

Economic Research Institute for ASEAN and East Asia. *SME Policy Index: ASEAN 2018 Boosting Competitiveness and Inclusive Growth: Boosting Competitiveness and Inclusive Growth*. OECD Publishing, 2018.

Farza, Khouloud, Zied Ftiti, Zaineb Hlioui, Waël Louhichi, and Abdelwahed Omri. "Does It Pay to Go Green? The Environmental Innovation Effect on Corporate Financial Performance." *Journal of Environmental Management* 300 (2021): 113695.

Frederiksen, Tomas. "Political Settlements, the Mining Industry and Corporate Social Responsibility in Developing Countries." *The Extractive Industries and Society* 6, no. 1 (2019): 162–170.

Ghadimi, Pezhman, Chao Wang, and Ming K. Lim. "Sustainable Supply Chain Modeling and Analysis: Past Debate, Present Problems, and Future Challenges." *Resources, Conservation and Recycling* 140 (2019): 72–84.

Greening, Daniel W., and Daniel B. Turban. "Corporate Social Performance as a Competitive Advantage in Attracting a Quality Workforce." *Business & Society* 39, no. 3 (2000): 254–280.

Günerergin, M., Ş. Penbek, and D. Zaptçıoğlu. "Exploring the Problems and Advantages of Turkish SMEs for Sustainability." *Procedia-Social and Behavioral Sciences* 58 (2012): 244–251.

Hansen, Ulrich Elmer, Ivan Nygaard, Henny Romijn, Anna Wieczorek, Linda M. Kamp, and Laurens Klerkx. "Sustainability Transitions in Developing Countries: Stocktaking, New Contributions, and a Research Agenda." *Environmental Science & Policy* 84 (2018): 198–203.

Hart, Stuart L., and Gautam Ahuja. "Does It Pay To Be Green? An Empirical Examination of the Relationship Between Emission Reduction and Firm Performance." *Business Strategy and the Environment* 5, no. 1 (1996): 30–37.

Holt, Diane, and Abby Ghobadian. "An Study of Supply Chain Management Practices Amongst UK Manufacturers." *Journal of Manufacturing Technology Management* 20, no. 7 (2009): 933–956.

Huang, Jui-Wen, and Yen-Hui Li. "Green Innovation and Performance: The View of Organizational Capability and Social Reciprocity." *Journal of Business Ethics* 145 (2017): 309–324.

Jahanger, Abdul, David Balsalobre-Lorente, Muhammad Ali, Anwar Samour, Shujaat Abbas, Turan Tursoy, and Fatoumatta Joof. "Going Away or Going Green in ASEAN Countries: Testing the Impact of Green Financing and Energy on Environmental Sustainability." *Energy & Environment* (2023): 0958305X231171346.

Jamali, Dima, and Charlotte Karam. "Corporate Social Responsibility in Developing Countries as an Emerging Field of Study." *International Journal of Management Reviews* 20, no. 1 (2018): 32–61.

Jansson, Johan, Jens Nilsson, Fredrik Modig, and Gudrun Hed Vall. "Commitment to Sustainability in Small and Medium-Sized Enterprises: The Influence of Strategic Orientations and Management Values." *Business Strategy and the Environment* 26, no. 1 (2017): 69–83.

Karagiannopoulou, Sofia, Nikolaos Sariannidis, Katerina Ragazou, Ioannis Passas, and Alexandros Garefalakis. "Corporate Social Responsibility: A Business Strategy that Promotes Energy Environmental Transition and Combats Volatility in the Post-Pandemic World." *Energies* 16, no. 3 (2023): 1102.

Khare, Vishnu Kant, Subash Raghuwanshi, Amit Vashisht, Priyanka Verma, and Richa Chauhan. "The Importance of Green Management and Its Implication in Creating Sustainability Performance on the Small-Scale Industries in India." *Journal of Law and Sustainable Development* 11, no. 5 (2023): e699–e699.

Lee, Sang M., Suntae Kim, and Dae-Soo Choi. "Green Supply Chain Management and Organizational Performance." *Industrial Management & Data Systems* 112, no. 8 (2012): 1148–1180.

Leonidou, Constantinos N., Constantine S. Katsikeas, and Neil A. Morgan. "'Greening' the Marketing Mix: Do Firms Do It and Does It Pay Off?" *Journal of the Academy of Marketing Science* 41 (2013): 151–170.

Lima, Cláudia O., and João Bonetti. "Bibliometric Analysis of the Scientific Production on Coastal Communities' Social Vulnerability to Climate Change and to the Impact of Extreme Events." *Natural Hazards* 102 (2020): 1589–1610.

Linnenluecke, Martina K., and Andrew Griffiths. "Corporate Sustainability and Organizational Culture." *Journal of World Business* 45, no. 4 (2010): 357–366.

Linton, Jonathan D., Robert Klassen, and Vaidyanathan Jayaraman. "Sustainable Supply Chains: An Introduction." *Journal of Operations Management* 25, no. 6 (2007): 1075–1082.

Lloret, Antonio. "Modeling Corporate Sustainability Strategy." *Journal of Business Research* 69, no. 2 (2016): 418–425.

Lu, Jun, and Jianfeng Wang. "Corporate Governance, Law, Culture, Environmental Performance, and CSR Disclosure: A Global Perspective." *Journal of International Financial Markets, Institutions and Money* 70 (2021): 101264.

Luchs, Michael G., Rebecca W. Naylor, Julie R. Irwin, and Rajagopal Raghunathan. "The Sustainability Liability: Potential Negative Effects of Ethicality on Product Preference." *Journal of Marketing* 74, no. 5 (2010): 18–31.

Luo, Xueming, and C. B. Bhattacharya. "Corporate Social Responsibility, Customer Satisfaction, and Market Value." *Journal of Marketing* 70, no. 4 (2006): 1–18.

Melnyk, Steven A., Robert P. Sroufe, and Roger Calantone. "Assessing the Impact of Environmental Management Systems on Corporate and Environmental Performance." *Journal of Operations Management* 21, no. 3 (2003): 329–351.

Mongeon, Philippe, and Adèle Paul-Hus. "The Journal Coverage of Web of Science and Scopus: A Comparative Analysis." *Scientometrics* 106 (2016): 213–228.

Montabon, Frank, Robert Sroufe, and Rachna Narasimhan. "An Examination of Corporate Reporting, Environmental Management Practices and Firm Performance." *Journal of Operations Management* 25, no. 5 (2007): 998–1014.

Mustapha, Mohd A., Zainuddin A. Manan, and Siti R. W. Alwi. "Sustainable Green Management System (SGMS)–An Integrated Approach towards Organisational Sustainability." *Journal of Cleaner Production* 146 (2017): 158–172.

Nowak, Anna, and Armand Kasztelan. "Economic Competitiveness vs. Green Competitiveness of Agriculture in the European Union Countries." *Oeconomia Copernicana* 13, no. 2 (2022): 379–405.

Olawumi, Tolulope O., and Daniel W. Chan. "A Scientometric Review of Global Research on Sustainability and Sustainable Development." *Journal of Cleaner Production* 183 (2018): 231–250.

Pagell, Mark, and David Gobeli. "How Plant Managers' Experiences and Attitudes toward Sustainability Relate to Operational Performance." *Production and Operations Management* 18, no. 3 (2009): 278–299.

Pham-Duc, Binh, Hieu Nguyen, Hai Phan, and Quang Tran-Anh. "Trends and Applications of Google Earth Engine in Remote Sensing and Earth Science Research: A Bibliometric Analysis Using Scopus Database." *Earth Science Informatics* 16, no. 3 (2023): 2355–2371.

Phillis, Yannis A., Evangelos Grigoroudis, and Vassilis S. Kouikoglou. "Sustainability Ranking and Improvement of Countries." *Ecological Economics* 70, no. 3 (2011): 542–553.

Pinzaru, Florina, Alina Mihaela Dima, Alexandra Zbuchea, and Zoltan Veres. "Adopting Sustainability and Digital Transformation in Business in Romania: A Multifaceted Approach in the Context of the Just Transition." *Amfiteatru Economic* 24, no. 59 (2022): 28.

Porter, Michael E., and Claas van der Linde. "Toward a New Conception of the Environment-Competitiveness Relationship." *Journal of Economic Perspectives* 9, no. 4 (1995): 97–118.

Pranckutė, Rasa. "Web of Science (WoS) and Scopus: The Titans of Bibliographic Information in Today's Academic World." *Publications* 9, no. 1 (2021): 12.

Raharjo, Kuswantoro. "The Role of Green Management in Creating Sustainability Performance on the Small and Medium Enterprises." *Management of Environmental Quality: An International Journal* 30, no. 3 (2019): 557–577.

Revell, Andrea, David Stokes, and Hsin Chen. "Small Businesses and the Environment: Turning over a New Leaf?" *Business Strategy and the Environment* 19, no. 5 (2010): 273–288.

Rossetto, Daniel E., Ricardo C. Bernardes, Felipe M. Borini, and Cláudio C. Gattaz. "Structure and Evolution of Innovation Research in the Last 60 Years: Review and Future Trends in the Field of Business through the Citations and Co-citations Analysis." *Scientometrics* 115, no. 3 (2018): 1329–1363.

Shrivastava, Paul. "The Role of Corporations in Achieving Ecological Sustainability." *Academy of Management Review* 20, no. 4 (1995): 936–960.

Torres, Lorena, Diego Ripa, Ashutosh Jain, Julio Herrero, and Stavroula Leka. "The Potential of Responsible Business to Promote Sustainable Work–An Analysis of CSR/ESG Instruments." *Safety Science* 164 (2023): 106151.

Trzaska, Rafał, Adam Sulich, Michał Organa, Jerzy Niemczyk, and Bartosz Jasiński. 2021. "Digitalization Business Strategies in the Energy Sector: Solving Problems with Uncertainty under Industry 4.0 Conditions." *Energies* 14, no. 23: 1–21.

Van Hoek, Remko, and Mark Johnson. "Sustainability and Energy Efficiency: Research Implications from an Academic Roundtable and Two Case Examples." *International Journal of Physical Distribution & Logistics Management* 40, no. 1/2 (2010): 148–158.

Van Raan, Anthony F. J. "Fatal Attraction: Conceptual and Methodological Problems in the Ranking of Universities by Bibliometric Methods." *Scientometrics* 62 (2005): 133–143.

Verrier, Baptiste, Benoît Rose, Elisabeth Caillaud, and Hamid Remita. "Combining Organizational Performance with Sustainable Development Issues: The Lean and Green Project Benchmarking Repository." *Journal of Cleaner Production* 85 (2014): 83–93.

Waltman, Ludo, Nees Jan van Eck, and Ed C. Noyons. "A Unified Approach to Mapping and Clustering of Bibliometric Networks." *Journal of Informetrics* 4, no. 4 (2010): 629–635.

Wu, Chien-Hua, and David Barnes. "An Integrated Model for Green Partner Selection and Supply Chain Construction." *Journal of Cleaner Production* 112 (2016): 2114–2132.

Wu, Shih-I, and Yung-Cheng Wu. "The Influence of Enterprisers' Green Management Awareness on Green Management Strategy and Organizational Performance." *International Journal of Quality & Reliability Management* 31, no. 4 (2014): 455–476.

Yang, Jie, Feng Zhang, Xiaojun Jiang, and Wenlong Sun. "Strategic Flexibility, Green Management, and Firm Competitiveness in an Emerging Economy." *Technological Forecasting and Social Change* 101 (2015): 347–356.

Yang, Meng, Daniela Vladimirova, and Steve Evans. "Creating and Capturing Value Through Sustainability: The Sustainable Value Analysis Tool." *Research-Technology Management* 60, no. 3 (2017): 30–39.

Zhang, Jun. "Delivering Environmentally Sustainable Economic Growth: The Case of China." *Asia Society Report* (2012): 2–25.

Zhu, Qinghua, and Joseph Sarkis. "An Inter-sectoral Comparison of Green Supply Chain Management in China: Drivers and Practices." *Journal of Cleaner Production* 14, no. 5 (2006): 472–486.

Zhu, Qinghua, Joseph Sarkis, and Kee-Hung Lai. "Green Supply Chain Management Implications for "Closing the Loop." *Transportation Research Part E: Logistics and Transportation Review* 44, no. 1 (2008): 1–18.

Zhu, Qinghua, Joseph Sarkis, and Kee-Hung Lai. "Institutional-Based Antecedents and Performance Outcomes of Internal and External Green Supply Chain Management Practices." *Journal of Purchasing and Supply Management* 19, no. 2 (2013): 106–117.

9 Green Purchase Behaviour

A Pathway to Corporate Sustainability through Bibliometric Lens

Divya Gogia, Nirmaljeet Virk, and Vinita Choudhary

9.1 INTRODUCTION

In a world characterised by persistent business activities and rising production, the consequential pollution of the natural environment is a well-acknowledged concern, leading to adverse impacts on people, wildlife, and crops (Elliott, 2003; Mohajan, 2018). The inconsistency of unlimited human needs against the backdrop of finite resources necessitates marketers to navigate the challenge of utilising limited resources efficiently (Laszlo, 1994). Green marketing emerges as a strategic approach, advocating for the adoption of environment-friendly products, for example, organic foods, recyclable products, and ozone-friendly items (Singh and Sharma, 2016; Gogia and Rathi, 2022). This eco-centric marketing philosophy not only aims to satisfy customer needs, wants, and demands but also endeavours to protect the environment and society in the process (Welford, 2013; Martinez et al., 2020).

The surge in consumer needs related to green products raises critical questions about consumer perceptions and the efficacy of marketing strategies employed by various companies. Researchers have delved into consumers' green purchase behaviours, exploring aspects such as emotions, perceived consumer effectiveness, connection to nature, and situational contexts like environmental co-governance and social interaction (Singh and Sharma, 2016; Maksudunov, 2020). A huge amount of literature has been conducted on green marketing, but an extensive research gap persists, especially in understanding the behaviour of consumers towards green marketing through bibliometric analysis. This research endeavours to address the noticeable gap in the examination of consumer behaviour towards green marketing, employing a quantitative analytical approach. Although numerous studies have explored sustainable consumption, green advertising, and the marketing mix within the broader domain of green marketing (Kartawinata et al., 2020; Le, 2021; Nekmahmud and Fekete-Farkas, 2020), a distinct void exists in

DOI: 10.1201/9781003626527-9

dedicated investigations into the intricate dynamics of consumer behaviour within green marketing initiatives. This gap hampers a comprehensive understanding of how consumers engage and respond to green marketing strategies, underscoring the need for targeted investigations.

To connect this gap and contribute to the growth of more efficient and persistent green marketing strategies, the present study utilises bibliometric analysis. By evaluating the existing literature and identifying key future research directions, the study seeks to answer several critical research questions:

1. What are the current study trends on consumers' purchasing behaviours in relation to green marketing?
2. In the body of current literature, who are the principal authors and contributors who have made a significant impact?
3. Which authors, co-words, and papers have had the most citations in the field of consumer behaviour research on green marketing?
4. In which countries is the majority of research in this domain carried out?

In order to address the research questions, this study uses a variety of bibliometric procedures, such as concurrences and co-authorship investigation, citation analysis, co-words analysis, Scopus-analysed search outcomes, and bibliographic couplings. The objective of this method is to extract insights that will lead to a deeper comprehension of customer behaviour in environmentally conscious marketing and provide guidance for the creation of efficient and focused green marketing campaigns.

9.2 RESEARCH METHODOLOGY AND STUDY DESIGN

This study aims to explore the progressions occurring in the domain of green marketing. In pursuit of this objective, a bibliometric analysis was executed on the papers in this domain that were listed in Scopus. This study's main focus is the Scopus database, which is acknowledged as one of the biggest sources of articles and biographical data. Figure 9.1 is a summary of the analytical methods and the bibliometric investigation outcomes. The details of search terms, withdrawal processes, and exclusion and inclusion criteria are covered in detail in the following sections.

9.2.1 Data Extraction Process and Search Strategy

The complete search process from the Scopus database, including the terms used, is shown in Figure 9.2. The search terms "green marketing" OR "consumer behaviour" gave access to publications published between 2014 and 2023 that were indexed in the Scopus database in the "Arts and Humanities, Business, Management and Accounting, Psychology, Social Sciences, Economics, Econometrics and Finance, and Multidisciplinary fields." Table 9.1 provides an explanation of the retrieval of meta-data. Prior to filtering, roughly 1046 broad topic papers were extracted on this

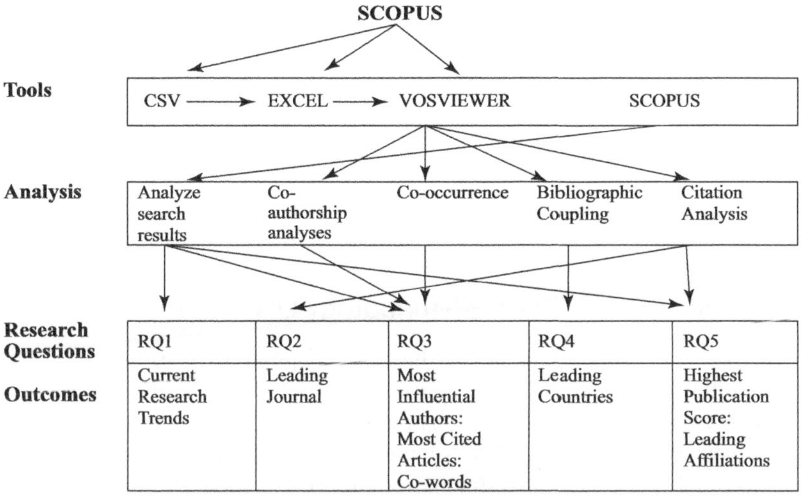

FIGURE 9.1 Analytical techniques and bibliometric study outcomes Source: Adapted from Sharma and Khushwaha (2023).

TABLE 9.1
Bibliography Data Retrieval Process

Criteria	Scopus Exclude	Scopus Include
Search date: November, 22 2023		1048
Database: Scopus		
Search Term: ("sustainability" OR "promotion") AND ("green" OR "marketing") AND ("sustainable" OR "green initiatives") AND ("company*" OR "programmes") AND ("consumer*" OR "behaviour")		
Year: 2014 to present	225	823
Subject areas: Social Sciences, Business, Management and Accounting, Arts and Humanities, Economics, Econometrics and Finance, Psychology, Multidisciplinary	227	594
Document types: Articles, book chapters, and reviews	38	556
Language: English	11	545
Erroneous records refinement: Duplication	2	543
Details missing	3	540
Content screening	42	498

subject. Following the application of the publication year, document type, and subject filter, 545 results were obtained. After refining erroneous records, addressing missing details, and conducting content screening, we obtained 498 results, and all subsequent analyses were performed on this database (Figure 9.2).

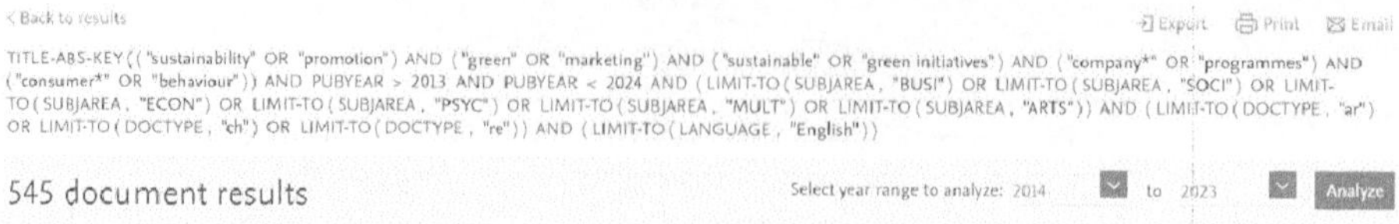

FIGURE 9.2 Capture of Scopus database search procedure.

9.2.2 Bibliometrics Research Methodology Development

In recent years, bibliometric analysis has gained popularity in the research community (Kumar et al., 2021b). The major reason for its popularity is the advancement of research databases like Scopus and Web of Science, and applications such as VOSviewer and Biblioshiny. The second reason is the expansion of bibliometric analysis to other fields, such as information theory and business, including business strategy (Kumar et al., 2021a), management studies (Ellegaard and Wallin, 2015), marketing research (Donthu et al., 2021), and others where the application of bibliometrics ranges from investigating publishing patterns to probing the intellectual arrangement of a research topic. In this study, we employed Excel, VOSviewer, and Scopus analysis to concentrate on primary analysis. Excel CSV is utilised to extract data files, while VOSviewer is used for bibliometric analysis and science mapping display. We analyse bibliographic couplings (BCs), concurrences and co-authorship analysis, co-words analysis, citation analysis, and co-citation, which are examples of science mapping analysis.

9.2.3 The Role of Data Analytics in Green Marketing Research

Scholars of green marketing have embraced data analytics as a cornerstone of their research, as it makes it possible to systematically analyse large volumes of data (Salhab, 2024). Therefore, the present study has used data tools for generating bibliometric data from the Scopus database on various trends, prominent works, and potential topics in the area of green marketing. Using analytical tools like co-word analysis and citation mapping, scholars can ascertain the important patterns and linkages within the field. Furthermore, data analytics helps to assess journal performance indicators, such as SNIP, SJR, and CiteScore, which are essential for evaluating the impact and topicality of the released materials. The inclusion of the big data approach also improves the reliability of the findings and helps organisations further their corporate sustainability agenda by connecting green marketing initiatives with wider overall sustainability objectives.

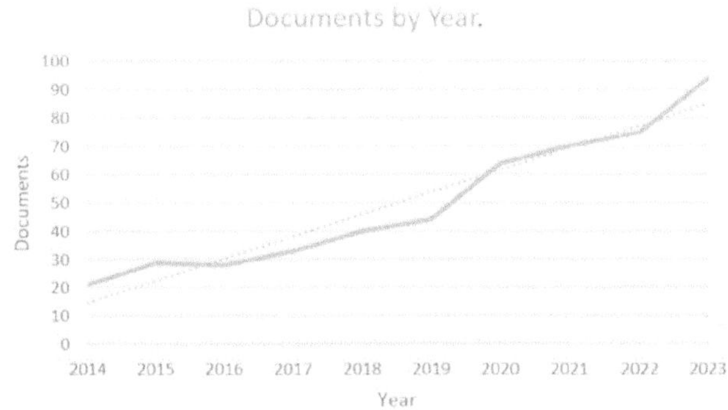

FIGURE 9.3 Documents by year.

9.3 RESULTS AND DISCUSSION

9.3.1 Recent Research Trends

The changing research trends seen in Scopus articles over the last ten years are depicted in Figure 9.3. According to the information in Table 9.2, studies looking at the study outcomes that were published after 2013 using the Scopus-extracted year of publication showed a significant increase in research activity. The research shows a steady rise in the quantity of published papers, indicating an ongoing expansion in scholarly discourse related to green marketing.

9.3.2 Top-Rated, Effective, and Significant Sources

The top journal is *Sustainability*, which has published 92 papers on green marketing with 1622 citations (Table 9.2). The *Journal of Cleaner Production* and *Business Strategy and the Environment*, with 48 and 8 documents, respectively, come next.

TABLE 9.2
Top Ten Leading Journals on Green Marketing

Source	Documents	Citations
Sustainability (Switzerland)	92	1622
Journal of Cleaner Production	48	2617
Business Strategy and the Environment	8	156
Resources, Conservation and Recycling	6	309
Environment, Development and Sustainability	5	57
International Journal of Sustainability in Higher Education	6	138
Developments in Marketing Science: Proceedings of the Academy of Marketing Science	17	8
World Sustainability Series	7	17

TABLE 9.3
Co-Authorship of Top Nine Authors in the Research Area

Authors	Documents	Citations
Amatulli and Cesare	5	181
De Angelis and Matteo	5	181
El-bassiouny and Noha	3	26
Oh and Kyungwha	4	113
Scarpato and Debora	3	68
Sogari and Giovanni	3	217

The subject's research over the past decade has been influenced by the top ten papers. Surprisingly, the *World Sustainability Series* has very few research papers on green marketing.

9.3.3 The Most Successful Authors in Green Marketing

Table 9.3 presents co-authorship analysis, which investigates how academics work together on a certain topic. Understanding how co-authors participate intellectually is important since co-authorship provides a formal means for researchers to do so (Acedo et al., 2006; Cisneros, Ibanescu, et al., 2018). On five papers, Amatulli and Cesare, and De Angelis and Matteo worked together. Oh and Kyungwha have collaborated on four articles.

9.3.4 Most Influential Articles

Based on citation analysis, Table 9.4 lists the top ten most significant papers, their publishing years, and the total number of citations. De Merdeiros et al.'s (2014) "Success Factors for Environmentally Sustainable Product Innovation: A Systematic Literature Review" is identified as the most influential paper, accredited with the highest number of citations (389), followed by Dangelico and Vocalleli's (2017) "Green Marketing: An Analysis of Definitions, Strategy Steps, and Tools through a Systematic Review of the Literature" with 329, and Bocken and Short's (2016) article "Towards a Sufficiency-Driven Business Model: Experiences and Opportunities," which has 262 total citations.

9.3.5 Co-Occurrence Author Keywords (Co-Words)

Co-word analysis phrases are typically obtained from author keywords, but also from article titles, abstracts, and full texts (Donthu et al., 2021; Baker et al., 2020). Co-word analysis shows a thematic relationship between commonly occurring keywords. Co-word analysis can be used to forecast future research in the subject when pertinent words from the publication's connotations and future study goals are incorporated in the analysis.

TABLE 9.4
Most Influential Articles

Article Title	Authors	Year	Total Citations
"Success factors for environmentally sustainable product innovation: A systematic literature review"	De Merdeiros, Ribeiro, and Cortimiglia	2014	389
"Green Marketing: An analysis of definitions, strategy steps, and tools through a systematic review of the literature"	Dangelico and Vocalleli	2017	329
"Towards a sufficiency-driven business model: Experiences and opportunities"	Bocken and Short	2016	262
"Consequences of "greenwashing": Consumers' reactions to hotels' green initiatives"	Rahman, Park, and Chi	2015	180
"Sustainable fashion supply chain: Lessons from H&M"	Shen	2014	175
"Factors affecting consumers' choices concerning sustainable packaging during product purchase and recycling"	Martinho, Portela, and Fonseca	2015	165
"Corporate investments in supply chain sustainability: Selecting instruments in the agri-food industry"	Rueda, Garrett, and Lambin	2017	140
"Public awareness of nature and the environment during the COVID-19 crisis"	Rousseau and Deschacht	2020	138
"Sustainability and branding: An integrated perspective"	Kumar and Christodoulopoulou	2014	131
Sustainable campus: Engaging the community in sustainability	Too and Bajracharya	2015	122

Table 9.5 displays the top 40 keywords of the researchers, which is a record of the most frequently used terms in the body of current literature on green marketing. The term "sustainability" ranks first with 168 occurrences, followed by "sustainable consumption" with 29. It indicates that most previous studies on the authors' keywords have been performed in the field of business management and accounting, i.e., sustainable consumption, sustainability, and corporate social responsibility. Table 9.5 also presents the total connection strength between keywords, indicating that sustainability, green marketing, sustainable development, sustainable consumption, and corporate social responsibility have strong network relations, implying that these five keywords have been often used together in research and review papers over the last ten years.

An overlay visualisation of forty of the authors' keywords produced using the VOSviewer programme is displayed in Figure 9.4. Throughout the course of a year, this network proposes keyword trends. The most recent keywords used in this study domain are sustainability, corporate social responsibility, and sustainable

TABLE 9.5
List of Top 40 Keywords of Authors in Green Marketing Domain

Keywords	Occurrences	Total Strength Link
"Sustainability"	168	156
"Green Marketing"	22	37
"Corporate Social Responsibility"	27	30
"Sustainable Development"	23	28
"Sustainable Consumption"	29	27
"Consumer Behavior"	16	25
"Sustainable Marketing"	10	23
"Consumer Behaviour"	12	22
"Marketing"	15	21
"Green Products"	7	16
"Innovation"	11	16
"Environment"	13	15
"Fast Fashion"	7	14
"Social Marketing"	9	14
"Circular Economy"	10	13
"Environmental Sustainability"	8	11
"Green Washing"	10	11
"Sustainable Fashion"	8	11
"Retail"	5	10
"Business Model"	6	9
"CSR"	7	9
"Green"	7	9
"Purchase Intention"	8	9
"Climate Change"	7	8
"Communication"	6	8
"Fashion"	6	8
"Green Product"	5	8
"Social Media"	6	8
"Supply Chain"	5	8
"Green Consumerism"	5	7
"Higher Education"	7	7
"Sustainable Products"	6	7
"Green Advertising"	6	5
"Green Consumption"	6	5
"Sustainable Tourism"	8	5
"COVID-19"	6	4
"Social Responsibility"	5	4
"Theory of Planned Behaviour"	6	4
"Corporate Social Responsibility (CSR)"	6	3
"Environmental Concern"	5	3

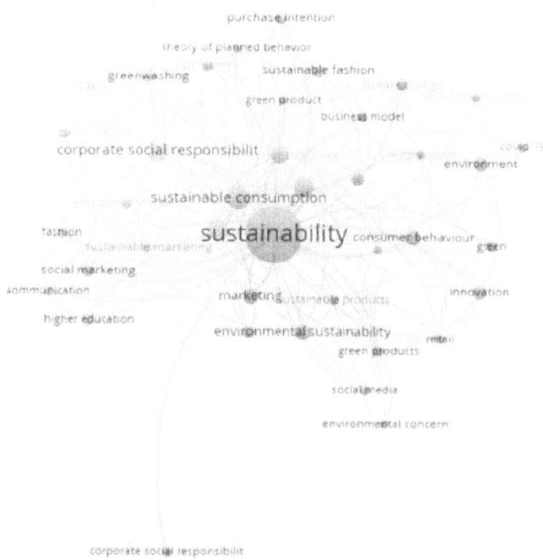

FIGURE 9.4 Network of authors' keywords trending in the green marketing research area.

consumption. Green marketing, sustainable development, and consumer behaviour are the next most often used terms. Future researchers will consequently be able to use these keywords in their studies. Most significantly, it is found that the most recent study issues being investigated at this moment include consumer behaviour, sustainability, and green marketing consumption.

9.3.6 Bibliographic Coupling of Countries

A scientific mapping technique called bibliographic coupling assumes that two papers with identical references also have similar content (Kessler, 1963; Weinberg, 1974). The bibliographic coupling networks of the top 33 nations are shown in Figure 9.5. There are five clusters visible on the network. Ten of the most common items are found in the first and largest cluster.

The United States and Italy, which are included in the second cluster, are the countries with the highest number of citations (Table 9.6).

9.3.7 Sources with Highest Publications

Table 9.7 lists the top ten most favoured journals with over 190 articles, along with the SJR, SNIP, and CiteScore for each. Of all the articles, nearly half on green marketing are published in the journals mentioned in Table 9.7, which constitute 39.56% of all publications. The journal *Sustainability* is the preferred publication for authors, followed by the *Journal of Cleaner Production* and *Business Strategy and the Environment*, which are based in Switzerland and the United Kingdom. The CiteScore, SJR, and SNIP values for the top 10 journals in 2023 are also included in Table 9.7. The total number of citations a journal achieved each year from papers

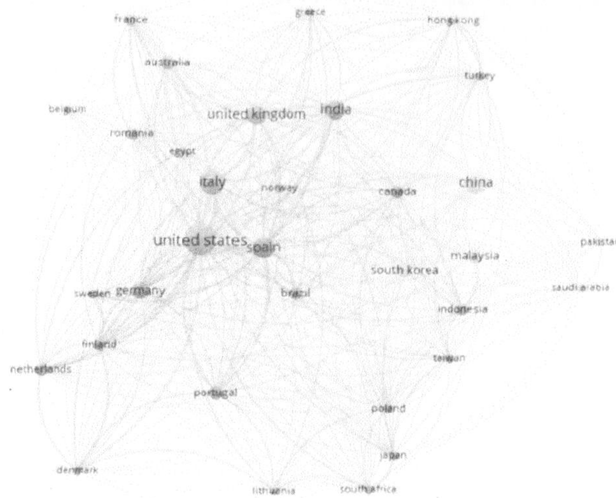

FIGURE 9.5 Network of bibliographic coupling of countries.

published in the three years prior is divided by the total number of publications indexed in Scopus during the same four years to arrive at the CiteScore. The provision of quantitative, statistical data derived from citation data requires a methodical, absolute approach to assessing the most esteemed journals globally. Out of these ten publications, *Resources Conservation and Recycling*, the *Journal of Cleaner Production*, and *Business Strategy and the Environment* had the highest CiteScore (20.3, 18.5, and 17.8, respectively).

A prominent metric known as the SCImago Journal Rank (SJR) is founded on the idea that "not all citations are treated equally." Based on the overall number of citations in a subject field, the topic field, journal quality, and reputation all directly affect the citation value, as per SJR. A higher SJR indicates that the journal is more prestigious. The *Journal of Business Research*, *Business Strategy and the Environment*, and Resources, Conservation and Recycling are the top three journals according to SJR, with 2.89, 2.87, and 2.68 citations, respectively. SNIP (Source Normalized Impact per Paper) assesses citations according to the overall number of citations within a topic field to ascertain the contextual citation efficacy of a source. It enables you to compare directly from sources on a variety of subjects. With SNIP values of 3.23, 3.01, and 2.93, respectively, the top three journals by SNIP values are the *Journal of Business Research*, *Technological Forecasting and Social Change*, and *Corporate Social Responsibility and Environmental Management*.

9.4 LIMITATIONS AND FUTURE RESEARCH

In this section, we provide limitations and offer suggestions for further study on green marketing that should be done in the future. Firstly, even if the Scopus database is

TABLE 9.6
Top 33 Countries across the Five Bibliographic Clusters

Country	Documents	Cluster	Citations
Australia	17	2	565
Belgium	5	2	278
Brazil	13	1	616
Canada	13	1	351
China	43	4	868
Denmark	7	3	145
Egypt	5	2	31
Finland	11	3	442
France	13	2	211
Germany	27	3	418
Greece	6	5	125
Hong Kong	6	5	108
India	31	5	282
Indonesia	12	1	61
Italy	53	2	1496
Japan	8	1	135
Lithuania	5	1	62
Malaysia	18	4	248
Netherlands	14	3	493
Norway	8	2	86
Pakistan	6	4	116
Poland	9	1	174
Portugal	17	1	355
Romania	15	2	273
Saudi Arabia	8	4	48
South Africa	7	1	225
South Korea	20	4	314
Spain	38	1	529
Sweden	8	3	181
Taiwan	8	1	67
Turkey	7	5	63
United Kingdom	36	2	656
United States	86	2	2108

meticulously used, it does not entirely cover the whole selection of relevant literature in the field of green marketing, which is a potential source of bias. This emphasises how important it is that future research projects incorporate several databases such as the Web of Science in order to provide a more thorough examination of the academic field.

Moreover, the temporal limitation imposed by concentrating only on articles released between 2014 and 2023 could introduce a temporal bias by leaving out

TABLE 9.7
The Top Ten Journals According to SNIP, SJR, and CiteScore

S.NO.	Journal	Country	Number of Papers	SNIP (2023)	SJR (2023)	CiteScore
1	*Sustainability (Switzerland)*	Switzerland	98	1.19	0.67	5.8
2	*Journal of Cleaner Production*	United Kingdom	55	1.98	1.98	18.5
3	*Business Strategy and the Environment*	United Kingdom	8	2.75	2.87	17.8
4	*International Journal of Sustainability in Higher Education*	United Kingdom	6	1.31	0.75	6.1
5	*Environment Development and Sustainability*	Netherlands	6	1.29	0.83	7.2
6	*Resources Conservation and Recycling*	Netherlands	6	2.77	2.68	20.3
7	*PLOS One*	United States	5	1.23	0.88	6
8	*Technological Forecasting and Social Change*	United States	5	3.01	2.64	17.2
9	*Corporate Social Responsibility and Environmental Management*	United Kingdom	4	2.93	2.13	15.6
10	*Journal of Business Research*	United States	4	3.23	2.89	16
	Total papers in top 10 journals		**197**			
	Total extracted papers		**498**			
	Share of top 10 journals in total extracted papers		**39.56**			

influential works that were written before this time and possibly ignoring fundamental understandings of the development of green marketing and consumer behaviour. Finally, despite methodological rigour, the inherent subjectivity of co-word analysis imposes a subtle constraint in the interpretive aspects of the results.

Ongoing dynamic analysis, cross-database comparisons, and inquiries into the multidisciplinary facets of green consumer behaviour are necessary as the subject of green marketing develops. In addition to adding to the body of knowledge on green marketing in academia, the study offers useful advice to companies looking to develop more targeted and effective green marketing campaigns.

This study acts as a springboard for more investigation, inspiring academics and industry professionals to delve deeper into the intricacies of consumer behaviour in the context of green marketing campaigns and encouraging the establishment of sustainable company practices.

REFERENCES

Baker, H. K., N. Pandey, S. Kumar, and A. Haldar. "A bibliometric analysis of board diversity: Current status, development, and future research directions." *Journal of Business Research* 108 (2020): 232–246.

Donthu, Naveen, Satish Kumar, DebiduttaPattnaik, and Weng M. Lim. "A bibliometric retrospection of marketing from the lens of psychology: Insights from Psychology & Marketing." *Psychology & Marketing* 38, no. 5 (2021): 834–865.

Ellegaard, Ole, and Johan A. Wallin. "The bibliometric analysis of scholarly production: How great is the impact?." *Scientometrics* 105 (2015): 1809–1831.

Elliott, Michael. "Biological pollutants and biological pollution—an increasing cause for concern." *Marine Pollution Bulletin* 46, no. 3 (2003): 275–280.

Gogia, Divya. and Rathi, Priya. "Determinants of organic food buying behaviour during crisis: A case of India." *International Journal of Research and Analytical Reviews* 9, no. 2 (2022): 445–453.

Kartawinata, Budi Rustandi, Dyah Maharani, Mahir Pradana, and Hilda Maulida Amani. "The role of customer attitude in mediating the effect of green marketing mix on green product purchase intention in love beauty and planet products in Indonesia." In *Proceedings of the International Conference on Industrial Engineering and Operations Management*, vol. 1, pp. 3023–3033, 2020.

Kessler, M. M. "Bibliographic coupling between scientific papers." *American Documentation* 14, no. 1 (1963): 10–25.

Kumar, Satish, Riya Sureka, Weng Marc Lim, Sachin Kumar Mangla, and Nisha Goyal. "What do we know about business strategy and environmental research? Insights from business strategy and the environment." *Business Strategy and the Environment* 30, no. 8 (2021a): 3454–3469.

Kumar, Satish, Weng Marc Lim, Nitesh Pandey, and J. Christopher Westland. "20 years of electronic commerce research." *Electronic Commerce Research* 21 (2021b): 1–40.

Laszlo, Ervin. *Macroshift: Navigating the Transformation to a Sustainable World*. Berrett-Koehler Publishers, 1994.

Le, Quang Hung. "Factors affecting consumer purchasing behavior: A green marketing perspective in Vietnam." *The Journal of Asian Finance, Economics and Business* 8, no. 5 (2021): 433–444.

Maksudunov, Azamat, and Murat Avci. "The color of the future in marketing is green." *Contemporary Issues in Strategic Marketing* (2020): 225–254.

Martínez, Marta Pagán, Camila Pires Cremasco, Luís Roberto Almeida Gabriel Filho, Sergio Silva Braga Junior, Adriano Viana Bednaski, Filipe Quevedo-Silva, Caroline Miranda Correa, Dirceu da Silva, and Rosamaria Cox Moura-Leite Padgett. "Fuzzy inference system to study the behavior of the green consumer facing the perception of greenwashing." *Journal of cleaner production* 242 (2020): 116064.

Mohajan, Haradhan. "Acid rain is a local environment pollution but global concern." *Open Science Journal of Analytical Chemistry* 3 (2018): 47–55.

Nekmahmud, Md, and Maria Fekete-Farkas. "Why not green marketing? Determinates of consumers' intention to green purchase decision in a new developing nation." *Sustainability* 12, no. 19 (2020): 7880.

Salhab, Hanadi A. "The use of data analytics in digital marketing for sustainable business growth." *Journal of Infrastructure, Policy and Development* 8, no. 8 (2024): 4894.

Sharma, S., and Khushwaha, B. "A retrospective overview of organic products in the personal care category: Science mapping, themes, and future research scope through bibliometric analysis." *International Journal of Business Excellence* (2023). https://doi.org/10.1504/IJBEX.2022.10049926.

Singh, Priyanka, Rajendra Singh, and Sanjay Sharma. "Emergence of green marketing strategies and sustainable development in India." *Journal of Commerce and Management Thought* 7, no. 4 (2016): 693–710.

Tomasovic, Brian. "Tradeoffs in environmental law." *Journal of Land Use & Environmental Law* 34 (2018): 93.

Weinberg, B. H. "Bibliographic coupling: A review." *Information Storage and Retrieval* 10, no. 5–6 (1974): 189–196.

Welford, Richard, ed. *Hijacking Environmentalism: Corporate Responses to Sustainable Development*. Routledge, 2013.

10 The Role of AI, ML, and Data Analytics in Overcoming Greenwashing

Enhancing Proactive Environmental Strategies for Improved Firm Performance

Vandana Gaur, Sonal Jauhari, and Vijayta Taneja

10.1 INTRODUCTION

A proactive environmental strategy is one in which organizations make a purposeful and future-oriented attempt to respond to environmental problems as well as promote sustainability. It goes beyond mere observation of regulations, instead actively identifying and implementing programs aimed at reducing impacts on the environment, conserving resources and enhancing environmental performance. Ambitious goal setting, investment in environmentally friendly technologies, stakeholder involvement, integration of sustainability principles into core business practices and decision-making processes are some of the characteristics of proactive environmental strategies. This approach is not only geared toward reducing adverse effects on the environment but also helps organizations create long-term value for themselves while contributing to wider environmental and social objectives.

Greenwashing refers to making false or misleading claims about a company's efforts to achieve a greener production process or its products that have environmentally beneficial attributes (refer to definitions of greenwashing in Tables 10.1, 10.2 and 10.3). It involves dishonest marketing that showcases the company as being more ecologically responsive than it actually is. Greenwashing includes making unsubstantiated claims about products' environmental attributes, overstating environmental initiatives and selective emphasis on positive aspects to downplay negative environmental impacts.

Companies greenwash for several reasons such as bolstering their public image, attracting environmentally aware consumers, or diverting attention from their real

DOI: 10.1201/9781003626527-10

TABLE 10.1
Definitions of Greenwashing

Reference	Definition
Seele, P., and Gatti, L. (2015)	Greenwashing involves misleading consumers about the environmental practices of a company or the environmental benefits of a product or service.
Leonidou, C. N., and Skarmeas, D. (2015)	Greenwashing is the practice of companies disingenuously spinning their products and policies as environmentally friendly.
Parguel, B., Benoît-Moreau, F., and Larceneux, F. (2011)	Greenwashing refers to the deceptive use of green marketing to give the impression that a company is environmentally friendly when it is not.
Marquis, C., and Toffel, M. W. (2012)	Greenwashing is the act of misleading stakeholders about the environmental practices of a company or the environmental benefits of a product.
Walker, K., and Wan, F. (2012)	Greenwashing involves the use of deceptive marketing practices by organizations to present an environmentally responsible public image.
Nyilasy, G., Gangadharbatla, H., and Paladino, A. (2014)	Greenwashing is the act of misleading consumers about the environmental practices of a company or the environmental benefits of a product or service.
Chen, Y., and Chang, C. (2012)	Greenwashing refers to the practice of companies exaggerating their environmental efforts to present a false image of environmental responsibility.
Delmas, M. A., and Burbano, V. C. (2011)	Greenwashing involves misleading consumers regarding the environmental practices of a company or the environmental benefits of a product or service.
Laufer, W. S. (2003)	Greenwashing is the practice of companies disingenuously spinning their products and policies as environmentally friendly.
TerraChoice Environmental Marketing (2009)	Greenwashing is defined as misleading consumers about the environmental practices of a company or the environmental benefits of a product or service.

environmental performance or controversial activities. Nevertheless, this practice is detrimental to customer trust; it undermines the integrity of the brand and acts as a barrier to genuine efforts to tackle environmental problems. Regulators, customers and advocacy groups are becoming more sensitive about companies' ecological assertions in the light of growing awareness regarding environmental issues, thereby making them responsible for deceitful greenwashing.

Proactive environmental strategy is an important part of corporate sustainability. Corporate sustainability has become central to contemporary business strategies due to its imperative role in ensuring ethical responsibility, long-term viability and competitive advantage. Companies can protect their operations and long-term

TABLE 10.2
Definition of Greenwashing by Organizations

Source	Definition
Environmental Protection Agency (EPA)	"Greenwashing is the practice of making an unsubstantiated or misleading claim about the environmental benefits of a product, service, technology, or company practice."
Greenpeace	"Greenwashing is used to describe the act of misleading consumers regarding the environmental practices of a company or the environmental benefits of a product or service."
World Wildlife Fund (WWF)	"Greenwashing refers to the deceptive use of green marketing to promote the perception that an organization's products, aims, or policies are environmentally friendly."
Corporate Finance Institute (CFI)	"Greenwashing is when a company purports to be environmentally conscious for marketing purposes but actually isn't making any notable sustainability efforts."
Bloomberg	"Greenwashing is the act of misleading consumers about the environmental practices of a company or the environmental benefits of a product or service."
Stanford Social Innovation Review (SSIR)	"Greenwashing is when a company spends more time and money claiming to be green through advertising and marketing than actually implementing business practices that minimize environmental impact."

TABLE 10.3
Definition of Greenwashing in Dictionares

Source	Definition
Oxford English Dictionary	"Greenwashing is disinformation disseminated by an organization so as to present an environmentally responsible public image."
Cambridge Dictionary	"Greenwashing is making people believe that your company is doing more to protect the environment than it really is."
Merriam-Webster	"Greenwashing is the act or practice of making a product, policy, activity, etc., appear to be more environmentally friendly than it really is."
Business Dictionary	"Greenwashing is the deceptive use of green marketing to give the impression that a company is environmentally friendly when it is not."
Encyclopedia Britannica	"Greenwashing is a form of marketing spin in which green PR and green marketing are deceptively used to persuade the public that an organization's products, aims, and policies are environmentally friendly."
Investopedia	"Greenwashing is the process of conveying a false impression or providing misleading information about how a company's products are more environmentally sound."

sustainability by foreseeing and resolving environmental issues before they become worse. Sustainable business methods help in the management of risks related to resource shortages, environmental deterioration and effects of climate change. Additionally, they can reduce any effect on their supply chains and operations through genuine sustainable practices rather than resorting to greenwashing.

In this context, emerging technologies such as artificial intelligence (AI), machine learning (ML), data analytics and automation offer options to reduce the danger of greenwashing by delivering more transparent, accurate and verifiable data regarding business environmental activities. This study examines how these technologies can augment a firm's capacity to execute authentic sustainability goals and enhance overall company performance. We examine the role of AI, ML and data analytics in enhancing accountability, improving decision-making and optimizing resource management to achieve sustainable business objectives.

10.1.1 Prevalence of Green Washing in India: Theoretical Framework

Increased market demand for sustainable products and greater awareness of environmental concerns are the main reasons for the prevalence of greenwashing prevalence in India (refer to Figure 10.1). The growing need for environmentally friendly products has fueled competition in the Indian consumer market, leading some companies to make green claims, sometimes without strong evidence. Even if people are becoming more conscious of sustainability, there is still a dearth of in-depth knowledge, making it easier for businesses to make false statements without encountering serious criticism. The regulatory framework in India is not as strict in comparison with other countries with respect to making environmental claims. This lack of strong regulation and enforcement allows companies to make false green claims with less risk of penalties. Companies use green marketing as a strategy to enhance their brand image and attract eco-conscious consumers. Additionally, they do not provide transparent information about their environmental practices, making it difficult for consumers to verify the authenticity of companies' green claims. Sectors like fast-moving consumer goods (FMCG), textiles and apparel, automobiles, and real estate have been engaged in greenwashing.

Greenwashing is a significant issue in India, fueled by an increasing demand for sustainable products, a less stringent regulatory environment and aggressive marketing strategies by companies. Addressing this issue requires stronger regulations, awareness of consumers, increased use of third-party certifications and higher transparency in corporate practices.

Thanh Tiep Le (2022) investigated the correlation between the environmental strategies used by small and medium-sized enterprises (SMEs) and their sustainable business performance. It specifically focuses on how corporate social responsibility and green innovation act as mediators in this relationship within the setting of a rising economy. This result indicates that incorporating environmental, social and economic sustainability into the business decision-making process can drive corporate social responsibility practices and green innovations for products, processes and services that ultimately lead to sustainable performance.

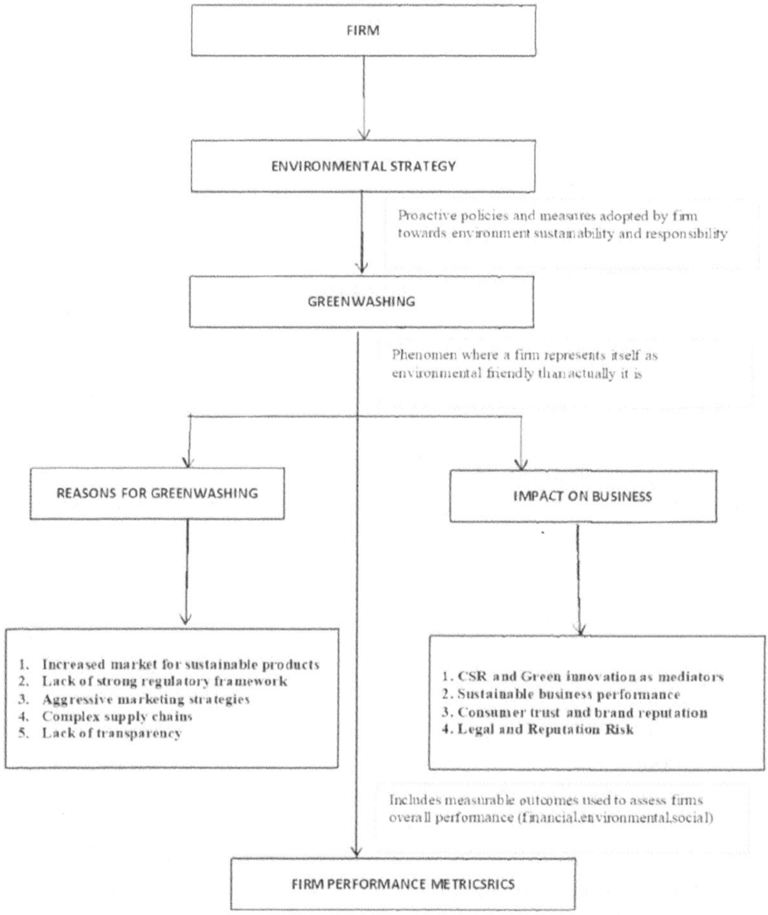

FIGURE 10.1 Greenwashing and environmental strategy: Theoretical framework.

10.1.2 History and Origin of Greenwashing

The concept of "greenwashing" was first introduced by Jay Westerveld, an environmentalist, in one of his essays. His criticism revolved around the hotel industry through which he condemned the placing of placards on bedsides encouraging the reuse of towels because it saves the environment. He thought that it was a cost-saving measure disguised as an environmental initiative when little else had been done to enhance overall environmental performance, if at all. Increasingly, many businesses began advertising their products as environmentally friendly due to the growth of ecological awareness. Nevertheless, some companies went overboard or even made false declarations. For instance, in the mid-1980s, Chevron launched its "People Do" campaign with pictures of its environmental commitments while facing several counts of environmental violations. From this example emerged the initial and most recognized instances of greenwashing. There has been a rise in products labeled "eco-friendly," "biodegradable" and "natural." Many of these claims are not substantiated and lead to greenwashing on a large scale without the necessary controls. This has led to the emergence of regulatory bodies and institutions in response to increasing instances of greenwashing. For example, the Federal Trade Commission (FTC) in the US issued the Green Guides that give companies guidelines on how they can make honest and non-deceptive environmental claims. Other countries also adopted similar guidelines to address deceitful claims relating to environmental protection (Federal Trade Commission, 2012). The 2010s witnessed a number of high-profile greenwashing scandals that put greenwashing under the spotlight. As social media became popular in the 2020s and made information more accessible, consumers became wiser about environmental advertisements. This development resulted in increased scrutiny for business practices as well as a call for transparency and accountability (White et al., 2019. However, companies still face the dilemma between genuine sustainability efforts and marketing, hence greenwashing remains a major concern even today. There has been some mitigation of this problem through third-party certifications, expanded regulatory oversight and increased consumer awareness, but it is still hotly debated within corporate environmental responsibility circles as an issue (Lyon et al., 2015).

10.1.3 Seven Sins of Greenwashing

Seven sins of greenwashing have been identified by the TerraChoice Group. Various tricks and methods are employed in order to create an impression that a company or product is environmentally responsible when it is not.

1. Sin of hidden trade-offs: This is where the positives are being advertised while the negatives are forgotten. For example, they can only be made from recycled materials but do not mention the pollution during the manufacturing process (TerraChoice Environmental Marketing, 2009).
2. No proof: This refers to making environmental claims without providing accessible evidence or a reliable third-party certification. For example, claiming a product as "eco-friendly" without any certification or proof of environmentally beneficial practices (Delmas et al., 2011).
3. Vagueness: This refers to using vague terms that are poorly defined or broad in meaning, such as when describing a product as "all natural," which can mislead people since many naturally occurring substances are harmful (Carlson et al., 1993).
4. Irrelevance: Irrelevance refers to making claims that may be true but do not matter to consumers who want environment-friendly products. For instance, saying that a particular product is "CFC-Free" when CFCs are already banned by law (TerraChoice, 2009).
5. Worshiping false labels: Conversely, there are products that mislead their consumers into thinking that they have undergone a legitimate green

certification process through false suggestions or certification-like images. An example is a paper towel whose packaging has a certification-like image that makes a claim that the product "fights global warming." Other examples include green jargon such as "eco-safe" and "eco-preferred."

6. The lesser of two evils: This simply means claiming to be greener than other products in a category that is itself environmentally unfriendly. For example, organic cigarettes might be less harmful than regular cigarettes but are still harmful overall (Parguel et al., 2011).
7. Stretching the truth: When a product is "certified organic" when it is actually not misleads consumers. Nowadays, there is huge demand for organic products, and consumers, in order to buy environmentally friendly products, trust products with an "organic" label. But companies sometimes resort to greenwashing by certifying products as organic when they are not.

10.2 THE ROLE OF AI, ML AND DATA ANALYTICS IN ADDRESSING GREENWASHING

In order to evaluate nonfinancial risks and allocate capital to businesses that manage themselves in a "socially responsible" manner and uphold their social contract, risk managers and investors have been looking for more high-quality environment, social and governance (ESG) data. Artificial intelligence, machine learning and data analytics can play a crucial role in mitigating greenwashing by enhancing transparency and delivering precise, real-time facts regarding a company's environmental impact. The issue is that there is a dearth of high-quality firm-level data that may be used as crucial inputs in evaluating a company's risks and compliance with ESG standards since there are no established standards for corporations to follow when reporting on sustainability issues. The solution for ESG rankings and ratings is big data, which was created using state-of-the-art statistical models, AI and natural language processing (NLP) in dozens of languages. It can also help fight self-reported bias and greenwashing and give high-quality data (Antonic, 2020).

10.2.1 Assessing and Tracking Sustainability Indicators

Machine learning algorithms can discern patterns and trends in data that allow firms to enhance procedures and minimize their environmental impact. Companies can monitor environmental data like carbon emissions, water usage, trash generation and energy consumption throughout their whole value chain by utilizing sensors, IoT devices and AI-driven analytics systems.

10.2.2 Authenticate Assertions Using Empirical Evidence

By incorporating machine learning models with external databases, firms may validate their sustainability claims against verifiable third-party data, thereby confirming the accuracy of their environmental declarations. This diminishes the probability of erroneous or deceptive assertions and enhances stakeholder trust.

10.2.3 Transparency in Supply Chain

Artificial intelligence can help in tracing the source of raw material supplied, ensuring that corporations source exclusively from providers. This reduces the risk of greenwashing within the supply chain and substantiates claims of responsible sourcing.

10.2.4 Predictive Analytics for Proactive Strategy Formulation

AI-driven predictive analytics can discern emerging environmental hazards and possibilities, allowing firms to adopt a proactive stance on sustainability. By utilizing historical data and market patterns, companies can predict legislative changes, alterations in customer preferences, and technical innovations that may affect their sustainability initiatives.

10.3 SIGNIFICANCE OF PROACTIVE ENVIRONMENTAL STRATEGY IN CORPORATE SUSTAINABILITY EFFORTS

Proactive environmental strategies are critical for corporate survival and can provide a variety of important benefits. These strategies require them to go beyond mere compliance with environmental regulations and actively reduce their environmental impact through innovation, resource efficiency and sustainable practices, reducing waste and disposal, enhancing social reputation, improving consumer preferences and creating new innovation capabilities.

Wijethilake (2017) examined the extent to which companies use sustainability control systems (SCS) to translate proactive sustainability strategy into corporate sustainability performance. The study concluded that although SCS plays an important role in facilitating proactive sustainability into corporate sustainability performance strategies, their partial mediation indicates the need for a more comprehensive approach. Corporations should integrate sustainability into management control systems and align internal functional departments to effectively implement sustainability strategies and achieve performance goals.

Sharma et al. (1998) examined the relationship between environmental policy and firm competitiveness, challenging the simplistic notion of "green is gold" through a resource-based perspective. It is shown by empirical evidence that environmental strategies can provide unique organizational capabilities such as stakeholder cohesion, higher levels of learning and continuous innovation. Statistical analysis shows that such strategies explain more than 20% of the variance. Overall, the findings suggest that proactive environmental strategies can indeed provide competitive advantage by strategies described in the resource-based view of the firm.

Li Yu et al. (2017), within the framework of environmental management strategies (GO framework), the resource-based view (RBV) and the development of specific strategic competitive capabilities, identified environmental performance as a key factor facilitating this relationship. While the study provides valuable insights into the capability–strategy interaction and the role firm-level conditions play in achieving competitive advantage, it recognizes several constraints these include focusing on specific sectors, limited coverage of green areas, a simple measure of supply chain variability, possible common-method bias in data collection and the use of cross-sectional data, limiting causal inferences. Despite these constraints, the findings hold significant implications for both theory and practice, suggesting avenues for future research, such as exploring regional variations and employing longitudinal data for causal analysis.

Benitez-Amado (2012) presented a pioneering conceptual and empirical analysis of the interplay between information technology (IT), corporate environmental management and corporate performance. It brings the concept of corporate environmental proactiveness into information systems research, emphasizing the role of IT in enabling this capability. By extending the interface between IT and business strategy to environmental strategy, it provides valuable insights for IT professionals. The study also contributes to the literature on the natural environment by highlighting the critical role of the IT department in shaping environmental policies. Conceptually, it shows that IT is able to positively impact environmental emergency management, leading to direct business benefits. Analytically, it highlights the positive relationship between proactive environmental strategies and firm performance, with IT capabilities being an important determinant of such strategies. Moreover, it finds a positive predictive relationship between IT and superior firm performance through the use of environmental strategies. These findings suggest that IT managers can contribute to environmental sustainability, ultimately leading to business benefits.

Villabruna et al. (2024) examine the factors that lead to active environmental strategies in the service sector, focusing on environmental consulting firms in particular in Spain. Through hierarchical regression analysis using data from 41 managers, the study finds significant relationships between managerial perceptions of sustainable development, short-term corporate efficiency, strategies adopted by companies for environmental passage and environmental strategies.

S. Zhang et al. (2019) conducted an empirical study with 260 companies from 11 countries/regions and examined the impact of a company's active environmental policies on its environmental performance. The findings included that the impact of green human resource management (GHRM) on the environment is fully mediated by the effective implementation of green organizational practices. Environmental leadership does not have a statistically significant direct or indirect effect on environmental performance but has a significant effect on temperature.

Shu et al. (2020) considered the differences in corporate business activities, focusing on two proactive environmental factors (PESs): labor production green hard work and green political influence. Analyzing a multi-indicator dataset, the study shows that proactive green management has a significant positive impact on firm performance as compared to green political influence. Also, international performance increases the positive effects of active green management on firm performance, while domestic performance strengthens the relationship between green political influence and firm performance. This study contributed to the development of a broader framework for aligning corporate business activities with PESs.

10.4 EFFECT OF GREENWASHING ON FIRM PERFORMANCE: THEORETICAL PERSPECTIVE

10.4.1 Legitimacy Theory

Legitimacy theory suggests that firms seek to operate within the norms and expectations of society. Greenwashing can be seen as one of the strategies by which organizations gain legitimacy from their stakeholders who perceive them to be responsible with regard to environmental concerns.

Initially, however, greenwashing could improve a firm's legitimacy, thereby having positive impacts on short-term performance through increased consumer approval and investor interest. Nonetheless, exposure may severely tarnish the company's reputation, leading to loss of trust, legal penalties and reduced sales (Suchman, 1995).

10.4.2 Signaling Theory

Signaling theory argues that signals are used by companies to inform stakeholders. Greenwashing can be viewed as a deceitful signal that intends to provide an erroneous impression of environmental accountability. The message behind it is tricky and can offer a competitive advantage if greenwashers take care of those who wish their purchases and investments to bear in mind sustainable principles. However, when the truth about greenwashing leaks out, it tends to lose its credibility and cause reputation damage as well as break or ruin trust, which is harmful for long-term performance (Connelly et al., 2011).

10.4.3 Resource-Based View (RBV)

RBV posits that a firm's competitive advantage originates from its unique resources plus capabilities. To this end, authentic environmental sustainability is seen as an asset that enhances organizational efficiency. Pseudo-resources like greenwashing may give a company some short-term positive results because it makes people believe in being sustainable. However, this does not equate to true environmental practices and consequently lacks genuine competitive advantage, possibly leading to future competitive disadvantages once this deception is unearthed (Barney, 1991).

10.4.4 Stakeholder Theory

According to stakeholder theory, any organization must satisfy its customers, employees, investors and members of the community. In addition, greenwashing can meet initial demands for environmentally responsible conduct from stakeholders, thus enhancing firm performance until they realize that they have been deceived by such claims. However, once stakeholders recognize the deception, the firm faces backlash, loss of stakeholder trust and potential boycotts, harming long-term performance (Freeman 1984).

10.4.5 Institutional Theory

Institutional theory focuses on the influence of societal values, norms and expectations on organizational behavior. Companies may engage in greenwashing in line with institutional pressures for environmental sustainability. Conformity through greenwashing can provide short-term profits to meet organizational expectations. However, exposure to greenwashing practices can lead to regulatory restrictions, legal challenges and reputational damage, negatively affecting long-term performance (DiMaggio and Powell, 1983).

10.4.6 Agency Theory

Agency theory addresses conflicts between principals (owners) and agents (managers). Managers may engage in greenwashing to satisfy short-term performance considerations or personal gain at the expense of long-term corporate value. Greenwashing can also improve short-term economic performance and managerial compensation. However, it poses significant risks to long-term performance due to potential legal ramifications, loss of trust and erosion of shareholder value once greenwashing is exposed.

These theoretical perspectives provide a comprehensive understanding of the complex relationship between greenwashing and firm performance, highlighting both the potential short-term benefits and the significant long-term risks (Table 10.4).

10.5 GREENWASHING AS A BARRIER TO PROACTIVE STRATEGIES

The act of misleading consumers about the environmental benefits of a product or company can have a significant impact on actions, efforts and attitudes toward proactive environmental strategies.

Lyon and Montgomery (2015) described how greenwashing can erode consumer trust in environmental issues, resulting in a more cynical public perception of corporate sustainability efforts. Greenwashing can damage consumer confidence. When companies misrepresent or exaggerate environmental claims, it not only causes outrage but can also raise doubts about the true sustainability of businesses.

Cherry and Sneirson (2011) highlighted the negative effects of greenwashing on consumer confidence and the wider implications for the market for green products.

Laufer (2003) examined how greenwashing undermines genuine sustainability efforts, where true and false information is indistinguishable, thereby displacing the effectiveness of all preemptive strategies.

Parguel, Benoît-Moreau and Larceneux (2011) examined the legal and reputational consequences of greenwashing, finding that companies found guilty of greenwashing experience long-term negative effects on their brand image and consumer loyalty. The spread of greenwashing can dilute the impact of genuine proactive environmental strategies. Consumers can be sensitive to environmental issues, making it harder for truly sustainable companies to stand out and gain a competitive advantage.

TABLE 10.4
Theories of Greenwashing

Author	Concept	Theory
Freeman, R. E. (1984)	Stakeholder theory posits that organizations must consider the interests of all stakeholders, not just shareholders, in their decision-making processes.	When companies engage in greenwashing, they mislead stakeholders (e.g., consumers, employees, communities) about their environmental performance, thereby violating stakeholder trust and potentially harming long-term relationships.
Spence, M. (1973)	Greenwashing can be seen as a form of false signaling, where companies use marketing and communication strategies to falsely signal environmental responsibility, thus gaining a competitive advantage or enhancing their reputation without substantiating their claims.	Signaling theory suggests that organizations send signals to the market to convey information about their quality and attributes.
DiMaggio, P. J., and Powell, W. W. (1983)	Institutional theory examines how institutional environments influence the behavior and practices of organizations.	Companies may engage in greenwashing to conform to institutional pressures and norms around sustainability. This behavior is often driven by the desire to appear legitimate in the eyes of regulators, consumers and other stakeholders.
Suchman, M. C. (1995)	Legitimacy theory suggests that organizations seek to operate within the bounds of societal norms and values to gain legitimacy.	By engaging in greenwashing, organizations attempt to align themselves with societal expectations for environmental responsibility, even if their actual practices do not support these claims.
Goffman, E. (1959)	Impression management theory, developed by Erving Goffman, explores how individuals and organizations attempt to control the perceptions others have of them. This theory suggests that entities engage in strategic behaviors to shape their public image and maintain a favorable impression.	Greenwashing can be seen as a form of impression management where companies manipulate their environmental image to appear more eco-friendly than they actually are. This is done to gain social approval, attract customers and avoid criticism.

Delmas and Burbano (2011) discussed the challenges faced by genuine sustainable companies, highlighting the challenges of authentic communication efforts. Greenwashing can lead to disinvestment and loss of investor confidence.

Kim and Lyon (2015) examined the importance of transparency to maintain investor confidence and how it affects consumers when it comes to light. Greenwashing can trigger strong regulatory measures as governments and regulatory agencies respond to public outcry and the need to create consumer protection. This can increase the compliance burden on all companies, including those with genuine sustainability practices.

Lyon and Maxwell (2011) discussed the possibility of a regulatory response to greenwashing and highlight how increased regulatory scrutiny can affect all companies within an industry. The impact of greenwashing on factors influencing active environmental strategies is vast and multifaceted. Greenwashing weakens genuine environmental efforts and creates consumer suspicion and distrust in green issues. This uncertainty can hinder proactive environmental initiatives as consumers may become cynical and detached from real sustainability efforts. Moreover, it can damage both brand equity and company reputation. Consumers who feel cheated by greenwashing are less likely to trust a company's future sustainability claims, which in turn can negatively affect the company's market performance and competitiveness. To maintain their credibility and support for active environmental strategies, companies must ensure that they are transparent and authentic in their green marketing efforts.

10.6 MITIGATING GREENWASHING

Reducing greenwashing requires the implementation of strategies and practices that promote transparency, accountability and real sustainability in corporate environmental claims. There are many ways to effectively reduce greenwashing.

- Strengthening legal systems: Implement and enforce rigorous regulations that define clear standards for environmental issues, for example, the Federal Trade Commission (FTC) green guidelines.
- Third-party testimonials and audits: Encourage the use of third-party testimonials and routine audits for environmental assurance. For example, LEED Certification for buildings, and Certified B Corporation for businesses meeting high social and environmental standards.
- Transparency and full disclosure: Companies must provide comprehensive and transparent sustainability reporting following recognized standards, such as the Global Reporting Initiative (GRI).
- Consumer education and awareness: Inform consumers about greenwashing and create awareness of how to know about true sustainability claims.
- Robust environmental, social and governance (ESG) criteria: Integrate rigorous ESG criteria into investment decisions and hold companies accountable for sustainability practices.

- Advocacy and watchdog organizations: NGOs and watchdog organizations are empowered to track corporate environmental issues and expose greenwashing. The Environmental Working Group (EWG) provides consumer guidelines and reports on greenwashing.
- Implementing life cycle assessments (LCA): Use LCA to assess the environmental impact of production from manufacturing to disposal ensures that sustainability claims have been validated.
- Corporate accountability and governance: Establish robust internal control systems that mandate accurate and detailed environmental reporting. For example, a code of conduct for environmental marketing claims.
- Enhanced stakeholder engagement: Liaise with stakeholders, including customers, employees and community groups, to ensure their concerns and ideas are incorporated into the company's development plans.
- Legal and financial penalties: Implementation of legal and financial penalties for companies found guilty of greenwashing in order to deter such practices.

Mitigating greenwashing requires a multifaceted approach involving regulatory measures, independent verification, transparency, consumer education, robust ESG criteria, advocacy, life cycle assessments, corporate governance, stakeholder engagement and legal penalties. By implementing these strategies, we can promote genuine sustainability and hold companies accountable for their environmental claims.

10.7 CASE STUDIES

10.7.1 Case Study 1: Volkswagen

The Volkswagen greenwashing scandal is an example of deceptive environmental practices. It highlights the need for strict regulatory frameworks, independent verification of corporate claims and a genuine commitment to sustainability. It's a subject of study for both companies and regulators who can work toward more transparent and accountable practices that support environmental protection and corporate responsibility.

10.7.1.1 Background

Volkswagen (VW), one of the largest automobile manufacturing companies, came into the limelight in 2015 when the Environmental Protection Agency (EPA) broke the news that the company had installed software in their diesel cars that had the potential to give positive results under controlled conditions and cheat emissions tests. The scam, widely known as "Dieselgate," brought to light the greenwashing practices of the automobile giant and its impact on the industry, environmental regulations and corporate governance.

10.7.1.2 Discovery and Impact

The devices installed in VW's vehicles were initially found by independent researchers at West Virginia University who conducted independent emissions tests on several VW diesel models. The discrepancies between laboratory results and real-world emissions prompted further investigation by the EPA and the California Air Resources Board (CARB).

After the revelations, Volkswagen admitted that approximately 11 million vehicles worldwide were equipped with the defeat devices. This led to the imposition of financial penalties on VW. The company agreed to a $14.7 billion settlement in the US alone, covering vehicle buybacks, environmental remediation and compensation to affected consumers. Globally, the total cost of fines, settlements and recalls exceeded $30 billion. In addition to the penalties, Volkswagen's reputation was severely damaged. It had to face widespread public backlash and a significant loss of consumer trust. Volkswagen's greenwashing scandal underscored the importance of transparency, integrity and accountability in corporate environmental claims.

10.7.1.3 The Deception

The main marketing strategy for VW diesel vehicles was to promote them as "clean diesel," which means that the company was complying with the stringent emission standards and was a good alternative to traditional gasoline engines. These vehicles were heavily promoted in the United States and Europe, claiming low emissions, fuel efficiency and high performance. It was discovered that VW had installed a "defeat device"—a kind of sophisticated software that was programmed to detect when the car was undergoing emissions testing. During the tests, the software activated full emissions controls, ensuring that the vehicles met the required standards. In normal driving conditions, however, the software reduced the effectiveness of these controls, which resulted in the cars emitting pollutants such as nitrogen oxides (NOx) at levels up to 40 times higher than the legal limits.

10.7.2 Case Study 2: British Petroleum

Rebranding of British Petroleum as BP, "Beyond Petroleum," aimed at positioning the company as a leader in renewable energy and environmental sustainability. The rebranding strategized with a logo resembling a green and yellow sunburst, which symbolized a commitment toward energy sources and environmental sustainability. The largest oil and gas giant was accused of greenwashing practices, as its misleading advertisements highlighted the company's investment in renewable sources of energy like solar and wind, indicating a significant shift from traditional sources, i.e., fossil fuels. However, the reality was that these investments accounted for a very small fraction of BP's overall budget while the vast majority of its capital expenditures were directed toward oil and gas exploration and production. The initiatives toward sustainability were also projected in corporate reports, and much information was omitted with respect to heavy reliance on fossil fuels and its contribution to greenhouse gas emissions. Hence, the scale of BP's renewable initiatives was insufficient to substantiate the company's claims of a fundamental transformation toward sustainability.

This has led to big companies accusing BP of misleading the public and failing to make meaningful progress in reducing carbon emissions.

Moreover, these greenwashing allegations severely damaged its reputation as a responsible and sustainable energy company. The company's credibility suffered further after the Deepwater Horizon oil spill in 2010, one of the largest environmental disasters in history. The spill highlighted ongoing risks to the environment emphasizing the losses associated with BP's core business activities, and as a result, it faced substantial fines, lawsuits and legislative changes aimed at improving safety and environmental standards in the oil and gas industry.

The BP greenwashing case reveals the pitfalls of trying to enhance corporate images through deceptive environmental marketing. It emphasizes the importance of real, critical action toward sustainable development, transparency in corporate communications and the critical role of accountability. When companies are taking on environmental responsibilities, the lessons learned from BP's greenwashing practices can point to more realistic and impactful sustainability strategies.

10.7.3 Case Study 3: Unilever

Unilever is one of the largest consumer goods companies and one of the few companies that have adopted AI and digital technologies in its operations. AI is helping Unilever to be faster and smarter while remaining dedicated to sustainability. It has incorporated AI in various areas like product design, research and development, supply chain and manufacturing, customer engagement and accessibility, recruiting and training, and sustainability.

Unilever's use of AI in its business shows how technology can drive innovation, increase efficiency and promote sustainability. Unilever is using AI to address present customer expectations while also paving the road for a more sustainable future.

10.8 CONCLUSION

Greenwashing poses serious threats to corporate trust, consumer confidence and environmental integrity. In order to promote true sustainability and avoid greenwashing, companies should adopt comprehensive policies and procedures that promote transparency, accountability and environmental realism.

Strengthening the regulatory framework is required, including stringent legislation that defines and enforces clear standards for environmental claims and establishes substantial penalties for non-compliance. Mandatory third-party audits should be conducted to confirm environmental claims, and the results should be made publicly available to ensure environmental protection. Standardized reporting frameworks such as the Global Reporting Initiative (GRI) and the Sustainability Accounting Standards Board (SASB) are being adopted to ensure comparability and consistency in environmental reporting. In addition, enhanced consumer protection legislation is needed to combat misleading environmental claims and make it easier for consumers to obtain verified information.

At the policy level, companies should commit to full transparency by releasing continuous improvement reports that include progress and areas for improvement and implement open data policies. There should be a focus on continuous improvement with regular reviews of environmental policies and practices and ongoing staff training on sustainability and greenwashing techniques. Stakeholder engagement is essential, involving them in developing and implementing sustainable strategies and conducting priority assessments to manage environmental issues.

A good corporate governance framework should provide accurate environmental reporting supported by accountability mechanisms like sustainability committees and management oversight. Life cycle assessment techniques should be used to assess environmental impacts from production to disposal and expose transparent processes for stakeholders. Digital tools such as interactive reporting and data visualization should be used to make complex environmental information accessible and understandable. Finally, industry cooperation is crucial for the development and promotion of environmental reporting, standards to prevent greenwashing and sharing best practices.

These ethics and recommendations can promote green waves, increase their environmental reliability and help regulators continue to monitor the beneficiary's training.

As companies are currently facing increasing pressure to adopt sustainable business practices, AI, ML and data analytics are saviors in this regard as they help increase the accuracy, transparency and effectiveness of environmental strategies.

Furthermore, embracing data-driven sustainability strategies not only benefits the environment but also offers substantial long-term value creation for businesses, strengthening their reputation, reducing costs and fostering innovation.

With the use of these technologies, firms can reduce the risk of greenwashing, improve operational efficiency and gain a competitive edge in the marketplace.

REFERENCES

Antoncic, M. "Uncovering Hidden Signals for Sustainable Investing Using Big Data: Artificial Intelligence, Machine Learning and Natural Language Processing." *Journal of Risk Management in Financial Institutions* 13, no. 2 (2020): 106–13.

Barney, Jay. "Firm Resources and Sustained Competitive Advantage." *Journal of Management* 17, no. 1 (1991): 99–120.

Benitez-Amado, Jose, and Rita M. Walczuch. "Information Technology, the Organizational Capability of Proactive Corporate Environmental Strategy and Firm Performance: A Resource-Based Analysis." *European Journal of Information Systems* 21, no. 6 (November 1, 2012): 664–79. https://doi.org/10.1057/ejis.2012.14.

Carlson, Les, Stephen J. Grove, and Norman Kangun. "A Content Analysis of Environmental Advertising Claims: A Matrix Method Approach." *Journal of Advertising* 22, no. 3 (1993): 27–39.

Chen, Yu-Shan, and Ching-Hsun Chang. "Greenwash and Green Trust: The Mediation Effects of Green Consumer Confusion and Green Perceived Risk." *Journal of Business Ethics* 114, no. 3 (June 10, 2012): 489–500. https://doi.org/10.1007/s10551-012-1360-0.

Cherry, M., and J. Sneirson. "Beyond Profit: Rethinking Corporate Social Responsibility and Greenwashing after the BP Oil Disaster." *Tulane Law Review* 85, no. 98 (2011): 1030.

Connelly, Brian L., S. Trevis Certo, R. Duane Ireland, and Christopher R. Reutzel. "Signaling Theory: A Review and Assessment." *Journal of Management* 37, no. 1 (2011): 39–67.

Delmas, Magali A., and Vanessa Cuerel Burbano. "The Drivers of Greenwashing." *California Management Review* 54, no. 1 (October 1, 2011): 64–87. https://doi.org/10.1525/cmr.2011.54.1.64.

DiMaggio, Paul J., and Walter W. Powell. "The Iron Cage Revisited: Institutional Isomorphism and Collective Rationality in Organizational Fields." *American Sociological Review* 48, no. 2 (1983): 147–60.

Federal Trade Commission. Green Guides, 2012. Retrieved October 17, 2023, from https://www.ftc.gov/sites/default/files/attachments/press-releases/ftc-issues-revised-green-guides/greenguides.pdf.

Freeman, R. E. *Strategic Management*. Cambridge: Cambridge University Press, 2010.

Freeman, R. E. "Strategic Management: A Stakeholder Theory." *Journal of Management Studies* 39, no. 1 (1984): 1–21.

Jensen, Michael C., and William H. Meckling. "Theory of the Firm: Managerial Behavior, Agency Costs and Ownership Structure." In *Corporate Governance*, pp. 77–132. Gower, 2019.

Goffman, Erving. *The Presentation of Self in Everyday Life*. Anchor, 1959.

Khan, Shamus. "Erving Goffman, The Presentation of Self in Everyday Life (1959)." *Public Culture* 32, no. 2 (May 1, 2020): 397–404. https://doi.org/10.1215/08992363-8090145.

Kim, E. H., and T. P. Lyon. "Greenwash vs. Brownwash: Exaggeration and Undue Modesty in Corporate Sustainability Disclosure." *Organization Science* 26 no. 3 (2015): 705–23.

Laufer, William S. "No Title." *Journal of Business Ethics* 43, no. 3 (January 1, 2003): 253–61. https://doi.org/10.1023/a:1022962719299.

Le, Thanh Tiep. "How Do Corporate Social Responsibility and Green Innovation Transform Corporate Green Strategy into Sustainable Firm Performance?." *Journal of Cleaner Production* 362 (2022): 132228.

Leonidou, Constantinos N., and Dionysis Skarmeas. "Gray Shades of Green: Causes and Consequences of Green Skepticism." *Journal of Business Ethics* 144, no. 2 (August 26, 2015): 401–15. https://doi.org/10.1007/s10551-015-2829-4.

Liu, Yang, Qinghua Zhu, and Stefan Seuring. "Linking Capabilities to Green Operations Strategies: The Moderating Role of Corporate Environmental Proactivity." *International Journal of Production Economics* 187 (May 1, 2017): 182–95. https://doi.org/10.1016/j.ijpe.2017.03.007.

Lyon, Thomas P., and A. Wren Montgomery. "The Means and End of Greenwash." *Organization & Environment* 28, no. 2 (2015): 223–49. https://www.jstor.org/stable/26164732.

Lyon, Thomas P., and J. W. Maxwell. "Greenwash: Corporate Environmental Disclosure Under Threat of Audit." *Journal of Economics & Management Strategy* 20, no. 1 (2011): 3–41.

Marquis, Christopher, and Michael W. Toffel. When Do Firms Greenwash?: *Corporate Visibility, Civil Society Scrutiny, and Environmental Disclosure*. Boston, MA: Harvard Business School, 2012.

Nyilasy, Gergely, Harsha Gangadharbatla, and Angela Paladino. "Perceived Greenwashing: The Interactive Effects of Green Advertising and Corporate Environmental Performance on Consumer Reactions." *Journal of Business Ethics* 125, no. 4 (November 6, 2013): 693–707. https://doi.org/10.1007/s10551-013-1944-3.

Parguel, Béatrice, Florence Benoît-Moreau, and Fabrice Larceneux. "How Sustainability Ratings Might Deter 'Greenwashing': A Closer Look at Ethical Corporate Communication." *Journal of Business Ethics* 102, no. 1 (June 22, 2011): 15–28. https://doi.org/10.1007/s10551-011-0901-2.

Qayyum, Abdul, Raja Ahmed Jamil, and Amnah Sehar. "Impact of Green Marketing, Greenwashing and Green Confusion on Green Brand Equity." *Spanish Journal of Marketing-ESIC* 27, no. 3 (July 19, 2022): 286–305. https://doi.org/10.1108/sjme-03-2022-0032.

Seele, Peter, and Lucia Gatti. "Greenwashing Revisited: In Search of a Typology and Accusation-Based Definition Incorporating Legitimacy Strategies." *Business Strategy and the Environment* 26, no. 2 (November 24, 2015): 239–52. https://doi.org/10.1002/bse.1912.

Sharma, Sanjay, and Harrie Vredenburg. "Proactive Corporate Environmental Strategy and the Development of Competitively Valuable Organizational Capabilities." *Strategic Management Journal* 19 (1998): 729–53.

Spence, Michael. "Job Market Signaling." In *Uncertainty in Economics*. New Yrork: Academic Press, 1973, pp. 281–306.

Suchman, Mark C. "Managing Legitimacy: Strategic and Institutional Approaches." *Academy of Management Review* 20, no. 3 (1995): 571–610.

"Shu, Chengli, Jinxin Liu, Mengli Zhao, and Per Davidsson. "Proactive Environmental Strategy and Firm Performance: The Moderating Role of Corporate Venturing." *International Small Business Journal* 38, no. 7 (June 8, 2020): 654–76. https://doi.org/10.1177/0266242620923897.

TerraChoice Environmental Marketing. *The Seven Sins of Greenwashing: Environmental Claims in Consumer Markets London*. TerraChoice Environmental Marketing 2009. http://sinsofgreenwashing.org/findings/greenwashing-report-2009/.

"The 'Six Sins of Greenwashing™' A Study of Environmental Claims in North American Consumer Markets," 2007. https://sustainability.usask.ca/documents/Six_Sins_of_Greenwashing_nov2007.pdf.

Villabruna, Vinicius Elias, Cleiton Hluszko, Daiane Rossi, Murillo Vetroni Barros, Jasmine Siu Lee Lam, and Fernando Henrique Lermen. "Barriers and Strategies for Green Investments in Environmental, Social and Governance: A Seaport Companies' Study." *Management of Environmental Quality*, February 19, 2024. https://doi.org/10.1108/meq-07-2023-0222.

Walker, Kent, and Fang Wan. "The Harm of Symbolic Actions and Green-Washing: Corporate Actions and Communications on Environmental Performance and Their Financial Implications." *Journal of Business Ethics* 109, no. 2 (November 26, 2011): 227–42. https://doi.org/10.1007/s10551-011-1122-4.

White, K., R. Habib, and D. J. Hardisty. "How to SHIFT Consumer Behaviors to Be More Sustainable: A Literature Review and Guiding Framework." *Journal of Marketing* 83, no. 3 (2019): 22–49.

Wijethilake, Chaminda. "Proactive Sustainability Strategy and Corporate Sustainability Performance: The Mediating Effect of Sustainability Control Systems." *Journal of Environmental Management* 196 (July 1, 2017): 569–82. https://doi.org/10.1016/j.jenvman.2017.03.057.

Zhang, D. "The Pathway to Curb Greenwashing in Sustainable Growth: The Role of Artificial Intelligence." *Energy Economics* 133 (2024): 107562.

Zhang, Shanshan, Zhiqiang Wang, and Xiande Zhao. "Effects of Proactive Environmental Strategy on Environmental Performance: Mediation and Moderation Analyses." *Journal of Cleaner Production* 235 (October 1, 2019): 1438–49. https://doi.org/10.1016/j.jclepro.2019.06.220.

11 Artificial Intelligence in Business Management and Its Impact on Sustainable Finance

A Bibliometric Overview of Past, Present, and Future Trends

Jyoti Verma and Anjali Sharma

11.1 INTRODUCTION

Sustainable finance (or green finance) refers to the integration of environmental, social, and governance (ESG) factors into investment decisions, with the aim of generating long-term value while promoting sustainability (Schumacher et al., 2020). Both current and future generations are at risk due to social inequality, increasing environmental degradation, and economic expansion at the cost of our environment (Alshater et al., 2021; Purnomo et al., 2021). Earlier, people believed that economic expansion would fix any kind of problem (Trivedi et al., 2024). In the 1960s, the idea of sustainable development drew a lot of attention among people and successfully integrated itself into government and business policies (Cunha et al., 2021). As part of sustainable finance, proactive environmental measures help companies achieve long-term financial success and satisfy environmental objectives (Choudhary & Trivedi, 2023; Trivedi, Choudhary, & Kamboj, 2024). Studies on investments in sustainable practices and technologies reveal that they result in better financial outcomes, therefore underlining the important relationship between business performance and environmental responsibility (Keskin et al., 2020; Kuo et al., 2022). Sustainable development is promoted through investments in fields such as clean technology (Rajkumar et al., 2022), environmentally friendly agriculture, and affordable housing (Choudhary, Kamboj, & Trivedi, 2023). These investments also promote economic growth, social progress, and environmental stewardship (Trivedi & Malik, 2021). ESG aspects are taken into account when making investment decisions, which protect long-term financial performance by helping identify

DOI: 10.1201/9781003626527-11

and manage environmental and social risks (Schoenmaker, 2017). Businesses that embrace sustainability and exhibit appropriate environmental practices may improve their brand image, attract clients, and foster consumer loyalty over the long term (Trivedi & Begde, 2024). Investors are increasingly emphasising sustainability in their investing strategies, seeking both financial gains and beneficial effects on the environment and society (Fatima et al., 2022; Kashi & Shah, 2023; Linnenluecke et al., 2020). These changing investor choices align with investing in a greener future (Trivedi, Aggarwal, & Singh, 2023). Thus, sustainability and climate change have received a lot of attention around the world (Zhang et al., 2019). Sustainable finance offers a mechanism for redirecting funds towards projects that encourage a green future as the globe faces environmental concerns such as resource depletion, climate change (Leichenko et al., 2009), and biodiversity loss (Saeed & Karim, 2022). Today, the concept of sustainable finance is quite important. It is essential in assisting the funding of measures to reduce greenhouse gas emissions, building energy-efficient infrastructure, and supporting the move towards a low-carbon economy (Morano et al., 2020; Saeed & Karim, 2022; Sibanda, 2013).

11.1.1 Growth and Proportion of Global Sustainable Investment

According to the *Global Sustainable Investment Review 2020*, these investments reached a peak of USD 35.3 trillion, a rise of 15% over the previous two years (2018–2020) and 55% over the previous four years (2016–2020) (Trivedi, Mehta, & Sharma, 2021). Table 11.1 represents the growth of sustainable investment assets by region in local currency for the period 2014 to 2020. Here, the asset values are expressed in billions. With the exception of Europe, where the data seem to indicate a decline due to significant changes in how sustainable investment is defined in this region under

TABLE 11.1
Growth of Sustainable Investing Assets by Region in Local Currency (in Billions) (2014–2020)

Year	Canada (CAD)	United States (USD)	Europe (EUR)	Japan (JPY)	Australia (AUD)
2014	$1,011	$6,572	€9,885	¥840	$203
2016	$1,505	$8,723	€11,045	¥57,056	$707
2018	$2,132	$11,995	€12,306	¥231,952	$1,033
2020	$3,166	$17,081	€10,730	¥310,039	$1,295
Growth rate 2014–2016	49%	33%	12%	6692%	248%
Growth rate 2016–2018	42%	38%	11%	307%	46%
Growth rate 2018–2020	48%	42%	−13%	34%	25%

Source: Global Sustainable Investment Review 2020. Retrieved from https://www.gsi-alliance.org/wp-content/uploads/2021/08/GSIR-20201.pdf.

European Union legislation, comparisons with previous versions of this report are quite challenging (Trivedi, 2024). It is abundantly clear from the data that sustainable investment assets are increasing all over the world (Trivedi, 2024). As shown in Table 11.1, sustainable managed assets increased by more than 48% in Canada over the previous two years, which was the highest increase (Trivedi, 2024). Canada was closely followed by the United States, which had growth of 42%, and Japan, which saw growth of 34% between 2018 and 2020 (Trivedi & Malik, 2022). Sustainable asset growth in Australia persisted, although at a slower rate than between 2016 and 2018 (46% increase from 2016 to 2018 to 25% growth from 2018 to 2020) (Trivedi & Malik, 2022). Due to a modified measuring technique from which the data for this year's report was derived, Europe shows a 13% decrease in the growth of sustainable investment assets for the period 2018 to 2020 (Trivedi & Malik, 2022).

11.1.2 Challenges and Strategies of Sustainable Finance

Investing in a green future through sustainable finance is most important for addressing environmental issues, promoting sustainable development, and meeting the expectations of investors and society (Hira, 2012; La et al., 2019; Macchiavello & Siri, 2022). By integrating ESG factors into investment decisions, financial resources can be directed towards environmentally friendly projects, fostering a transition to a sustainable and resilient global economy (Contreras et al., 2019; Soppe, 2009). Collaboration among many stakeholders, legislative support, standardisation of procedures and practices, and the availability of trustworthy data are crucial for sustainable finance to reach its extent potential (Trivedi, 2023). Choosing to invest in a greener future is not only the right thing to do morally, it also provides the possibility to gain money over the long term and contribute to the development of a more sustainable world (Akomea et al., 2022; Mohamed et al., 2023). The following are some benefits of sustainable finance.

- Access to information: To make wise investment decisions, investors need accurate and thorough information on ESG aspects. Sustainable financial practices can be facilitated by improving data accessibility, quality, and availability (Mohamed et al., 2023).
- Measurement and reporting: Accurate evaluation and comparison of the sustainability performance of investments depend on standardisation of ESG criteria and reporting standards. Efforts towards harmonisation and transparency are necessary (Trivedi, 2023).
- Risk assessment and valuation: It is essential to create trustworthy methods for assessing the financial effects of ESG risks and opportunities in order to facilitate the efficient pricing and valuation of sustainable investments (Chami et al., 2002; Zadek & Robins, 2018).
- Policy and regulatory support: Governments and regulatory bodies can play a pivotal role in fostering an environment that is conducive to sustainable finance by implementing supportive policies, incentives, and regulations (Purnamawati, 2021).

- Collaboration and partnerships: To accelerate the adoption of sustainable finance principles, stakeholders in the finance sector, including investors, financial institutions, businesses, and civil society organisations, must cooperate and exchange best practices (Trivedi, 2023).

11.1.3 AI Solutions to Sustainable Finance Challenges

- Enhancing data collection and analysis: Artificial intelligence (AI) can analyse large quantities of information from different sources (like social media, financial statements, satellite images, etc.) to offer precise and up-to-date ESG data.
- Navigating regulatory changes: AI systems can easily track regulatory developments, enabling financial institutions to stay updated and compliant with new sustainability standards.
- Advanced risk modelling: AI-driven predictive models can simulate long-term sustainability risks, such as climate-related financial risks, helping investors better understand the potential impacts and make informed decisions.

11.2 RESEARCH METHODOLOGY

The objective of this research is to conduct a bibliographic survey of global research in the field of sustainable finance to identify and analyse trends and patterns, and thus provide information about publications, organisations, and countries, as well as the authors who have received the most citations.

The following research questions have been formulated to conduct the analysis:

1. What are the latest trends in scholarly publications on sustainable finance?
2. What organisations, countries, and authors contribute most frequently in the context of sustainable finance?
3. What are the latest issues and challenges in sustainable finance?

A total of 2556 publications (published articles) made up this investigation, which were obtained from 482,906 articles by filtering out data based on the publication type, language, and time period. To perform the network analysis, data was obtained from the Dimensions database and entered into the VOSviewer application. In order to examine the data and gather information for these study objectives, network maps were used.

The following are the retrieval techniques used for the current study:

- Database: Dimensions
- Topic: Sustainable finance
- Study period: 2000–2022

11.3 BIBLIOMETRIC ANALYSIS

The term 'sustainable finance' was investigated using a bibliometric analysis of various research articles produced between 2000 and 2022 (only published papers). A collection of techniques called bibliometric analysis is used to analyse research articles in a quantitative and qualitative way (Annand, 2015; Ikpaahindi, 1985; Muhmad & Muhamad, 2021; Shashi et al., 2021). This technique evaluates and analyses previously published research articles using a quantitative tool. Such analysis provides a comprehensive overview of the most beneficial contributions made by authors in publication houses. The data have been further analysed on two aspects: (1) type of analysis and (2) unit of analysis. The 'Citation' criteria are used in the type of analysis, and 'Countries', 'Organisations', and 'Authors' have been selected in the unit of analysis, which is explained in this research. Bibliometric analysis techniques have been used to examine the collected data based on the following key information:

- Publication trends over time
- Geographic distribution
- Authorship patterns

11.3.1 Publication Trends over Time

Here, the growth of publications related to sustainable finance was analysed. Figure 11.1 shows the number of publications for the years 2000–2022. The increase in scholarly interest shows that sustainability concepts are becoming increasingly important in financial practices and decision-making. It is not only beneficial for businesses, government agencies, academicians, and researchers, but is also important for every individual on the planet Earth.

According to this analysis, 56,971 documents were published in 2022, followed by 54,422 in 2021 and 45,461 in 2020. According to this growing publication trend, the number of publications on sustainable finance has increased tremendously and will continue to grow in the future. In particular, publications have been discovered containing 'sustainable finance' in the title, abstract, and keywords. Since there is a

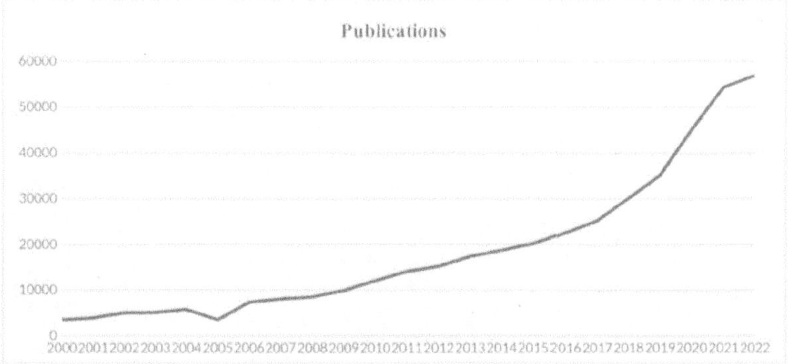

FIGURE 11.1 Number of publications for 2000–2022.

lot of information available on this topic, we exclusively used published publications to gain a thorough understanding of the topic.

11.3.2 Geographic Distribution

Here, the author evaluated the countries and organisations with the highest concentration of the most significant studies on sustainable finance. Including restrictions such as minimum number of documents in a country (20) and a minimum number of citations of a country (20) leads to further filtering of our data to 37 countries out of 129 that met this threshold. As a consequence, the 37 countries that are considered pertinent for the current study constitute the final sample. The results clearly show that the United Kingdom and China are in the top position to publish papers on this topic (Table 11.2 and Figure 11.2).

TABLE 11.2
Top 10 Countries on the Basis of Most Publications

Country	Documents	Citations	Total Link Strength
China	1166	13588	565
United Kingdom	264	3422	398
United States	249	2587	317
Pakistan	228	4279	327
Australia	132	2435	198
Malaysia	131	2851	209
India	117	1360	144
Germany	79	830	123
Spain	71	599	68
Italy	70	636	83

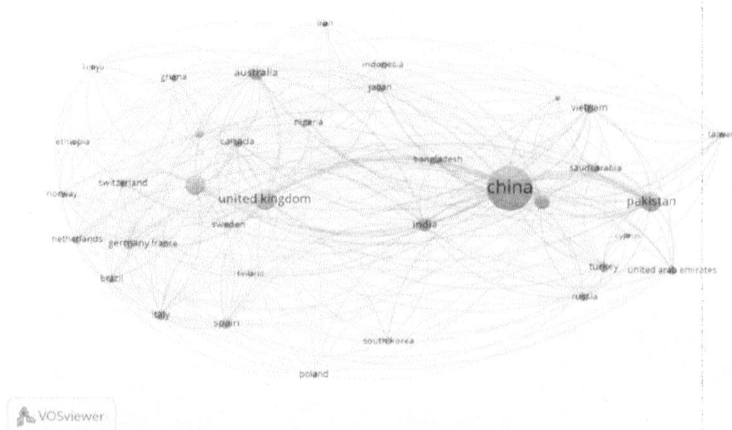

FIGURE 11.2 Top 37 countries.

TABLE 11.3
Top 23 organisations

Organisation	Documents	Citations
Jiangsu University	55	1351
City University of Macau	28	1189
Fuzhou University	30	1096
University of Economics Ho Chi Minh City	30	958
Northern University of Malaysia	21	952
Beijing Institute of Technology	27	839
Anhui University of Finance and Economics	32	729
Southwestern University of Finance and Economics	47	720
Central South University	21	484
Shanghai University of Finance and Economics	40	415
Ilma University	28	383
Xi'an Jiaotong University	26	351
Huazhong University of Science and Technology	21	347
Shandong University of Finance and Economics	28	336
University of Oxford	26	329
Shenzhen University	26	305
Shandong University	26	212
University College London	20	198
University of Queensland	20	195
Central University of Finance and Economics	21	118
London School of Hygiene & Tropical Medicine	22	102
Chongqing University	22	98
Jiangxi University of Finance and Economics	22	97

Furthermore, including restrictions such as minimum number of documents in an organisation of (20) and a minimum number of citations of an organisation (20) leads to filtering our data to 23 organisations out of 2266 that met this threshold, as shown in Table 11.3 and Figure 11.3. As a consequence, the 23 organisations that are considered pertinent for the current study constitute the final sample.

It is quite clear from the bibliometric analysis that sustainable finance has wider aspects and its publication trend has also been accelerating. Also, China has reached a remarkable position for publishing highly cited documents.

11.3.3 Authorship Patterns

The analysis identified prolific authors and their collaborations. Further including restrictions such as minimum number of documents by an author (10) and a minimum number of citations by an author (5) leads to further filtering of our data to 11 authors out of 8327 that met this threshold, as shown in Table 11.4. As a consequence, the 67 countries that are considered pertinent for the current study constitute the final sample.

FIGURE 11.3 Top 23 organisations.

TABLE 11.4
Top 11 Authors

Author	Documents	Citations
Sadiq, Muhammad	14	1013
Chien, Fengsheng	18	941
Mohsin, Muhammad	14	697
Kirikkaleli, Dervis	15	541
Ahmad, Munir	10	361
Murshed, Muntasir	15	347
Adebayo, Tomiwa Sunday	10	286
Naeem, Muhammad Abubakr	13	252
Mahmood, Haider	11	243
Ahmed, Zahoor	10	180
Karim, Sitara	11	167

It is quite clear from the bibliometric analysis that the publications on the topic of sustainable finance reached a higher level of citations per document. Table 11.4 shows the list of authors, their publications, and citations that have been the most influential contributions in existing literature. From our analysis, it is noted that Muhammad Sadiq and Fengsheng Chien published papers that have over 900 citations.

11.4 RECOMMENDATION

This comprehensive bibliometric research on sustainable finance provides invaluable insights into global research patterns. By identifying key contributors, organisations, and countries, the study lays a foundation for the promotion of sustainable financial practices throughout the world. The study concludes with a recommendation for increased research to address environmental risks, underscoring the ongoing need for scholarly inquiry and knowledge dissemination in the field of sustainable finance.

11.5 CONCLUSION

The bibliometric survey on sustainable finance research reveals several key findings that shed light on the current state of the field and its global trends and patterns. First, the survey identifies a significant increase in the number of publications related to sustainable finance from the years 2000 to 2022. This increase in scholarly interest shows that sustainability concepts are becoming increasingly important in financial practices and decision-making. Another key finding is the dominant focus on climate-related topics within sustainable finance research. This suggests that climate change and its implications for financial systems have emerged as a primary concern among researchers and practitioners. The emphasis on climate concerns is an indication of how crucial it is to manage environmental hazards and coordinate financial activities with climate goals such as those outlined in the Paris Agreement. Globally, sustainable finance research is gaining popularity, and this study finds differences in research output between nations. China has reached a remarkable position for publishing highly cited documents and publications. This may be explained by the varied levels of global awareness, legislative support, and institutional incentives for sustainable finance research. To foster sustainable finance practices around the world, continued research, international cooperation, and policy support are essential to address environmental risks. In addition, AI effectively addresses key challenges in sustainable finance, driving better decision-making and promoting long-term sustainability. The growing interest around the world in sustainable finance shows how important it is to keep doing study, work together with other countries, and support policies that protect the environment and make sure that financial actions are in line with sustainability goals.

REFERENCES

Akomea-Frimpong, I., Adeabah, D., Ofosu, D., & Tenakwah, E. J. A review of studies on green finance of banks, research gaps and future directions. *Journal of Sustainable Finance & Investment*, 12(4), 1241–1264, 2022.

Alshater, M. M., Atayah, O. F., & Hamdan, A. Journal of sustainable finance and investment: A bibliometric analysis. *Journal of Sustainable Finance & Investment*, 1–22, 2021.

Annand, D. Developing a sustainable financial model in higher education for open educational resources. *International Review of Research in Open and Distributed Learning*, 16(5), 1–15, 2015.

Chami, Ralph, Cosimano, Thomas F., and Fullenkamp, Connel. Managing ethical risk: How investing in ethics adds value. *Journal of Banking & Finance*, 26, 1697–718, 2002.

Chen, Xiaoyan, and Liu, Yisheng. Visualization analysis of high-speed railway research based on CiteSpace. *Transport Policy*, 85, 1–17, 2020.

Choudhary, Vinita, and Trivedi, Sonal. "The role of internal audit in risk management with a perspective of sustainability development." *The Impact of Climate Change and Sustainability Standards on the Insurance Market*, 287–315, 2023.

Choudhary, Vinita, Kamboj, Neha, and Trivedi, Sonal. "Scoping review of literature on the application of blockchain in finance." *Perspectives on Blockchain Technology and Responsible Investing*, 1–28, 2023.

Contreras, G., Bos, J. W., & Kleimeier, S. Self-regulation in sustainable finance: The adoption of the Equator principles. *World Development*, 122, 306–324, 2019.

Cunha, F. A. F. D. S., Meira, E., & Orsato, R. J. Sustainable finance and investment: Review and research agenda. *Business Strategy and the Environment*, 30(8), 3821–3838, 2021.

Fatima, R., Boitan, I. A., & Carè, R. How many shades are there in sustainable finance?: A bibliometric review. *Handbook of Research on Global Aspects of Sustainable Finance in Times of Crises*, 1–26, 2022.

Global Sustainable Investment Review (2020). Retrieved from: https://www.gsi-alliance.org/wp-content/uploads/2021/08/GSIR-20201.pdf

Hira, T. K. Promoting sustainable financial behaviour: Implications for education and research. *International Journal of Consumer Studies*, 36(5), 502–507, 2012.

Ikpaahindi, L. An overview of bibliometrics: its measurements, laws and their applications. *Libri*, 35, 163, 1985.

Kashi, A., & Shah, M. E. Bibliometric review on sustainable finance. *Sustainability*, 15(9), 7119, 2023.

Keskin, A. İ., Dincer, B., & Dinçer, C. Exploring the impact of sustainability on corporate financial performance using discriminant analysis. *Sustainability*, 12(6), 2346, 2020.

Khan, A., Goodell, J. W., Hassan, M. K., & Paltrinieri, A. A bibliometric review of finance bibliometric papers. *Finance Research Letters*, 47, 102520, 2022.

Kuo, K. C., Yu, H. Y., Lu, W. M., & Le, T. T. Sustainability and corporate performance: Moderating role of environmental, social, and governance investments in the transportation sector. *Sustainability*, 14(7), 4095,2022.

La Torre, M., Trotta, A., Chiappini, H., & Rizzello, A. Business models for sustainable finance: The case study of social impact bonds. *Sustainability*, 11(7), 1887, 2019.

Leichenko, Robin, Karen O'Brien, and William Solecki. Climate change and the global financial crisis. *GECHS Synthesis Human Security in an Era of Global Change*, 25, 25–52, 2009.

Linnenluecke, M. K., Marrone, M., & Singh, A. K. Sixty years of Accounting & Finance: A bibliometric analysis of major research themes and contributions. *Accounting & Finance*, 60(4), 3217–3251, 2020.

Macchiavello, E., & Siri, M. Sustainable finance and fintech: Can technology contribute to achieving environmental goals? A preliminary assessment of 'green fintech' and 'sustainable digital finance'. *European Company and Financial Law Review*, 19(1), 128–174, 2022.

Mohamed Buallay, A., Al Marri, M., Nasrallah, N., Hamdan, A., Barone, E., & Zureigat, Q. Sustainability reporting in banking and financial services sector: A regional analysis. *Journal of Sustainable Finance & Investment*, 13(1), 776–801, 2023.

Morano, Pierluigi, Tajani, Francesco, and Anelli, Debora. A decision support model for investment through the social impact bonds. The case of the city of Bari (Italy). *Valori e Valutazioni*, 24, 163–78, 2020.

Muhmad, S. N., & Muhamad, R. Sustainable business practices and financial performance during pre-and post-SDG adoption periods: A systematic review. *Journal of Sustainable Finance & Investment*, 11(4), 291–309, 2021.

Purnamawati, I. G. A. Sustainable finance for promoting inclusive growth. *JIA (Jurnal Ilmiah Akuntansi)*, 6(2), 435–454, 2021.

Purnomo, A., Sari, A. K., Susanti, T., Mannan, S. S. A., & Lumentut, T. M. B. A. Sustainable finance study of bibliometric overview. In *IOP Conference Series: Earth and Environmental Science* (Vol. 729, No. 1, p. 012124). IOP Publishing, 2021.

Rajkumar, K., Shenbagharaman, A., Malik, R., Nagarjuna, P., & Kuchhal, S. The recent advancements in the wireless security and smart grid technology in protecting the ecological aspects for better future generation—analysis through structural equation modeling. In *2022 2nd International Conference on Innovative Practices in Technology and Management (ICIPTM)* (Vol. 2, pp. 90–94). IEEE, 2022.

Saeed Meo, M., & Karim, M.Z.A. The role of green finance in reducing CO2 emissions: An empirical analysis. *Borsa Istanbul Review*, 22, 169–178, 2022. https://dx.doi.org/10.1016/j.bir.2021.03.002.

Schoenmaker, D. From risk to opportunity: A framework for sustainable finance. *RSM Series on Positive Change*, 2, 2017.

Schumacher, K., Chenet, H., & Volz, U. Sustainable finance in Japan. *Journal of Sustainable Finance & Investment*, 10(2), 213–246, 2020.

Shashi, P., Cerchione, R., & Mittal, A. Managing sustainability in luxury industry to pursue circular economy strategies. *Business Strategy and the Environment*, 30(1), Art. no. 1, 2021. doi:10.1002/bse.2630.

Sibanda, Mabutho. Financialization of green capital: A panacea? *Mediterranean Journal of Social Sciences*, 4, 371–71, 2013.

Soppe, A. Sustainable finance as a connection between corporate social responsibility and social responsible investing. *Indian School of Business WP Indian Management Research Journal*, 1(3), 13–23, 2009.

Trivedi, Sonal, and Prasad Begde. "Venture capital: Trends, investment strategies, and impact." In *Fostering Innovation in Venture Capital and Startup Ecosystems* (pp. 171–190). IGI Global, 2024.

Trivedi, Sonal, and Malik, Reena. "Blockchain technology as an emerging technology in the insurance market." In *Big Data: A Game Changer for Insurance Industry* (pp. 81–100). Emerald Publishing Limited, 2022.

Trivedi, Sonal, and Malik, Reena. "Role and significance of data protection in risk management practices in the insurance market." In *Big Data Analytics in the Insurance Market* (pp. 263–273). Emerald Publishing Limited, 2022.

Trivedi, Sonal, and Malik, Reena. "Role of cryptocurrency in digital marketing." In *Blockchain Technology and Applications for Digital Marketing* (pp. 175–185). IGI Global, 2021.

Trivedi, Sonal, and Reena Malik. "Social media marketing as new marketing tool." In *Research Anthology on Social Media Advertising and Building Consumer Relationships* (pp. 18–33). IGI Global, 2022.

Trivedi, Sonal, Kiran Mehta, and Renuka Sharma. "Systematic literature review on application of blockchain technology in E-finance and financial services." *Journal of Technology Management & Innovation*, 16(3), 89–102, 2021.

Trivedi, Sonal, Aggarwal, Rashmi, and Singh, Gurmeet, eds. *Perspectives on Blockchain Technology and Responsible Investing*. IGI Global, 2023.

Trivedi, Sonal, Choudhary, Vinita, and Kamboj, Neha. "Bridging discovery to market: A case study exploring successes and failures in basic science and commercial integration." In *Unleashing the Power of Basic Science in Business* (pp. 42–54). IGI Global, 2024.

Trivedi, Sonal, Choudhary, Vinita, Kamboj, Neha, and Virk, Nirmaljeet Kaur. "New avenues of opportunities and challenges for start-ups, MSMEs, the Indian financial sector, and the indian insurance sector." *Business Drivers in Promoting Digital Detoxification*, 226–250, 2024.

Trivedi, Sonal. "An umbrella review of the literature on blockchain and distributed ledger technology and their roles in future banking." *Perspectives on Blockchain Technology and Responsible Investing*, 29–57, 2023.

Trivedi, Sonal. "Blockchain framework for insurance industry." *International Journal of Innovation and Technology Management*, 20(6), 2350034, 2023.

Trivedi, Sonal. "Blockchain technology and structural change in financial services." In *The Application of Emerging Technology and Blockchain in the Insurance Industry* (pp. 249–263). River Publishers, 2024.

Trivedi, Sonal. "Convention of blockchain in financial services: A state-of-the-art literature review." *Revolutionizing Financial Services and Markets Through FinTech and Blockchain*, 237–250, 2023.

Trivedi, Sonal. "Explainable artificial intelligence in consumer-centric business practices and approaches." *AI Impacts in Digital Consumer Behavior*, 36–55, 2024.

Trivedi, Sonal. "Pedagogical approaches and innovative teaching: Simulation activity for new product design using osborn checklist (SCAMPER)." In *Fostering Pedagogical Innovation Through Effective Instructional Design* (pp. 245–261). IGI Global, 2024.

Wang, Xinxin, Xu, Zeshui, and Škare, Marinko. A bibliometric analysis of Economic Research-Ekonomska Istra zivanja (2007–2019). *Economic research-Ekonomska istraživanja*, 33, 865–86, 2020.

Zadek, S., & Robins, N. Aligning the financial system with sustainable development. *Financing Sustainable Development*, 102, 2018.

Zhang, Dayong, Zhang, Zhiwei, and Managi, Shunsuke. A bibliometric analysis on green finance: Current status, development, and future directions. *Finance Research Letters*, 29, 425–30, 2019.

12 Consumer Attitudes toward Greenwashing

Exploring Skepticism and Green Fatigue

K. Vinodha Devi, J. Malarvizhi, Ashwitha Shetty, and Priyanka Ghosh

12.1 INTRODUCTION

Greenwashing, a term that describes the practice of companies falsely marketing their products as environmentally friendly, has become a significant issue in recent years. As consumer awareness about environmental issues has grown, so too has the sophistication of greenwashing tactics, leading to increased skepticism and what is now termed "green fatigue" among consumers. When consumers perceive that companies are not genuinely committed to sustainable practices, they become skeptical of all green claims, leading to a broader erosion of trust in the market for eco-friendly products (Chen and Chang, 2013). Consumers often doubt whether the higher costs are justified or merely a ploy for profit under the guise of environmental concern (Bowen and Aragón-Correa, 2014). Sociocultural factors also contribute significantly; social media and peer influences can either amplify or mitigate skepticism, depending on the credibility of the sources (Nguyen et al., 2018). Psychological factors are equally important. Cognitive dissonance, where consumers experience a conflict between their environmental values and their perception of corporate practices, often leads to green fatigue (Gleim et al., 2013). The perception of authenticity in green marketing messages is crucial; when consumers feel that companies are genuine, their trust increases, reducing skepticism and fatigue (Nyilasy et al., 2014). The implications of greenwashing extend beyond consumer attitudes. Companies that engage in greenwashing not only risk their reputation but also undermine the broader movement toward sustainable consumption. Policymakers and marketers need to address these issues by promoting transparency and authenticity in green marketing efforts to rebuild consumer trust and encourage sustainable consumption. This research aims to analyze economic, sociocultural, and psychological factors to understand how greenwashing influences consumer attitudes.

DOI: 10.1201/9781003626527-12

12.2 DESIGN AND METHODS

The primary objective of the study is to analyze the impact of greenwashing on consumer attitudes toward marketing eco-friendly products. The study applied an exploratory data analysis method to study the problem of greenwashing. Data was collected from 232 sample respondents through a pre-tested questionnaire. The data are confined to six months of the financial year from July to December 2023. Data accuracy was ensured by using multiple cross-checks and a reliability test. The reliability of the scale in the questionnaire is assessed using Cronbach's α, which measures internal consistency or how closely related a set of items are as a group (Table 12.1).

With a Cronbach's α of 0.823, the scale demonstrates good internal consistency, meaning that the items on the scale are reliably measuring the same underlying construct. This suggests that the scale is a reliable tool for assessing. The mean of 3.80 suggests that, on average, respondents' ratings are relatively high on the scale, assuming the scale is appropriately centered. A standard deviation of 0.523 indicates moderate variability in the respondents' ratings.

12.2.1 Socioeconomic Profile of the Respondents

Understanding the socioeconomic profile of the sample respondents is important to analyze the relationship between demographic factors and their attitude toward greenwashing. It provides a foundation for analyzing any research problem as there is always an association between demographic factors and the attitude of the population. This study tries to examine various socioeconomic characteristics, such as age, gender, education, occupation, income levels, and family size, to gain valuable insights into the diverse backgrounds of the respondents and how these factors are associated with the particular research problem. This section aims to provide a detailed overview of the socioeconomic characteristics of the respondents in our sample.

Table 12.2 provides an overview of the demographic characteristics of respondents in the study. The table reveals that sample respondents are evenly split between male (50%) and female (50%) respondents. The majority of respondents have a Master's degree (32.8%), followed by those with a Professionals (27.6%). A smaller portion has a Bachelor's degree (8.2%) or others (21.1%). Regarding age-wise classification, most respondents fall within the 20–40 age group (51.3%), with a significant number in the 40–60 age group (39.7%). Nearly half of the respondents (48.3%) are regular

TABLE 12.1
Scale Reliability Statistics

	Mean	SD	Cronbach's α
Scale	3.80	0.523	0.823

TABLE 12.2
Socioeconomic Profile of the Sample Respondents

Variable	Category	Number of Respondents	Percentage
Gender	Male	116	50
	Female	116	50
Educational level	High school	24	10.3
	Bachelor's degree	19	8.2
	Master's degree	76	32.8
	Professional	64	27.6
	Others	49	21.1
Age	Less than 20	6	2.6
	20–40	119	51.3
	40–60	92	39.7
	More than 60	15	6.5
Occupation	Self-employed	41	17.7
	Wage earner	56	24.1
	Regular earner	112	48.3
	Retired	11	4.7
	Dependent	12	5.2
Social group	General	141	60.8
	BC/OBC	51	22
	SC/ST	40	17.2
Monthly income (rupees)	Less than 30,000	41	17.7
	30,000–60,000	34	14.7
	60,000–90,000	40	17.2
	90,000–120,000	88	37.9
	More than 120,000	29	12.5

earners, while 24.1% are wage earners, and 17.7% are self-employed. The table also reveals that the majority of respondents belong to the general category (60.8%), followed by backward classes and other backward classes (BC/OBC; 22%) and scheduled castes and scheduled tribes (SC/ST; 17.2%). A significant portion of respondents (37.9%) have a monthly income of Rs. 90,000 to Rs. 1,20,000. About 17.7% earned less than Rs. 30,000 monthly, and the remaining respondents are distributed across other income brackets.

12.3 GREEN CONSUMPTION BEHAVIOR

Recently, environmental issues have become major social concerns, and consumers are becoming increasingly aware of environmental matters; however, they remain hesitant to purchase eco-friendly products (Kim and Lee, 2023). Table 12.3 states how frequently the respondents buy eco-friendly products.

TABLE 12.3
Green Consumption Behavior

Variable	Category	Number of Respondents	Percentage
Purchasing behavior	Never (Not intentionally)	2	0.9
	Rarely	31	13.4
	Occasionally	79	34.1
	Frequently	56	24.1
	Always	64	27.6
Total		**232**	**100.0**

Table 12.3 on the purchasing behavior of the green products shows that a very few sample respondents, 0.9%, never intentionally buy eco-friendly products. Meanwhile, 13.4% of respondents rarely make such purchases. A notable proportion of the sample, 34.1%, occasionally buys eco-friendly products, while 24.1% frequently do so. Additionally, 27.6% of respondents always opt for eco-friendly products. The majority of respondents either purchase these products occasionally or always, with 51.7% frequently or always making eco-friendly choices. This suggests a generally positive trend toward the adoption of eco-friendly purchasing habits among the sample respondents.

12.4 RESULTS AND ANALYSIS

Understanding the determinants of health disparities is crucial for developing effective public health policies. Rural areas, characterized by their geographic isolation, limited access to healthcare services, and often lower socioeconomic status presents unique challenges in terms of health outcomes. Socioeconomic status (SES) encompasses various dimensions, including income, education, employment, and social class, all of which profoundly influence health. In rural areas, these issues are exacerbated by additional factors such as limited healthcare infrastructure, fewer educational opportunities, and economic instability.

12.4.1 Impact of Education Condition on Green Behavior

Higher education levels were associated with increased sustainability behaviors aimed at environmental preservation (Piao and Managi, 2023). This study tried to analyze the association between the educational level and the frequency of buying eco-friendly products.

Null hypothesis (H0): There is no significant association between education and green behavior of the sample respondents.

Alternative hypothesis (H1): There is a significant association between education and green behavior of the sample respondents.

TABLE 12.4
Contingency Tables: Education and Green Behavior

Frequently Buy EFP	High School	Bachelor's Degree	Master's Degree	Professionals	Others	Total
Never	0	0	1	0	1	2
Rarely	0	10	14	0	7	31
Occasionally	5	4	16	43	11	79
Frequently	1	2	37	8	8	56
Always	18	3	8	13	22	64
Total	24	19	76	64	49	232

TABLE 12.5
Chi-Square Analysis

	Value	df	p
χ^2	130	16	<.001
N	232		

Table 12.4 tries to analyze the association between the educational level and green behavior of the respondents. A contingency analysis of education levels and green behavior, specifically focusing on the frequency of purchasing eco-friendly products (EFP), was conducted. The results show that occasionally purchasing green products has the highest value. In this category, the majority are professionals (43 respondents), followed by those with a master's degree (16 respondents). Additionally, 64 respondents always purchase eco-friendly products, with the highest numbers in the high school category (18 respondents). Fifty-six respondents frequently purchase eco-friendly products, with the majority being master's degree holders (37 respondents).

The chi-square test results ($\chi^2 = 130$, df = 16, $p < 0.001$) indicate a statistically significant association between education level and the frequency of purchasing eco-friendly products (Table 12.5). The study rejects the null hypothesis that there is no significant association between education and green behavior of the sample respondents.

12.4.2 Impact of Income on Green Consumption Behaviour

An association between higher educational attainment and sustainable consumption patterns has been repeatedly reported in the literature. The study tried to analyze the relationship between the income level and green consumption behavior of the respondents.

Null hypothesis (H0): There is no significant difference in the attitudes of consumers based on income level toward eco-friendly products.

Alternative hypothesis (H1): There is a significant difference in the attitudes of consumer based on income level toward eco-friendly products.

Table 12.6 indicates that there is a statistical significance in the differences between different income groups related to eco-friendly products. The F value of "Environmental protection is important to ensure a sustainable future" is 9.141. It states that there is a significant difference between groups regarding the importance of environmental protection for a sustainable future. The F value of "Human activities are significantly contributing to climate change" is 9.856, which shows significant differences exist between groups in their beliefs about human activities contributing to climate change. The value of F for "Action to mitigate environmental damage" is 9.275. It shows that there is a significant difference between groups on the necessity of actions to mitigate environmental damage. The value of F for "Aware of the environmental impact of the products I purchase" is 8.597. It clearly shows that there

TABLE 12.6
ANOVA

Consumer Attitude		Sum of Squares	df	Mean Square	F	Sig.
Environmental protection is important to ensure a sustainable future	Between groups	18.531	4	4.633	9.141	0.000
	Within groups	113.522	224	0.507		
	Total	132.052	228			
Human activities are significantly contributing to climate change	Between groups	19.388	4	4.847	9.856	0.000
	Within groups	111.642	227	0.492		
	Total	131.030	231			
Individual action to mitigate environmental damage	Between groups	16.664	4	4.166	9.275	0.000
	Within groups	101.957	227	0.449		
	Total	118.621	231			
Aware of the environmental impact of the products I purchase	Between groups	15.610	4	3.903	8.597	0.000
	Within groups	103.041	227	0.454		
	Total	118.651	231			
Knowledgeable about eco-friendly products and their benefits	Between groups	16.560	4	4.140	11.683	0.000
	Within groups	80.436	227	0.354		
	Total	96.996	231			
Understood the importance of using eco-friendly products	Between groups	5.960	4	1.490	3.062	0.017
	Within groups	110.454	227	0.487		
	Total	116.414	231			

(Continued)

TABLE 12.6 (CONTINUED)
ANOVA

Consumer Attitude		Sum of Squares	df	Mean Square	F	Sig.
Always up-to-date with information	Between groups	9.078	4	2.270	3.850	0.005
about eco-friendly products	Within groups	133.819	227	0.590		
and practices	Total	142.897	231			
Eco-friendly products are	Between groups	11.289	4	2.822	4.520	0.002
generally of high	Within groups	141.728	227	0.624		
quality	Total	153.017	231			
Buy eco-friendly	Between groups	9.257	4	2.314	0.301	0.877
products even if	Within groups	1744.187	227	7.684		
they are more expensive	Total	1753.444	231			
Feel good about	Between groups	4.386	4	1.096	1.531	0.194
myself when I	Within groups	162.610	227	0.716		
buy eco-friendly products	Total	166.996	231			
Eco-friendly	Between groups	13.733	4	3.433	3.836	0.005
products have	Within groups	203.159	227	0.895		
influenced other aspects of my lifestyle	Total	216.892	231			
Eco-friendly	Between groups	4.499	4	1.125	1.782	0.133
products are	Within groups	143.289	227	0.631		
healthier for me and my family	Total	147.789	231			
Eco-friendly	Between groups	46.570	4	11.643	10.606	0.000
products are of	Within groups	249.184	227	1.098		
better quality than non-eco-friendly products	Total	295.754	231			
Overall lead to	Between groups	21.540	4	5.385	4.537	0.002
sustainable	Within groups	269.455	227	1.187		
consumption and production	Total	290.996	231			

is a significant difference in their awareness of the environmental impact of their purchases. The F value of "Knowledgeable about eco-friendly products and their benefits" is 11.683. This clearly shows that there is a significant difference between groups in their knowledge about eco-friendly products and their benefits. P values less than 0.001 indicate a very high level of statistical significance of the result. The

F value of "Understand the importance of using eco-friendly products" is 3.062. The p value is 0.017, and it shows that there is a significant difference between groups in their understanding of the importance of using eco-friendly products. The F value of "Keep up to date with information about eco-friendly products and practices" is 3.850, with a p value of 0.005, it shows that there are significant differences between groups in how well they keep up to date with information about eco-friendly products and practices. The F value of "Eco-friendly products are generally of high quality" is 4.520. The p value is equal to 0.002, which states that there is a significant difference between groups regarding the belief that eco-friendly products are generally of high quality. The F value of "Eco-friendly products are worth the extra cost" is 6.384, and the p value is 0.001. It states that the groups significantly differ in their belief that eco-friendly products are worth the extra cost. The F value of "Buy eco-friendly products even if they are more expensive" is 0.301, and the p value is 0.877. It states that no significant differences were found between groups regarding the willingness to buy eco-friendly products even if they are more expensive. The F value of "Feel good about myself when I buy eco-friendly products" is 1.531, and the p value is 0.194. It shows that there are no significant differences between groups in terms of feeling good about buying eco-friendly products. The F value of "Eco-friendly products have influenced other aspects of my lifestyle" is 3.836, and the p value is 0.005. It shows that there are significant differences between groups on the influence of eco-friendly products on other aspects of their lifestyle. The F value of "Eco-friendly products are healthier for me and my family" is 1.782, and the p value is 0.133. It shows that no significant differences were found between groups regarding the belief that eco-friendly products are healthier. The F value of "Eco-friendly products are of better quality than non-eco-friendly products" is 10.606 ($p < 0.001$). This states that significant differences exist between groups on the belief that eco-friendly products are of better quality than non-eco-friendly products. The F value of "Overall lead to sustainable consumption and production" is 4.537, and the p value is 0.002. It shows that there are significant differences between groups in the belief that eco-friendly products lead to sustainable consumption and production.

12.4.3 Consumer Attitudes toward Skepticism and Green Fatigue

Table 12.7 presents group statistics comparing male and female respondents' attitudes toward skepticism regarding various environmental issues.

Table 12.7 reveals males tend to be more skeptical about the severity of environmental problems compared to females, with a higher mean score of 3.91 compared to 3.57. The results also state that males have a higher mean score (3.89) than females (3.50) for the belief that many environmental threats are exaggerated for political or economic reasons, indicating greater skepticism among males. Males are more skeptical about the motives of those who advocate for environmental protection, with a significantly higher mean score (3.98) compared to females (3.12). Males also show higher skepticism regarding environmental regulations being unnecessary and overreaching, with a mean score of 3.69 compared to females' mean score of 3.16. Across all measured areas of skepticism regarding environmental issues, males exhibit higher mean scores compared to females.

TABLE. 12.7
Gender-Wise Group Statistics on Skepticism

Group Statistics

	Gender	N	Mean	Std. Deviation	Std. Error Mean
Skeptical about the severity of environmental problems	Male	116	3.91	0.987	0.092
	Female	116	3.57	1.144	0.106
Many environmental threats are exaggerated for political or economic reasons	Male	116	3.89	1.045	0.097
	Female	116	3.50	1.282	0.119
Skeptical about the motives of those who advocate for environmental protection	Male	116	3.98	0.834	0.077
	Female	116	3.12	1.056	0.098
Environmental regulations are often unnecessary and overreaching	Male	116	3.69	1.254	0.116
	Female	116	3.16	1.316	0.122

Table 12.8 states that both males and females report similar levels of feeling overwhelmed by constant messages about being environmentally friendly, with nearly identical mean scores of 3.69 and 3.66, respectively. Males are slightly more tired of hearing about the importance of recycling and reducing waste, with a mean score of 3.47 compared to 3.08 for females. Males feel more indifferent toward green initiatives because there are too many of them, with a higher mean score of 3.82 compared to females' mean score of 3.16. Males also find it more exhausting to keep up with

TABLE 12.8
Gender-Wise Group Statistics on Green Fatigue

Group Statistics

Green Fatigue	Gender	N	Mean	Std. Deviation	Std. Error Mean
Feel overwhelmed by the constant messages about being environmentally friendly	Male	116	3.69	0.715	0.066
	Female	116	3.66	0.933	0.087
Tired of hearing about the importance of recycling and reducing waste	Male	116	3.47	1.059	0.098
	Female	116	3.08	1.173	0.109
I feel indifferent towards green initiatives because there are too many of them	Male	116	3.82	0.956	0.089
	Female	116	3.16	1.087	0.101
I find it exhausting to keep up with all the environmental recommendations	Male	116	3.55	0.762	0.071
	Female	116	2.99	1.205	0.112
I ignore messages about being eco-friendly because they are repetitive	Male	116	3.58	0.846	0.079
	Female	115	2.95	1.091	0.102

TABLE 12.9
Kaiser–Meyer–Olkin Measure

KMO and Bartlett's Test		
Kaiser–Meyer–Olkin measure of sampling adequacy		0.727
Bartlett's test of sphericity	Approx. chi-square	2.150E3
	df	78
	Sig.	.000

all the environmental recommendations, with a mean score of 3.55 compared to females' mean score of 2.99. Males tend to ignore messages about being eco-friendly because they are repetitive more than females do, with a mean score of 3.58 compared to females' mean score of 2.95. The data indicates that males generally experience higher levels of green fatigue across all measured areas compared to females.

12.4.4 Factors Contributing to Skepticism and Green Fatigue

Table 12.9 reveals that the Kaiser–Meyer–Olkin (KMO) test measures the adequacy of sampling for factor analysis. A KMO value of 0.727 is considered "acceptable." The value indicates that there is a reasonable correlation between the variables and that the data is likely to produce reliable factors. The large chi-square value (2150) and the associated p value of 0.000 indicate that the correlation matrix is not an identity matrix. The results of Bartlett's test of sphericity confirm that data has significant intervariable correlations and that factor analysis is appropriate for identifying underlying factors.

Table 12.10 states that the first three components together account for 68.84% (47.86%, 11.937%, 9.036%) of the total variance, which is substantial. Additional components contribute less variance, suggesting that fewer components are needed to explain the majority of the variance.

The rotated component matrix (Table 12.11) provides the factor loadings of each variable after rotation. The rotation method used here is varimax with Kaiser normalization, which aims to simplify the factor structure by maximizing the variance of each factor's loadings. The table was broken into three components. The first component is Green Fatigue (GF); this factor is characterized by individuals feeling overwhelmed or indifferent toward environmental messages due to their frequency and repetitiveness. The second component is Environmental Skepticism (ES); this factor captures skepticism about the legitimacy and seriousness of environmental issues, influenced by conflicting information and perceived biases. The third component is Critical Skepticism (CS); this factor reflects doubts about the motives behind environmental regulations and the exaggeration of environmental threats. Overall, these components help to understand different dimensions of attitudes toward environmental

TABLE 12.10
Total Variance

	Total Variance Explained					
	Initial Eigenvalues			Rotation Sums of Squared Loadings		
Component	Total	% of Variance	Cumulative %	Total	% of Variance	Cumulative %
1	6.222	47.862	47.862	3.316	25.505	25.505
2	1.552	11.937	59.799	2.954	22.723	48.227
3	1.175	9.036	68.835	2.679	20.608	68.835
4	0.887	6.819	75.655			
5	0.721	5.550	81.204			
6	0.668	5.141	86.346			
7	0.491	3.778	90.124			
8	0.380	2.926	93.050			
9	0.367	2.821	95.870			
10	0.243	1.873	97.743			
11	0.129	0.994	98.737			
12	0.086	0.665	99.402			
13	0.078	0.598	100.000			

Extraction method: principal component analysis

TABLE 12.11
Rotated Component Matrix

Rotated Component Matrix[a]

	Component		
	GF	ES	CS
Tired of hearing about the importance of recycling and reducing waste	0.865		
I find it exhausting to keep up with all the environmental recommendations	0.798		
I feel indifferent towards green initiatives because there are too many of them	0.791		
Ignore messages about being eco-friendly because they are repetitive	0.762		
Conflicting scientific reports make me doubt the seriousness of environmental problems		0.774	
Political and economic interests often cloud the truth about environmental issues		0.765	
Personal experiences and observations lead me to question environmental claims		0.753	
Mixed messages in the media contribute to my skepticism about environmental issues		0.659	
Feel overwhelmed by the constant messages about being environmentally friendly		0.462	
Many environmental threats are exaggerated for political or economic reasons			0.795
Environmental regulations are often unnecessary and overreaching			0.772
Skeptical of motives of those who advocate for environmental protection			0.732
Skeptical about the severity of environmental problems			0.617

Extraction method: principal component analysis
Rotation method: varimax with Kaiser normalization

[a] Rotation converged in six iterations.

12.4.5 Factors Determining Consumer Attitude: Regression Analysis

The regression analysis results give insights into how different factors—Green Fatigue (GF), Environmental Skepticism (ES), and Critical Skepticism (CS), identified through factor analysis—influence consumer attitude toward eco-friendly products.

The R square (coefficient of determination) value of 0.705 indicates that approximately 70.5% of the variance in consumer attitude toward eco-friendly products can be explained by the model, which includes GF, ES, and CS (refer to Table 12.12). This is a relatively high percentage, suggesting that the model fits the data well. The

TABLE 12.12
Model Summary

Model	R	R Square	Adjusted R Square	Std. Error of the Estimate
1	0.798	0.705	0.703	1.40245

Predictors: (Constant), F3, F2, F1
Dependent variable: total of consumer attitudes toward eco-friendly products

TABLE 12.13
Consumer Attitudes toward Eco-Friendly Product

	Unstandardized Coefficients		Standardized Coefficients		
Model	B	Std. Error	Beta	t	Sig.
1	(Constant)	65.533	0.623		105.211
	GF	0.765	0.630	0.077	1.215
	ES	2.854	0.622	0.291	4.585
	CS	0.628	0.635	0.063	0.990

Dependent variable: total of consumer attitudes toward eco-friendly products

adjusted R square value of 0.703 accounts for the number of predictors in the model and provides a more accurate measure of goodness-of-fit.

The regression results show that the intercept (constant) of the model is 65.533 (refer to Table 12.13). This is the minimum value of consumer attitudes toward eco-friendly products when all predictors (GF, ES, CS) are zero. The unstandardized coefficient (B) of Green Fatigue is 0.765. It states that for each unit increase in Green Fatigue, the consumer attitude toward eco-friendly products increases by 0.765, holding other variables constant. The unstandardized coefficient (B) of Environmental Skepticism is 2.854, which shows that for every unit increase in Environmental Skepticism, the consumer attitude score increases by 2.854, holding other variables constant. The unstandardized coefficient (B) of Critical Skepticism is 0.628. It clearly indicates that for each unit increase in Critical Skepticism, the consumer attitude score increases by 0.628, holding other variables constant.

$$\textit{Consumer Attitude} = 65.533 + (0.765GF) + (2.854ES) + (0.628CS)$$

These results state that while all three factors play an important role in determining consumer attitudes toward eco-friendly products, Environmental Skepticism is the most influential factor, and Green Fatigue and Critical Skepticism have a relatively smaller impact.

12.4.6 Leveraging Artificial Intelligence to Combat Greenwashing

Factor analysis identifies three major dimensions affecting consumer attitudes—Green Fatigue, Environmental Skepticism, and Critical Skepticism—with Environmental Skepticism having the strongest impact on consumer attitudes, as confirmed by regression analysis. One of the strategies firms can apply is to use artificial intelligence (AI) to address these challenges and bridge the gap between consumer expectations and company practices. By leveraging AI, companies can predict consumer preferences and behaviors toward eco-friendly products, enabling them to design marketing strategies that directly address skepticism and fatigue.

12.5 DISCUSSION: INSIGHTS ON ADOPTING GREENWASHING AS A MARKETING STRATEGY

The socioeconomic profile of the respondents highlights that a significant portion of them with higher education levels (master's and professional degrees) frequently purchase eco-friendly products. This suggests that educated consumers are more inclined toward green consumption. It also states that respondents with higher income levels (Rs. 90,000 to Rs. 120,000) are more likely to purchase eco-friendly products. Higher income often correlates with the ability to afford premium-priced green products. Males tend to be more skeptical about environmental issues and green initiatives compared to females. Firms using greenwashing need to consider these gender differences. Both males and females experience green fatigue, with males exhibiting higher levels of fatigue across all measured areas. It states that the awareness level of males compares to their female counterparts. To counter skepticism and green fatigue, particularly among male consumers, firms should ensure transparency in their environmental claims. The chi-square test also confirms a significant association between education level and green behavior. Firms should invest in educating their consumers about their green initiatives and the benefits of their products through third-party certification. ANOVA results highlight significant differences in attitudes toward eco-friendly products across different income groups. This suggests the need for income-specific marketing strategies, where high-income consumers might respond better to premium green products, and lower-income consumers might need value propositions that emphasize cost savings from sustainable practices. The factor analysis reveals three distinct dimensions of attitudes towards environmental issues. Green Fatigue (GF) reflects feelings of being overwhelmed and indifferent toward environmental messages due to their frequency and repetitiveness. This suggests that excessive or repetitive environmental communication can lead to disengagement and fatigue among individuals. Environmental Skepticism (ES) states skepticism about the legitimacy and seriousness of environmental issues, influenced by conflicting information, perceived biases, and mixed media messages. This suggests that the credibility of environmental claims is questioned due to perceived inconsistencies and biases. Critical Skepticism (CS) reveals doubts about the motives behind environmental regulations and the perception of exaggeration in environmental threats. This indicates a critical perspective on the political and economic motives behind environmental protection efforts and the regulation of environmental issues. Regression was applied to study which factor out of the three has a highly significant relationship with consumer attitude. The result shows that the model explains a substantial portion (70.5%) of the variance in consumer attitudes toward eco-friendly products, suggesting a good fit. All three predictors (Green Fatigue, Environmental Skepticism, and Critical Skepticism) are statistically significant, with Environmental Skepticism having the strongest effect on consumer attitudes, followed by Green Fatigue and Critical Skepticism. Firms can use AI-powered supply chain tracking systems as a marketing strategy to enhance transparency by allowing consumers to verify the reliability of sustainability claims. Such tracking systems can be integrated into product packaging to address environmental skepticism by providing credible evidence of sustainable practices, thereby fostering trust and reducing greenwashing allegations. Beyond tracking systems, AI can optimize resource use and develop genuinely sustainable practices, reducing the need for misleading greenwashing campaigns. By combining insights from socio-economic analysis and AI-driven strategies, firms can create a long-run approach to keeping the trust of consumers in their products.

REFERENCES

Boulstridge, Elizabeth, and Matthew Carrigan. "Do Consumers Really Care About Corporate Social Responsibility? Highlighting the 'Greenwashing' Issue." *International Journal of Consumer Studies* 24, no. 3 (2000): 270–277.

Bowen, Frank, and José A. Aragón-Correa. "Greenwashing in Corporate Environmentalism Research and Practice: The Importance of What We Say and Do." *Organization & Environment* 27, no. 2 (2014): 107–112.

Chen, Yu-Shan, and Ching-Hsun Chang. "Greenwash and Green Trust: The Mediation Effects of Green Consumer Confusion and Green Perceived Risk." *Journal of Business Ethics* 114 (2013): 489–500.

Delmas, Magali A., and Vanessa C. Burbano. "The Drivers of Greenwashing." *California Management Review* 54, no. 1 (2011): 64–87.

Gleim, Mark R., Jeffery S. Smith, Demetra Andrews, and J. Joseph Cronin. "Against the Green: A Multi-Method Examination of the Barriers to Green Consumption." *Journal of Retailing* 89, no. 1 (2013): 44–61.

Kim, Namkyung, and Kihwan Lee. "Environmental Consciousness, Purchase Intention, and Actual Purchase Behavior of Eco-Friendly Products: The Moderating Impact of Situational Context." *International Journal of Environmental Research and Public Health* 20, no. 7 (2023): 5312.

Lyon, Thomas P., and John W. Maxwell. 2008. "Corporate Social Responsibility and the Greenwashing Problem." *Journal of Economics & Management Strategy* 17, no. 1: 21–37.

Nguyen, Hung, Nguyen Cuong, and Hoang Thoa. "Green Consumption: Closing the Intention-Behavior Gap." *Sustainable Development* 27 (2018). https://doi.org/10.1002/sd.1875.

Nyilasy, Gergely, Harsha Gangadharbatla, and Antonio Paladino. "Perceived Greenwashing: The Interactive Effects of Green Advertising and Corporate Environmental Performance on Consumer Reactions." *Journal of Business Ethics* 125 (2014): 693–707.

Piao, Xiaoyue, and Shunsuke Managi. "The International Role of Education in Sustainable Lifestyles and Economic Development." *Scientific Reports* 13 (2023): 8733.

Schaefer, Amanda, and Bruce A. Kahn. "The Influence of Corporate Greenwashing on Consumer Trust: An Empirical Study." *Journal of Environmental Management* 91, no. 2 (2010): 424–431.

Singh, Jitender, and Surabhi Singh. "Greenwashing and Its Impact on Brand Perception and Consumer Behavior." *Journal of Brand Management* 25, no. 6 (2018): 641–653.

Walker, Kent, and Joanne H. Wan. "The Impact of Greenwashing on Consumer Behavior." *Journal of Business Ethics* 109, no. 2 (2012): 227–242.

13 Green Finance and Artificial Intelligence

The Role of Higher Education Institutions in Building a Sustainable Future

Jaswinder Pal Singh and Shivani Inder

13.1 INTRODUCTION

Green finance is an evolving concept and has not been defined precisely in the literature. Green finance can be considered a broad term that suggests financial investments flowing into sustainable development projects and initiatives, environmental products, and policies that encourage the development of a more sustainable economy (Höhne, Khosla, Fekete, and Gilbert, 2012). It is considered as a subset of climate finance. Further, mitigation finance and adaptation finance are also subsets of climate finance. Mitigation finance refers to financial activities in projects that aim to reduce greenhouse emissions to help organizations adopt sustainable goals (Tsai, 2024). Adaptation finance refers to the flow of funds toward those objectives/organizations/projects that aim to reduce the vulnerability of persons to the negative effects of climate change (Hohne et al., 2012). Green finance is also a subset of sustainable finance. Sustainable finance has been referred to as "the process of taking environmental, social and governance considerations into account when making investment decisions in the financial sector, leading to more long term investments in sustainable economic activities and projects" (EU, 2024).

Green finance can help an organization to achieve green growth. Green growth suggests balanced economic development that favors sustainability (CİTİL, 2024). In order to achieve sustainable growth, organizations must maximize the positive changes resulting from ecological development, which requires the adoption of new models of environment-friendly economic growth and the creation of new opportunities (Sharma & Kautish, 2023). Green development involves further increasing productivity, reducing environmental pollution through the use of green technologies and knowledge, and expanding energy and resource use. Green finance plays a key role in achieving green growth. As part of corporate social responsibility, institutions should actively contribute to society, which involves participation in the management of the green environment, financing of green activities and making

DOI: 10.1201/9781003626527-13

their organizations more sustainable. Green finance is also considered a socially responsible activity for financial companies such as banks, cooperative funds and securities (Tolliver et al., 2019). These companies should seek green financing as part of their corporate social responsibility, even though they are not currently profitable. The aim of green finance is economic growth and improving the environment with the development of the financial sector (Mohsin & Jamaani, 2023; Ping & Shah, 2023). It includes the allocation of funds to green economic activities through targeted financing supported by the participation of public institutions in market operations. This approach, known as "target-based financing," effectively promotes green economic activities and removes restrictions on non-green activities (Lamptey et al., 2023; Vikas et al., 2022). Green finance is a strategic approach to encourage the financial sector in the transition to resource-efficient and low-carbon economies, especially in the context of adaptation to climate change. The financing of green initiatives in an economy depends to a large extent on three sources:

- Internal public finances: The government of a country provides money directly.
- International public finance: Financing by international organizations and development banks.
- Private sector financing: Financing by national and international non-governmental organizations.

The main task of the financial system is to ensure funding is directed to the most productive uses. Green finance can play an instrumental role in improving the productive use of funds while placing them in the best aspect of sustainability (Gutiérrez-Torrenova, 2021; G. Liu & Liang, 2024). Higher education institutions, especially universities, can help the transition toward being a low-carbon circular economy.

13.1.1 How Can Universities Offer a Connecting Link between Green Finance and Education?

Green finance can play an instrumental role in promoting sustainability in educational institutions, especially universities, by enabling and equipping them with funds for initiatives focusing on sustainable development, i.e., environmental, societal and governance (Chen & Xie, 2023). Studies have emphasized that green finance contributes to sustainable economic and environmental development, enabling universities to integrate sustainable goals, values and strategic intent into their operations (Ma, 2022). Additionally, green finance would enable universities to bring in greater transparency and accountability in their endeavors, supported by projects that use digitized technologies like blockchain.

Furthermore, the implementation of green finance in day-to-day operations and other operative mechanisms can lead to improvements in enhancing green total factor productivity within regions, leading to increases in technological innovation and industrial structure upgrading. Improving green factor productivity is an essential element for achieving the sustainable development goals of universities (Z. Liu

& Abu Hatab, 2023). By leveraging green finance, universities can improve their achievement targets for sustainable development goals and gain greater participation in sustainability rankings.

13.1.1.1 Addressing Environmental Challenges

The world is facing unprecedented environmental challenges, such as climate change, the reduction of biodiversity and over-exhaustion of resources. Universities are viewed as an obvious platform to raise green finance to handle environmental challenges like energy transition, raising awareness, mitigating environmental damage, and providing adequate information and resources for financing problems.

Universities can channel green finance instruments to engage different stakeholders. Higher education institutions have been perceived as a natural fit for raising green debt for a variety of reasons. Since 2021, the issuance of green debt by higher education institutions, especially universities, has risen. Universities with global repute like Macquarie University, Concordia University and the National University of Singapore have issued "green, social, and sustainable bonds" in the recent past (Capital Monitor, 2021). In 2021, Singapore's Nanyang Technological University sold sustainability-linked bonds amounting to $650 million. The university intends to use the proceeds from the bonds to achieve its goals for carbon neutrality and also to reduce carbon emission intensity by half. However, California's Stanford University was the first university to sell bonds with climate and sustainability objectives. Stanford aimed to use the proceeds for financing campus construction and renovation aligned with environmental objectives (Capital Monitor, 2024).

Additionally, universities can bring students in as active stakeholders in protecting the environment through green finance initiatives. Initially, a prompt and comprehensive response is warranted. One of the ways to respond is teaching environmental finance. With a background in environmental finance, students will know the necessary information and resources to address these problems and tackle them with financing solutions. They will be able to enact or promote new projects and become part of the global energy transition, thus mitigating environmental damage and enhancing environmental protection.

13.1.1.2 Enhancing Financial Literacy

Sustainable finance literacy is an extension of financial literacy. Financial literacy builds the capacity, skills and knowledge for an individual to make appropriate financial decisions. Anderson and Robinson (2022) found financial literacy to be a significant barrier in making choices related to investment or purchase of products and services with higher environmental values.

Research focusing on sustainable development goals has a different set of opportunities and challenges. Sustainable development is viewed as the culmination of three concepts, i.e., environment, society and economy (Giddings et al., 2002). Literature suggests that teachers need to voice their views and raise awareness levels about the critical theoretical concepts to view sustainability issues (Sund, 2016). Universities can also engage students in research activities to critically address the concerns and issues related to the relationships between humans and nature (Spannring, 2017).

Studies (Filippini et al., 2024 Bethlendi et al., 2022; Brent & Ward, 2018) also support lower levels of financial literacy related to sustainable products and sustainable investing among investors across different countries. Higher education institutions like universities have a critical role to play in this context by raising awareness and literacy levels of young generations.

Through the integration of green finance into education, students gain knowledge of conventional financial principles within the context of the development of sustainable development. It allows them to improve their total financial accountability and understanding of the long-term effects of their financial decisions on the environment and society. People who have literacy skills know how to manage their finances efficiently, make sound investments and actively further stable economic growth and development in the long term. Thus, the next generation of consumers and investors can be prepared to participate actively in green investments, products and services.

13.1.1.3 Fostering Responsible Citizenship

Higher education institutions play a pivotal role in preparing young people as society's leaders. Universities need to inculcate appropriate knowledge, capabilities and disposition among young minds so that they become capable of appreciating diversity and addressing social injustice.

Universities are evolving and developing their curricula to contribute toward the global knowledge economy. Supporting this idea, Borkovic et al. (2020) identify two major themes in university education: expression of internationalization policy and global citizenship as an obscured educational intention. Hadjichambis and Paraskeva-Hadjichambi (2020) discuss the role of universities in empowering students to become environmental citizens. By acting as environmental citizens, young people can respond to environmental issues and bring in sustainability. They discuss a pedagogical approach of education for environmental citizenship as a distinct, integrated and holistic approach.

Providing students with financial support for environmentally friendly initiatives helps in molding responsible and accountable citizens. Through these projects, students learn that money-related decisions have crucial repercussions that go beyond their personal circumstances. As a result, they develop a sense of responsibility for the planet. By gaining a true understanding of green finance, students may possibly become more sustainable and responsible consumers, as well as professional hopefuls, possibly assisting in the creation of a more sustainable and equitable planet.

13.1.1.4 Supporting the Achievement of Sustainable Development Goals (SDGs)

The United Nations Sustainable Development Goals (SDGs) provide a solid global basis for a better and more prosperous future (Baloch & Ulucak, 2023; Popkova et al., 2023). To take part in achieving these objectives, education is the main tool (Dai & Chen, 2023). Green finance education directly contributes to the following main f(SDGs): SDG 4 Quality Education, which means providing learning outcomes that might result in intelligent students becoming modern-day proficient professionals. Green finance education ensures SDG 7 Affordable and Clean Energy, which requires stimulating the global utilization of renewable sources of energy and skills in using green money to invest in green electricity projects and infrastructure. Easier access to affordable green energy leads, of course, to affordable living for entire families. SDG 13 Climate Action, in line with the criteria and subgoals for that SDG, green finance education aims to teach students to understand the financial driving mechanisms of the green world economy, provide them with green factors for investments, and empower them to undertake and defend their own green projects on climate protection.

13.1.1.5 Bridging the Skills Gap

With the rising focus on decarburizing economies, governments and corporations are prioritizing sustainability and the environment to achieve sustainable development goals. As a consequence, workers need to be equipped with skills that can support the green transition. However, the need for sustainability expertise is greater than the supply, which has created "the green skills gap." In other words, the green skill gap refers to the shortage of skill sets among workers who need to perform activities created by the green transition. According to World Economic Forum estimates, the green skills gap is growing each year. The green skills gap broadly points toward the requirement of skills in sustainability literacy, critical thinking, green retrofitting, digital skills linked to energy modeling, waste management, and reporting and compliance (University College of Estate Management, 2024).

Many countries have admitted to lagging in fulfilling the green skills gap. According to the LinkedIn Global Green Skills Report 2023, society is in urgent need to reimagine the future of work and transition toward green skills (Global Green Skills Report 2023 | LinkedIn, n.d.). Universities can play a significant role in addressing this situation. They can channel green funds to develop green skills at undergraduate and graduate levels. Specialized training programs are urgently needed to bridge the gap and enable the transformation of business models and practices toward a sustainable green industry.

For instance, American University's Kogod School of Business and Columbia Business School have considered major investments in the field of climate finance and are prioritizing sustainability in their courses, campus infrastructure and administrative processes. The higher focus on sustainable development by industry and government determines the increasing demand for qualified experts in the field of green finance (Han et al., 2024). The increasing inclusion of green finance in the curricula is expected to lessen this skills gap. As a result, specialists who acquire a profound understanding of sustainable methods in finance will be able to address the job market demands and contribute to the increase of sustainability in their specific realms.

13.1.1.6 Encouraging Innovation and Entrepreneurship

Green finance-focused education can instill students' proclivity for prompting innovations and projective reasoning. Simultaneously, such an approach propels their ambitious aims to address environmental issues by all means possible (Hou et al., 2022). This type of education helps students familiarize themselves with modern and sophisticated financial tools, such as green bonds and attractive and profitable investments. In effect, this new knowledge can inspire students to develop their sustainable company ideas and participate in facilitating the development of a green and vibrant economy (Popkova et al., 2023).

13.1.1.7 Promoting Long-Term Thinking

The essence of green finance education is that while "conventional financial education typically focuses on short-term gains and immediate financial successes, green finance education promotes the creation of a long-term mindset and strategic planning." Students attending such a course learn how to relate investment and financial decisions to their long-term environmental and social consequences. This is important for establishing a sustainable future. Green finance education within educational settings is essential as it provides students with the required tools and knowledge to understand and control environmentally friendly finance. As a result, the system of education forms a cohort of students who specialize in the fields of economics, environment and sustainable development. Such an approach benefits the students and also supports the ultimate purposes of world sustainable development and well-being.

13.1.2 Exploring the Niche Area: How Green Finance Can Help Libraries Sustain a Changing Environment

13.1.2.1 Funding for Sustainable Infrastructure

Green financing provides monetary assistance to libraries, which concretely implement programs oriented toward being environmentally safe. This tool is used with the help of bonds, green loans, and grants to the libraries for the construction of eco-buildings, installation of solar panels, and realization of environmentally safe construction approaches. After implementation, this tool allowed the relevant carbon emissions of the library to be reduced, which significantly reduced energy and maintenance costs. On the whole, libraries can efficiently perform basic functions with buildings that are safe in various situations.

13.1.2.2 Enhancing Energy Efficiency

Energy efficiency is of essential importance for the purposes of sustainable development. On top of that, energy-efficient technologies also allow libraries to use green financing to receive state-of-the-art equipment such as an optical-driven hydrocarbon lighting source and a facility management system of the latest generation. As a result, the harm done to the environment can be reduced, and the operating costs can be cut. The identified premises can also be used as a foundation for the guidelines to ensure that limited financial resources are used to improve the energy efficiency of libraries and expand their services and products.

13.1.2.3 Promoting Resource Conservation

Green funding is a way to support the preservation and safeguarding of library resources. It also addresses land designing to use water wisely, constructing devices for the collection of rainwater and creating methods to prevent waste. These measures enable the proper use of trees and to simultaneously prevent waste. Green funding is also involved in digitizing the library; a digital library requires no physical storage or paper for storage. This ensures that libraries can continue to operate.

13.1.2.4 Supporting Renewable Energy Adoption

An opportunity for a library in this regard is the use of green funding, which today can help a library switch to solar, wind or some other kind of renewable energy, such as geothermal energy. Renewable energy capability not only decreases reliance on fossil fuels but also provides a consistent, renewable source of energy. Moreover, these systems are capable of creating surplus electricity, which the library can either sell to the grid or use to operate nearby town software. This method, first, indicates the library's dedication to green growth, and, second, how a library can serve as an instructive example for local cities.

13.1.2.5 Funding Sustainable Programs and Initiatives

Green finance can be used to create and implement plans to promote and maintain awareness and sustainability among the members of the surrounding community. There are opportunities to invite potential participants into libraries to try out classes and to help fund their efforts toward the creation of awareness programs, cleanups and the promotion of bicycle use. Another opportunity for green finance is funding the running of the green space in the community, community action or any other sustainable effort that ultimately brings benefits to those living in the neighborhood. The library is changing as it takes on the role of an educational environment that is geared toward sustenance in an evolving community.

13.1.2.6 Strengthening Community Resilience

The assistance of libraries to local communities in adjusting to environmental change is vital. Therefore, green financing can provide resources to libraries to enhance their capacity to act as emergency responders, information providers, and disaster recovery and support centers during natural disasters and inclement weather. Implementing resources in sustainable core structures like energy storage systems and flood-proof designs empowers libraries and their services even under certain disasters. This greatly contributes to the development of local communities and makes them more powerful and confident.

13.1.2.7 Encouraging Green Practices among Patrons

Libraries have the potential to adopt environmentally friendly policies via green funding and, consequently, encourage sustainable health behavior throughout the community. This may include sustainable living resources, recycling software, and increased utilization of digital resources to reduce paper. Libraries should be an ideal example of sustainable procurement policy by adopting environmentally friendly products and reducing the use of plastics. Overall, the use of environmentally friendly practices by the libraries and advancing them among library users greatly increases the capacity for both health behaviors and sustainable development in libraries.

13.1.2.8 Enhancing Library Services and Outreach

The monies acquired from green initiatives in the library services and advocacy programs have the potential to be reinvested. By decreasing operational expenses, libraries can allocate additional resources to extending their collections, improving digital services, and increasing opportunities for educational and cultural initiatives. This contributes to the continued vitality and significance of libraries within local communities, drawing in a larger audience and fostering a culture of continuous learning and sustainable progress. The implementation of improved services and educational initiatives aimed at increasing knowledge about library resources assists in accommodating the evolving requirements and shifting demands of nearby communities.

13.1.2.9 Attracting New Funding Opportunities

Library funding opportunities can be increased by demonstrating a strong dedication to sustainable growth. Numerous scholarships and funding opportunities are available for projects aimed at advancing environmental sustainability and enhancing local community adaptation. Libraries may become sustainable leaders and attract more investment by utilizing green funds and promoting effective sustainable development activities. This will establish a beneficial cycle, whereby initial investments in green finance will enhance funding prospects and perpetuate endeavors toward sustainable growth.

13.2 STRATEGIES FOR GENERATING GREEN FINANCE THROUGH UNIVERSITIES

As the funds for green finance are short of the requirements, it is important to understand different options for generating green finance to finance sustainable development goals' achievement. The following sections cover different strategies that universities can adopt to support green finance.

13.2.1 Grants and Donations

- **Government grants**: Apply for federal, state and local government grants focused on sustainability, energy efficiency and green initiatives.
- **Private foundations and NGOs**: Seek funding from private foundations and non-governmental organizations that support environmental and educational projects.
- **Crowdfunding**: Launch crowdfunding campaigns to engage the community and gather small donations for specific green projects.

13.2.2 Leveraging Technology

- **Digital resources**: Develop and promote digital assets, such as e-books and online courses that specifically focus on green finance and sustainability, with the aim of generating additional income.

- **Virtual events**: Host online events, webinars and conferences centered on topics related to green finance, while charging participants fees to generate funds.

13.2.3 Partnerships and Collaborations

- **Corporate partnerships**: Collaborate with corporations that have corporate social responsibility (CSR) programs focused on sustainability. These partnerships can provide financial support and resources.
- **Academic and research institutions**: Partner with universities and research institutions to secure joint funding for green projects and educational programs.

13.2.4 Energy Efficiency Programs

- **Energy savings performance contracts (ESPCs)**: Engage with energy service companies to implement energy-saving measures. The savings generated can be used to repay the investment over time.
- **Utility rebates and incentives**: Take advantage of rebates and incentives offered by utility companies for energy-efficient upgrades and renewable energy installations.

13.2.5 Fundraising Events

- **Green fundraisers**: Organize events such as eco-fairs, green marathons and sustainability workshops to raise funds and awareness for green initiatives.
- **Membership programs**: Develop membership programs where patrons contribute annually to support the library's green projects.

13.2.6 Sponsorship and Naming Rights

- **Sponsor a green space**: Offer naming rights for green spaces or eco-friendly installations within the library to attract donations from individuals or businesses.
- **Sponsor programs and events**: Seek sponsorship for green programs, events and initiatives from local businesses and environmentally conscious organizations.

13.2.7 Revolving Green Fund

- **Establish a revolving green fund**: Create a fund that finances energy-efficient projects and reinvests the savings from reduced energy costs into future green projects.

13.2.8 Community Engagement and Support

- **Friends of the Library groups**: Mobilize Friends of the Library groups to advocate and raise funds specifically for green initiatives.
- **Volunteer programs**: Engage volunteers in sustainability projects, reducing labor costs and fostering community investment in green initiatives.

By utilizing these strategies, libraries can secure funding to support green finance initiatives, promoting sustainability and environmental responsibility within their communities.

13.3 THE INTERLINKAGES BETWEEN ARTIFICIAL INTELLIGENCE, GREEN FINANCE AND UNIVERSITIES

Artificial intelligence (AI) has revolutionized the financial industry, and green finance is not an exception. Educational institutions, especially universities, hold the potential to become centers of learning and innovation. This can be achieved by leveraging AI to play an instrumental role in promoting and devising innovative ways of making green finance central to managerial decisions and implementation plans. The following are a few ways through which AI can become a crucial part of university plans and execution.

13.3.1 Curriculum Design, Development and University Research

Universities hold the responsibility to generate awareness and extend knowledge for specialized fields in the society. Universities can offer specialized courses focusing on green finance, sustainable finance, achievement of sustainable development goals, climate finance, carbon credit analytics, etc., while incorporating artificial intelligence into the student curriculum. This would help universities act as catalysts in extending knowledge, awareness and discussion on the channels through which AI can influence young minds, academicians, professionals and researchers.

Further, universities can encourage research on the dimensions of climate finance, green finance and related areas like green bonds, investment strategies, usage of AI tools in measuring the carbon credits and gaining insights on the promotion of green finance. Universities act as an apt platform to invoke discussions on the usage of AI to promote green finance. Additionally, universities can promote interdisciplinary collaboration among different fields like finance, business administration, computer science and environmental science departments to develop innovative solutions.

13.3.2 Education and Training

Universities hold a responsibility to prepare a workforce that is adequately equipped with the skills needed in the upcoming time period. Universities are the connecting link between learning, skills and industry requirements. Educational institutions can utilize and develop AI-powered tools to enhance personalized learning, such as adaptive learning platforms and programs to incorporate green finance into the skill set among students belonging to different fields.

Universities can develop skills among their stakeholders, especially students, to employ AI, data science and machine learning to analyze complex financial data and make informed decisions. By offering continuing education programs for professionals in the industry, offering platforms for discussion like symposiums, conferences, round table discussions and chat rooms for discussing thought-provoking ideas on the usage of AI in green finance.

13.3.3 Campus Sustainability

Universities can plan and implement sustainable infrastructure projects, such as energy-efficient buildings and renewable energy installations, funded through green finance mechanisms. Universities should employ AI and data science to optimize energy consumption across campus buildings, aiming to reduce and optimize the consumption of energy as well as reduction of carbon emissions. Universities can employ AI-powered solutions to improve waste management practices, indulge in the usage of renewable energy resources and reduce the negative impact on the environment.

Further, universities can develop and employ AI-enabled solutions to improve waste management practices in order to reduce environmental impact.

13.3.4 Industry Partnerships and Innovation

Universities can call for partnerships with banks, asset management firms and insurance companies in order to develop AI-driven solutions for understanding better ways of employing green finance. Universities can support startups and entrepreneurs working on innovative green finance solutions. Also, universities can organize events that can encourage students and researchers to develop AI-based solutions for environmental and social challenges. Additionally, universities can advocate for policies that support green finance and sustainability at local, national and international levels, ensuring that higher education institutions are at the forefront of the transition to a sustainable economy.

Planning for the future involves recognizing the crucial role that green finance will play in shaping sustainable development in higher education institutions. Universities should prioritize integrating AI and green finance into their strategic planning processes, ensuring that sustainability goals are aligned with institutional missions and operations.

When considering the future, the higher education industry needs to give importance to incorporating green money responsibly within the necessity of achieving the SDGs. In order to transition to a net-zero, ecologically balanced economy, a more significant and transformative approach is needed, rather than just superficially incorporating green practices into existing business models. The successful implementation of the SDGs necessitates the meaningful integration of these goals into the fundamental operations of influential organizations that possess the credibility and influence, if not the direct ecological capacity, to bring about more than half of the necessary changes within the next ten years. This pertains to the subtle alteration of the creative, world-class, research-intensive, problem-solving curriculum provided by higher education institutions in the United States. This represents a significant and mostly untapped possibility. It also involves implementing environmentally responsible facilities management activities, forming collaborations across the entire campus, engaging in community service programs and developing finance methods. These efforts collectively represent the ambitious yet attainable road toward a responsible higher education ecosystem.

13.4 CONCLUSION

The field combining AI with green finance has the potential to greatly benefit higher education. Green finance provides colleges and universities with a means to attain sustainability and expand their impact in the quest for an education focused on sustainability. AI provides the right amount of leverage to green finance to achieve sustainable development goals. This proposition is exceptionally clear and evident. Colleges and green finance possess at least two fundamental characteristics in common. Green finance mostly revolves around monetary aspects, as indicated by the term "finance." It covers the full financial environment in which colleges and universities can directly or indirectly achieve their sustainability objectives and even have an impact on their communities. Combining green finance with AI helps universities to reap the benefits that can pave the way toward the achievement of sustainable development goals. Furthermore, both colleges and green finance share a common focus on the tangible and symbolic aspects of sustainability within their immediate surroundings.

REFERENCES

Anderson, Anders, and David T. Robinson. "Financial literacy in the age of green investment." *Review of Finance* 26, no. 6 (2022): 1551–1584.

Baloch, Muhammad Awais, Danish, and Zubeyde Senturk Ulucak. "Understanding the role of green finance and renewable energy consumption for sustainable development in ACI economies." *Climatic Change* 176, no. 11 (2023): 151.

Bethlendi, András, László Nagy, and András Póra. "Green finance: The neglected consumer demand." *Journal of Sustainable Finance & Investment* (2022): 1–19.

Borkovic, Shinead, Toula Nicolacopoulos, Dell Horey, and Tracy Fortune. "Students positioned as global citizens in Australian and New Zealand universities: A discourse analysis." *Higher Education Research & Development* 39, no. 6 (2020): 1106–1121.

Brent, Daniel A., and Michael B. Ward."Energy efficiency and financial literacy." *Journal of Environmental Economics and Management* 90 (2018): 181–216.

Capital Monitor (2024) https://www.capitalmonitor.ai/analysis/universities-are-waking-up-to-green-finance-but-say-it-quietly/?cf-view.

Chen, Shuyang, and Gang Xie. "Assessing the linkage among green finance, technology, and education expenditure: Evidence from G7 economies." *Environmental Science and Pollution Research* 30, no. 17 (2023): 50332–50345.

CİTİL, Mucahit. "Is green finance a prerequisite for green growth of G-20 economies." *Innovation and Green Development* 3, no. 4 (2024): 100170.

Dai, Yijia, and Xuanyuan Chen. "Evaluating green financing mechanisms for natural resource management: Implications for achieving sustainable development goals." *Resources Policy* 86 (2023): 104160.

EU (2024) https://op.europa.eu/webpub/com/general-report-2024/en/index.html.

Filippini, Massimo, Markus Leippold, and Tobias Wekhof. "Sustainable finance literacy and the determinants of sustainable investing." *Journal of Banking & Finance* 163 (2024): 107167.

Giddings, Bob, Bill Hopwood, and Geoff O'Brien. "Environment, economy and society: fitting them together into sustainable development." *Sustainable Development* 10, no. 4 (2002): 187–196.

Global Green Skills Report 2023 | LinkedIn. (n.d.). Retrieved August 6, 2024, from https://economicgraph.linkedin.com/research/global-green-skills-report.

Hadjichambis, Andreas Ch, and Demetra Paraskeva-Hadjichambi. "Education for environmental citizenship: The pedagogical approach." *Conceptualizing Environmental Citizenship for 21st Century Education* 4 (2020): 237–261.

Han, Jie, Wei Zhang, CemIşık, Wenju Zhao, Muhammad Anas, Qinglan Zheng, Danxi Xie, and Satar Bakhsh. "Sustainable development pathways: Exploring the impact of green finance on urban metabolic efficiency." *Sustainable Development* (2024).

Höhne, N., S. Khosla, H. Fekete, and A. Gilbert. *Mapping of green finance delivered by IDFC members in 2011.* Cologne: Ecofys, 2012.

Hou, Ruirui, Lijie Du, Syed Abdul Rehman Khan, Asif Razzaq, and Muhammad Ramzan." Assessing the role of green finance and education as new determinants to mitigate energy poverty." *Frontiers in Psychology* 13 (2022): 924544.

Lamptey, Richmond Odartey, Michael Zisuh Ngoasong, and Richard Blundel. "Green Finance Strategies in Africa: A Focus on Capital Market-Based Impact Investments in Small and Medium-Sized Enterprises in Ghana." In *The Routledge Handbook of Green Finance*, edited by Othmar M. Lehner, Theresia Harrer, Hanna Silvola, Olaf Weber, pp. 317–336. London: Routledge, 2023.

Liu, Guangwei, and Kaipeng Liang." The role of technological innovation in enhancing resource sustainability to achieve green recovery." *Resources Policy* 89 (2024): 104659.

Liu, Zhen, and Assem Abu Hatab."Assessing stakeholder engagement in public spending, green finance and sustainable economic recovery in the highest emitting economies." *Economic Change and Restructuring* 56, no. 5 (2023): 3015–3040.

Ma, Wenxuan. "Exploring the role of educational human capital and green finance in total-factor energy efficiency in the context of sustainable development." *Sustainability* 15, no. 1 (2022): 429.

Mohsin, Muhammad, and Fouad Jamaani. "Unfolding impact of natural resources, economic growth, and energy nexus on the sustainable environment: Guidelines for green finance goals in 10 Asian countries." *Resources Policy* 86 (2023): 104238.

Ping, Shen, and Syed Ahsan Ali Shah. "Green finance, renewable energy, financial development, FDI, and CO2 nexus under the impact of higher education." *Environmental Science and Pollution Research* 30, no. 12 (2023): 33524–33541.

Popkova, Elena, Natalia G. Vovchenko, and Olga V. Andreeva, eds. *Climate-Smart Innovation: Social Entrepreneurship and Sustainable Development in the Environmental Economy.* Vol. 1. Singapore: World Scientific, 2023.

Sharma, Aditya, and Pradeep Kautish. "Measuring green finance." In *Recent Developments in Green Finance, Green Growth and Carbon Neutrality*, edited by Suleiman O. Mamman, Kazi Sohag, pp. 171–191. Amsterdam: Elsevier, 2023.

Spannring, Reingard. "Animals in environmental education research." *Environmental Education Research* 23, no. 1 (2017): 63–74.

Sund, Louise. "Facing global sustainability issues: Teachers' experiences of their own practices in environmental and sustainability education." *Environmental Education Research* 22, no. 6 (2016): 788–805.

Tolliver, Clarence, Alexander Ryota Keeley, and Shunsuke Managi. "Green bonds for the Paris agreement and sustainable development goals." *Environmental Research Letters* 14, no. 6 (2019): 064009.

Torrenova, Manuel Gómez Gutiérrez. "Sustainability. The end of finance as it was." *Studies of Applied Economics* 39, no. 3 (2021). https://doi.org/10.25115/eea.v39i3.5535.

Tsai, Wen-Tien. "Green finance for mitigating greenhouse gases and promoting renewable energy development: Case study in Taiwan." *Green Finance* 6, no. 2 (2024): 249–264.

University College of Estate management (March 18, 2024). What is the green skills gap? https://www.ucem.ac.uk/whats-happening/articles/green-skills-gap/.

Vikas, Namita, Percy Venegas, and Sourajit Aiyer."Role of banks and other financial institutions in enhancing green digital finance." In *Green Digital Finance and Sustainable Development Goals*, edited by Farhad Taghizadeh-Hesary and Suk Hyun, pp. 329–352. Singapore: Springer Nature Singapore, 2022.

14 Integrating Data Analytics, AI, and Machine Learning to Optimize Environmental Impact and Foster Green Innovation

Isha Chhabra, Khushboo Gulati, Soumya Singhal, and Nand Dhameja

14.1 INTRODUCTION

In 2018, the United Nations urged governments and businesses around the world to consider sustainable development as be a key priority. Climate change has garnered significant attention due to its profound impact on the economy and society. This impact is characterized by the potential for substantial property damage and loss of life resulting from increasingly regular occurrences of extreme weather events and the rise in sea levels (Jin et al., 2019).

Such companies are being closely examined for their detrimental impact on the environment (Berrone et al., 2013). Corporations are experiencing mounting pressure to diminish their ecological impact and adhere to legislation concerning climate change and sustainability (Liu et al., 2023). The effectiveness of environmental legislation is a crucial determinant that impacts corporate green innovation (GI) (Liu et al., 2023). As there is a contradiction between environmental regulation and economic development, development strategies should be carefully framed to mitigate environmental damage. Environmental regulations aimed at rectifying environmental harm have been enforced to alleviate the negative impact on the environment. These policies can facilitate the transition toward a more environmentally friendly economy by encouraging the development and adoption of green technologies (Du, 2021).

Previous studies have extensively examined the relationship between environmental regulation and green innovation (Shao, 2019; Guo et al., 2021; Shu, 2016). Green innovation is a successful tactic used by businesses to achieve cleaner manufacturing

DOI: 10.1201/9781003626527-14

and increased competitiveness (Jaffe and Palmer, 1997; Tseng et al., 2013). The importance of green innovation and energy efficiency in increasing environmental performance is vital in both developed and emerging economies, highlighting the need for sustainable development (Singh, 2023).

Green innovation has been recommended as a strategic approach to accomplish environmental preservation and economic development by a considerable number of organizations and communities (Takalo & Tooranloo, 2021). Green innovation, also known as "eco-innovation," is the process of developing new products and technology with the intention of lowering environmental risks, such as pollution and the unfavorable effects of resource use (Castellacci and Lie, 2017).

The OECD (2011) highlights that green innovation is essential for transitioning to a sustainable economy. Firms that invest in green innovation can benefit from increased efficiency, reduced operational costs, and enhanced market opportunities. For example, renewable energy technologies, such as solar and wind power, have not only reduced environmental impacts but also created new business models and revenue streams (IRENA, 2019).

Moreover, green innovation can enhance a firm's reputation and brand value. Consumers and investors are increasingly prioritizing sustainability, and firms that demonstrate environmental leadership can attract environmentally conscious customers and investors. This shift in consumer and investor preferences underscores the importance of integrating green innovation into business. Previous researchers have found that adopting green innovation in businesses can lead to a positive impact on a firm's performance, investments, and risk mitigation (Liu, 2023; Nassani et al., 2023; Ahmed et al., 2021).

Despite the substantial research that has been done, there is room to show the effects of environment regulations-driven green innovation and its impact on the financial performance of firms. Since green innovation is a multifaceted and complex factor to comprehend, there exists a restricted comprehension of the factors driving green innovation. Therefore, the chapter also intends to shed light on the factors that drive green innovation.

Hence, by addressing these research gaps, the chapter contributes to enhancing the understanding of the factors that drive corporate green innovation and its impacts, including the mechanisms and contextual factors involved. This will enable policymakers to make more informed decisions and develop effective strategies to promote sustainable business practices.

14.2 LITERATURE REVIEW

This section discusses previous studies depicting the concept of green innovation, the relationship between environmental regulation and green innovation, how green innovation influences business performance, and the drivers that promote green innovation, as well as how artificial intelligence (AI), machine learning (ML), data analytics, and automation improve business sustainability.

14.2.1 Green Innovation

Tariq et al. (2017) stated that the concept of green innovation has gained popularity in recent decades as a means of addressing the problems posed by global warming and environmental issues. The word "green innovation" is a concept that is used interchangeably with "environmental innovation," "eco-innovation," and "ecological innovation" by many authors, in different contexts and countries. This concept goes beyond traditional innovation by integrating ecological concerns into the core of innovation activities, aiming to achieve sustainable development. Green innovation is crucial in addressing global environmental challenges such as climate change, resource depletion, and pollution, while also providing economic and social benefits.

14.2.1.1 Historical Context and Evolution

The concept of green innovation has evolved alongside the growing awareness of environmental issues. In the 1960s and 1970s, environmental concerns began to gain prominence with the rise of the environmental movement, leading to increased public awareness and the introduction of environmental regulations. In 1987, the term "sustainable development" received attention from policymakers around the globe with the Brundtland Report. This report emphasized that the world needs to redirect its activities so as not to compromise the needs of future generations for today's needs. Additionally, this report highlighted the importance of integrating environmental considerations into economic and social development. In the 1990s and early 2000s, the concept of green innovation began to take shape as businesses and policymakers recognized the potential of innovation to address environmental challenges. The rise of corporate social responsibility (CSR) and the increasing emphasis on sustainability in business practices further propelled the adoption of green innovation. Today, green innovation is seen as a key driver of sustainable development and a crucial element in the transition toward a low-carbon and resource-efficient economy. Green innovation improves both the environmental and financial performance of a company. Multiple academics have suggested different categories of green innovation, such as green product innovation (Tariq et al., 2017; Calza et al., 2017; Xie, Huo, & Zou, 2019), green process innovation (Xie, 2019), and green service innovation (Calabrese et al., 2018).

This section outlines several categories of green innovation, which encompass innovation in product development, innovation in process, organizational innovation, marketing innovation, and business model innovation (Figure 14.1).

14.2.1.1.1 Innovation in Product Development

Product innovation in the context of green innovation refers to the creation of new or greatly enhanced products that effectively decrease environmental harm during their entire lifespan. This style of innovation centers on improving the environmental attributes of products, typically by utilizing sustainable materials, expanding energy efficiency, or refining end-of-life management. Some instances of green product innovation include the following:

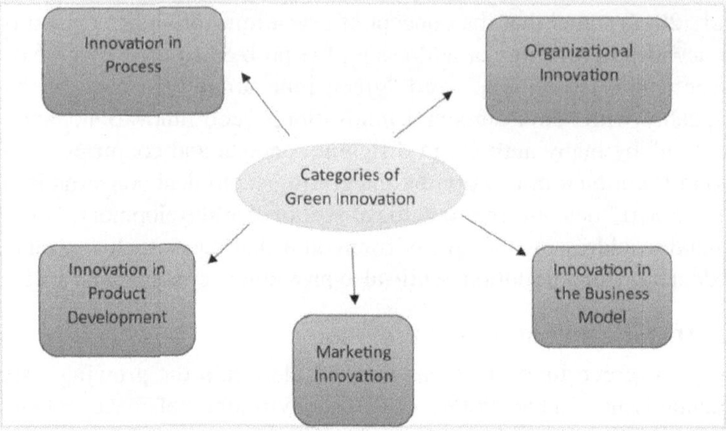

FIGURE 14.1 Different categories of green innovation.

- Energy-efficient appliances are devices that utilize minimal energy during their operational phase, resulting in reduced greenhouse gas emissions and lower energy expenses for consumers.
- Electric vehicles (EVs) are vehicles that are powered by electricity rather than fossil fuels. They have the advantage of producing lower emissions and reducing our reliance on non-renewable energy sources. Biodegradable packaging refers to materials used for packaging that have the ability to disintegrate organically, hence decreasing waste and minimizing pollution.
- Organic and eco-friendly items are goods that are manufactured utilizing environmentally sustainable methods, such as organic foods, natural personal care items, and eco-friendly clothes.

14.2.1.1.2 Innovation in Processes

Process innovation refers to the implementation of novel or enhanced techniques for production that improve environmental performance through increased resource efficiency, waste reduction, and emission minimization. This form of innovation centers on enhancing industrial processes to increase their sustainability. Notable instances of green process innovation include:

- Cleaner production technology: This refers to a range of techniques employed in industrial processes to reduce waste and emissions. These techniques include closed-loop systems, pollution control technology, and waste minimization strategies. Energy-efficient manufacturing refers to production methods that utilize less energy through the implementation of advanced automation, precision manufacturing, and energy recovery technologies. Recycling and trash management are procedures that facilitate the recycling and reutilization of materials, thereby decreasing the need for new resources and minimizing the amount of garbage sent to landfills.

- Water conservation techniques: Strategies that decrease water usage and improve water efficiency in industrial operations, including the implementation of water recycling systems and the utilization of modern filtration technology.

14.2.1.1.3 Organizational Innovation

Organizational innovation includes alterations in a company's framework, ethos, and managerial methodologies that foster sustainability and facilitate environmentally friendly innovation. This form of innovation centers on incorporating environmental factors into the fundamental operations and strategic orientation of the firm. Some instances of green organizational innovation are as follows:

- Environmental management systems (EMS) are structured frameworks that assist firms in effectively managing their environmental obligations, such as obtaining ISO 14001 certification.
- Sustainability reporting and transparency refer to the practices of openly sharing information on environmental performance and sustainability activities with stakeholders. These practices aim to increase accountability and promote engagement with stakeholders.
- Green corporate culture refers to the implementation of initiatives aimed at cultivating a sustainable mindset within the firm, motivating employees to embrace environmentally conscious behaviors and practices. Sustainable supply chain management refers to the implementation of strategies and practices that guarantee the ecologically responsible operation of the entire supply chain. This includes conducting evaluations to assess the sustainability of suppliers and adopting green procurement policies.

14.2.1.1.4 Marketing Innovation

Marketing innovation entails the creation of novel tactics and methodologies to advertise and market environmentally friendly products and services. This form of innovation centers on effectively conveying the ecological advantages of products and cultivating customer consciousness and desire for sustainable alternatives. Notable instances of innovative green marketing include:

- Eco-labeling refers to certification schemes that offer consumers information regarding the environmental attributes of items, enabling them to make well-informed choices when making purchases.
- Green branding refers to branding techniques that prioritize a company's dedication to sustainability and showcase the environmental advantages of its products.
- Sustainable marketing campaigns are marketing initiatives that highlight the environmental qualities of products and services, typically employing messaging that resonates with environmentally conscious consumers.

Consumer education and engagement involve programs that inform customers of the ecological consequences of their decisions and promote sustainable consumption habits.

14.2.1.1.5 Innovation in the Business Model

Business model innovation entails a thorough reconsideration of the fundamental methods by which a company generates, delivers, and acquires value, with a specific emphasis on sustainability. This form of innovation frequently results in the creation of wholly novel business models that prioritize environmental and social advantages in addition to commercial profits. Some instances of green business model innovation include:

- Product–service systems are models in which customers pay for the service offered by a product rather than the product itself. This approach promotes resource efficiency and helps to minimize waste. Circular economy models prioritize the reuse, repair, remanufacturing, and recycling of items and materials to establish closed-loop systems, thereby reducing waste and limiting resource consumption. Sharing economy platforms are the business models that enable the sharing and collaborative consumption of resources. These platforms include ride-sharing apps, peer-to-peer housing services, and equipment-sharing platforms.
- Zero-waste business models are strategies that focus on eradicating waste production by creating goods and procedures that guarantee the reuse, recycling, or composting of all resources.

14.2.1.2 Role of AI, ML, Data Analytics, and Automation Green Innovation

Automation, data analytics, AI, and ML are critical components of contemporary corporate management, especially when it pertains to promoting green innovation. These technologies improve decision-making, increase operational efficiency, and assist companies in achieving sustainability objectives.

AI and ML give companies data-driven insights that help with energy management, waste reduction, and resource allocation. For example, real-time energy optimization and environmental metric monitoring by AI algorithms can lower greenhouse gas emissions (Berg et al., 2020). Businesses can anticipate resource requirements with the use of predictive machine learning models, which reduce waste and increase productivity (Daugherty & Wilson, 2018a).

By identifying possible equipment problems before they become more serious, ML-powered predictive maintenance minimizes downtime and saves resources. By prolonging the lifespan of equipment and preventing resource-intensive repairs, this proactive strategy supports sustainability goals.

Data analytics provides information on waste management, supply chain effectiveness, and resource usage. Businesses can identify inefficiencies and switch to low-emission logistics or renewable resources by analyzing consumption data, which supports sustainability initiatives (Lacy & Rutqvist, 2015). Implementing circular economy principles, where product life cycles prioritize reuse and recycling to cut waste, also requires data analytics.

Businesses may quantify and lessen their environmental impact by evaluating real-time data, which makes data analytics a crucial component of green innovation (Bibri, 2018).

Automating repetitive processes lowers labor and resource expenses, especially when it comes to robotic process automation (RPA) and Internet of Things-enabled devices. RPA simplifies processes, lowering energy use and human error (Willcocks et al., 2015). Companies can modify processes for maximum efficiency and lower emissions by using smart automation systems that can track environmental parameters and energy consumption in real time (Daugherty & Wilson, 2018b).

When combined, these technologies increase profitability and promote sustainable business practices. For instance, automation simplifies resource usage throughout supply chains, and artificial intelligence and data analytics can assess providers' sustainability. This collaboration puts businesses in a position to comply with green laws and satisfy consumers' demands for environmentally friendly operations (Elkington, 2018).

14.2.1.3 Environmental Regulation and Green Innovation

Environmental regulation refers to the establishment of government laws to control the conduct of businesses with the goal of encouraging them to restructure and improve their business activities with the objective of mitigating environmental damage. Environmental regulation can incentivize firms to implement reforms and develop innovative pollution control technology at both the micro and macro levels (Shen et al., 2022). To ensure the preservation of the local environment, the government should implement environmental regulations, including restrictions on emission standards and technical requirements for businesses (Fan et al., 2021).

Researchers have found that environmental regulation-driven green innovation can positively impact energy savings and the reduction of emissions. Environmental regulation can promote green innovation, product innovation, and technological innovation (Cruz et al., 2006; Carrillo-Hermosilla et al., 2010). Scholars have developed some indicators to measure the intensity of environmental regulation based on research issues and the availability of data (Figure 14.1). The indicators are (a) the ratio of investment in pollution treatment to the amount of pollutant emissions (Ederington, 2003); (b) the ratio of the expenses incurred for pollution management in the industrial sector to the added value (Zhang, 2012); (c) pollutant removal rate indices; and (d) a weighted composite index that represents the combined elimination rate index of multiple contaminants (Fu, 2010).

Song et al. (2018) opined that the advantages of GI must outweigh the expenses associated with adhering to environmental regulation (ER) rules; this motivates enterprises to allocate resources to green innovation. Wang and Shen (2016) studied the relationship between GI and ER efficiency and found an inverted U-shaped relationship by analyzing data from a variety of industrial sectors in China. Zhang et al. (2020) also discovered an inverted U-shaped relationship between ER and GI efficacy in the city of Xi'an, China. This relationship is calculated using directional distance functions.

Green technology innovation encompasses a range of innovative actions undertaken during the production, operation, and sales processes to minimize pollution emissions, reduce energy consumption, and improve environmental competitiveness (Wang et al., 2018). Environmental laws pertain to the policies and tools implemented by the government to address environmental issues and promote economic growth (Peng & Li, 2016). The specific mechanisms of these regulations will be explored individually in three categories. Figure 14.2 shows the green technology mechanism encompassing three different environment regulations: public regulations, market regulations, and command regulations.

In contrast to this, other authors have argued that environmental regulation does not promote green innovation. The cost for firms may increase as a result of the reduction in emissions to the environment that involves environmental regulations.

Domazlicky and Weber (2004) argued the advantages of technological advancements resulting from environmental restrictions were insufficient to offset the higher expenses incurred by businesses. Li and Bi (2013) argued that environmental regulation hinders the technological advancement of firms and inhibits green innovation. There is an alternative perspective that suggests the effect of environmental legislation on the efficiency of green innovation is not clear. Kneller and Manderson (2012) contended that implementing compulsory environmental regulations would result in elevated expenses for reducing pollutants and conducting research and development

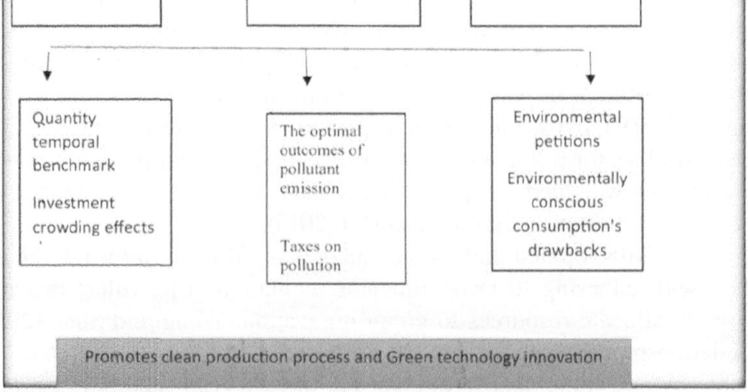

FIGURE 14.2 Impact of different environment regulations on green technology innovation.

(R&D) in the United Kingdom, while yielding minimal effects on overall capital accumulation. Peuckert (2014) highlighted that environmental legislation would constrain spending, impede short-term technical progress, and foster long-term development. Peng et al. (2017) found that the efficiency of green innovation was correlated with formal and informal environmental control policies in a U-shaped and inverted U-shaped manner, respectively.

The importance of the factors that encourage environmental innovation practices and green technological advancements has been underscored by numerous authors who have conducted research on the subject. Foreign direct investment inflows (Song et al., 2015), the intensity of market competition, environmental regulations, and investments in R&D (Song and Wang, 2018), and strategies to implement sustainable practices in supply chain management (Luthra and Mangla, 2018) may all influence the capacity to innovate and develop a green economy.

14.2.1.4 Green Innovation and Firm Performance

Tang (2018) stated that green innovation has been recognized as a catalyst for enhancing the performance of corporations. In the current highly competitive landscape, companies are striving hard to meet the business niche and for survival. Such a competitive environment has compelled companies to adopt green innovation policies and actively involve, construct, and enhance their relevance to the stakeholders. With the implementation of effective green innovation policies, companies have the opportunity to progress and establish themselves as global competitors (Lai et al., 2003; Antonionli et al., 2013).

Other scholars have stated that managerial environmental concern and absorptive capacity enhance firm performance. Managerial environmental concern is a representation of a company's leaders' attention to environmental issues. It was found that the level of concern that managers have for the environment can influence the innovation activities within a business, leading to improved performance (Qi et al., 2010).

On the contrast, absorptive capacity is the ability of a firm to recognize, assimilate, and apply new information for commercial purposes. It is considered a crucial aspect of a firm's dynamic capabilities, allowing them to learn from partners, access external information, and incorporate it into their existing knowledge base. There is a significant relationship between absorptive capacity and innovation, hence influencing firm performance. Executives of firms should prioritize the development of absorptive capacities to improve organizational activities. Practitioners should strive to integrate external knowledge that can enhance their existing expertise and address relevant concerns related to green innovation initiatives (Xue et al., 2019).

Jia, Huang and Zhang (2019) analyzed the performance of state-owned enterprises in China and found there is a significant positive relationship between green utility model innovation and firm performance. The author's study also supported the Porter hypothesis, as there is an improvement in sales growth and net profits of the firms. Albort-Morant et al. (2016) contended that enterprises can improve their competitive advantage by implementing green innovation and green management practices in their study. Furthermore, these companies may acquire additional advantages, including the capacity to increase their prices, stakeholder trust, and increased goodwill. This is especially true if the organization is the first to introduce the concept. Chen et al. (2018) suggests that the firm can enhance its performance and profitability by implementing green innovation in a manner that is both effective and efficient. Green innovation allows firms to include environmental considerations in their strategies, while also integrating an additional method, including its execution. Many firms fail to recognize the potential benefits of gaining a competitive advantage and do not make an effort to solve environmental concerns (Yin, Gong & Wang, 2018).

14.2.1.5 Drivers of Green Innovation

The implementation of green innovation in a corporation is influenced by both internal and external drivers of innovation, as well as their respective subelements. The strategic capacity of a firm to implement an innovative response in a dynamic environment is contingent upon the integration of the necessary organizational skills and resources to develop processes, new products, and organizational innovations (Bernauer et al., 2007).

Singh et al. (2022) contended that both internal and external stakeholders are essential in the advancement of green innovation practices. The market and the institutional environment are the primary sources of external environmental influences (Singh et al., 2022; Kawai et al., 2018). The market environment is significantly influenced by consumer demand and competitive pressure. The demand for environmentally sustainable products among consumers is increasing. In addition to satisfying consumer demand, green innovation methods have the potential to generate substantial revenues by establishing a positive green reputation for enterprises (Wang et al., 2020). Nevertheless, companies have been subjected to subliminal pressure as a result of competitors' emphasis on environmental concerns, and industrial pressure is a significant determinant of institutional environmental. Cai and Li (2018) and Huang et al. (2022) have emphasized the importance of companies implementing green innovation strategies to counteract their competitors' environmental ambitions.

Peng and Liu (2016) found that top managers play a pivotal role in shaping corporate strategy, and the degree to which they prioritize environmental issues has a considerable impact on the effective implementation of eco-innovation. Scholars widely acknowledge that senior management commitment and managerial concern can provide resource support for green innovation (Song et al., 2021).

Formal institutions mostly utilize written requests and explicit coercive action to compel enterprises to implement green innovation strategies. The main factor is the implementation of rigorous policies and regulations by the government. Companies are influenced by deterrence and use environmentally-friendly innovation approaches to meet legitimacy standards (Li et al., 2023). Furthermore, the government provides assistance and guidance to firms by implementing various policies, such as green subsidies and tax incentives, which successfully counterbalance the initial costs of adopting green innovation strategies.

Market-oriented environmental supervisory measures, such as resource taxes, pollution permits, and carbon trading rights, offer enterprises greater flexibility and allow them to choose green innovation models that align with their capabilities (Xie et al., 2017). Some researchers contend that stakeholder pressure positively influences corporate green innovation (Shahzad et al., 2020; Rui, 2021), whereas others argue that there is insufficient evidence to establish a link between corporate green practices and stakeholder pressure (Qi et al., 2010). One plausible reason for the disparity in these research findings is that experts analyze different sectors. The relevant government departments should implement targeted environmental legislation and regulations to encourage companies to embrace green innovation practices that align with the local economic and market conditions. In the future, additional research can be conducted to investigate the ideal circumstances under which environmental legislation might successfully stimulate green innovation in businesses. Furthermore, conducting comparative studies across other industries is essential for acquiring valuable insights.

14.3 CHALLENGES IMPLEMENTING GREEN INNOVATION

Innovation is the main factor that stimulates growth, propelling a corporation toward exceptional performance and guaranteeing a competitive advantage. Furthermore, it enhances environmental efficiency, therefore garnering the necessary social capital for future progress (Cillo et al., 2019). Although there are numerous advantages to adopting green innovation, the adoption of green innovation faces several barriers. Researchers in many industries have found several challenges to the application of green practices.

Government involvement, economic resources, consumer support, perception, and attitude are the primary barriers to the adoption of green innovation in the manufacturing industry. The primary challenges to the implementation of green technology in large Chinese automobile companies, as identified by the research conducted by Xia, Zhang, Yu, and Tu (2019), are the absence of advanced technologies, inadequate infrastructure, limited expertise and resources, and uncertain advantages. Chang et al. (2022) analyzed the suggestions put out by environmental leaders to facilitate the adoption of environmentally friendly legislation. A grand total of 26 hurdles were uncovered, with the most prominent ones including a dearth of demand for environmental merchandise, exorbitant costs, inadequate resources, and buyers' apathy toward paying for ecologically sustainable products. In their study, Jayant and Azhar (2014) examined the obstacles that impede the successful implementation of green supply chain management. A total of 20 barriers were found, with the most prominent ones being top management commitment, cost, fear of failure, market, and customer-related obstacles.

Giunipero et al. (2012) discovered that the implementation of green supply chain management is hindered by the costs associated with sustainability, environmental legislation, and a lack of alignment in top-level decision-making. The implementation of green supply chain activities in the expanding economy of Bangladesh has considerable hurdles such as resource scarcity, customer demand, financial constraints due to limited short-term financial benefits for enterprises, and a lack of government regulations. Gupta and Barua (2017) identified internal barriers to green innovation in manufacturing firms as financial, human, and resource constraints. Silva et al.

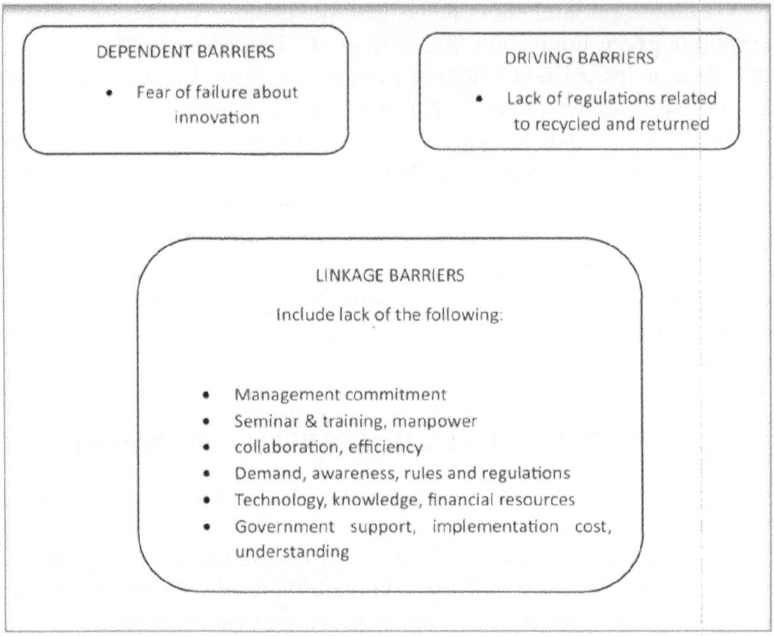

FIGURE 14.3 Challenges to implementing of green innovation.

(2019) examined substantial impediments to the implementation of greener production. The primary impediments were technological and economic; however, sustainability is promoted by government policies and customer demand.

A study conducted by Ullah et al. (2021) in the manufacturing sector categorized the barriers to innovation into three quadrants: dependent barriers, linkage barriers, and driving barriers.

- Dependent barriers: These factors have a low pushing force but strong dependence power. This category encompasses a solitary obstacle, specifically the apprehension of experiencing failure in relation to green innovation (Figure 14.3).
- Linkage barriers: These are the obstacles that possess significant driving force and reliance power. There are a total of 16 barriers that have been identified as mediating variables: insufficient commitment by the management, inadequate seminars and training, lack of manpower for green innovation, insufficient collaboration between the environment agencies and the government, no demand certainty, inadequate awareness about reverse logistics, lack of rules and regulations, ambiguity surrounding the efficiency of green innovation, the absence of novel technology, the dearth of knowledge on the market, insufficient financial or material resources, absence of governmental financial support, the implementation of a green innovation

system incurs a significant financial burden, inadequate understanding of green innovation and behaviors, government subsidy, and identification of environmental opportunity.

- Driving barriers: These obstacles possess a robust impetus but exhibit a feeble reliance and are situated at the lowermost part of the model. The absence of legally binding regulations pertaining to the return and recycling of products is the only barrier found as a driving barrier.

14.4 CONCLUSION

The concept of "going green" has been compelling globally active companies to continuously enhance their environmental capabilities, implement green initiatives to mitigate further environmental degradation, and enhance overall corporate performance. Implementing green environmental initiatives is a significant possibility for a company's expansion, which includes improved performance and long-term viability. The majority of companies in developed countries have profited from environmentally-friendly initiatives and have positioned themselves competitively on the global stage, to the disadvantage of companies from rising and developing nations.

Green innovation is a critical component of sustainable development, offering numerous environmental, economic, and social benefits. By addressing key drivers and overcoming challenges, firms can successfully adopt green innovation to enhance their performance and contribute to a more sustainable future. Understanding the concept of green innovation and its implications is essential for businesses, policymakers, and stakeholders striving to achieve sustainability goals. As the global focus on sustainability intensifies, green innovation will continue to play a pivotal role in shaping a greener and more resilient economy.

This chapter has applied an exploratory review to understand how environment-driven green regulation can influence the performance of firms, the factors that drive green innovation, and challenges of adopting green innovation. Companies must exhibit a strong dedication and backing toward enhancing environmental concerns. Management must prioritize environmental initiatives and consistently adhere to all environmental legislation and policies to achieve sustainable development. Indeed, giving priority to environmental operations is not just an expense but a viable strategy that can result in competitive advantages in the market. Therefore, it is advisable for practitioners to allocate resources toward environmental programs to improve the proficiency of their workforce. Consequently, this will enhance the performance of an organization by improving its green innovation efforts (Xue et al., 2019). The research explored the internal and external factors that drive green innovation, and the challenges of adopting green innovation have been categorized as dependent, linkage, and driving barriers.

Policymakers should persist in promoting awareness of environmental entrepreneurship. Strict environmental oversight of companies' operations should be implemented to encourage green innovation. Undoubtedly, implementing such policy recommendations might significantly contribute to fostering a favorable environmental climate conducive to achieving sustainable development. Policymakers should establish a conducive environment to promote corporate green innovation. Implementing efficient policies and innovative initiatives, such as providing grants, tax holidays, and rebates, can assist companies in mitigating environmental threats such as pollution and emissions.

REFERENCES

Ahmed, M., Guo, Q., Qureshi, M.A., and Salam, J. 2021. "Do Green HR Practices Enhance Green Motivation and Proactive Environmental Management Maturity in Hotel Industry?" *International Journal of Hospitality Management* 94: 102852.

Albort-Morant, G., Leal-Millán, A., and Cepeda-Carrión, G. 2016. "The Antecedents of Green Innovation Performance: A Model of Learning and Capabilities." *Journal of Business Research* 69, no. 11: 4912–4917.

Aldoseri, Abdulaziz, Al-Khalifa, Khalifa N., and Hamouda, Abdel Magid. 2024. "AI-Powered Innovation in Digital Transformation: Key Pillars and Industry Impact." *Sustainability* 16, no. 5: 1790. https://doi.org/10.3390/su16051790.

Antonioli, D., Mancinelli, S., and Mazzanti, M. 2013. "Is Environmental Innovation Embedded within High-Performance Organisational Changes? The Role of Human Resource Management and Complementarity in Green Business Strategies." *Resource Policy* 42: 975–988.

Barin Cruz, L., Ávila Pedrozo, E., and de Fátima Barros Estivalete, V. 2006. "Towards Sustainable Development Strategies: A Complex View Following the Contribution of Edgar Morin." *Management Decision* 44 (7): 871–891.

Bernauer, T., Engel, S., Kammerer, D., and Sejas Nogareda, J. 2007. "Explaining Green Innovation: Ten Years after Porter's Win-Win Proposition: How to Study the Effects of Regulation on Corporate Environmental Innovation?" *Politische Vierteljahresschrift* 39: 323–341.

Berg, H., M. Linnenluecke, and Griffiths, A. 2020. "Implementing Green Innovation in Business." *Journal of Cleaner Production* 45, no. 4: 357–366.

Berrone, P., Fosfuri, A., Gelabert, L., and Gomez-Mejia, L.R. 2013. "Necessity as the Mother of 'Green' Inventions: Institutional Pressures and Environmental Innovations." *Strategic Management Journal* 34, no. 8: 891–909.

Bibri, S.E. 2018. *Smart Sustainable Cities of the Future: The Untapped Potential of Big Data Analytics and Context-Aware Computing for Advancing Sustainability.* Springer.

Cai, W., and Li, G. 2018. "The Drivers of Eco-Innovation and Its Impact on Performance: Evidence from China." *Journal of Cleaner Production* 176: 110–118.

Calabrese, A., Castaldi, C., Forte, G., and Levialdi, N.G. 2018. "Sustainability-Oriented Service Innovation: An Emerging Research Field." *Journal of Cleaner Production* 193: 533–548.

Calza, F., Parmentola, A., and Tutore, I. 2017. "Types of Green Innovations: Ways of Implementation in a Non-Green Industry." *Sustainability* 9, no. 8: 1301.

Carrillo-Hermosilla, J., Del Río, P., and Könnölä, T. 2010. "Diversity of Eco-Innovations: Reflections from Selected Case Studies." *Journal of Cleaner Production* 18, no. 10: 1073–1083.

Castellacci, F., and Lie, C.M. 2017. "A Taxonomy of Green Innovators: Empirical Evidence from South Korea." *Journal of Cleaner Production* 143: 1036–1047.

Chang, T.W., Chen, K.H., Lee, Y.S., and Yen, S.J. 2022. "How Does Environmental Leadership Cause Adoption of Environmental Policy?" *Frontiers in Environmental Science* 10: 928463.

Chen, X., Yi, N., Zhang, L., and Li, D. 2018. "Does Institutional Pressure Foster Corporate Green Innovation? Evidence from China's Top 100 Companies." *Journal of Cleaner Production* 188: 304–311.

Cillo, V., Petruzzelli, A.M., Ardito, L., and Del Giudice, M. 2019. "Understanding Sustainable Innovation: A Systematic Literature Review Corporate Social Responsibility and Environment." *Management Review* 26: 1012–1025.

Daugherty, P.R., and Wilson, H.J. 2018a. "Book Highlight—The Self-Aware Factory Floor: AI in Production, Supply Chain, and Distribution." *Global Business & Organizational Excellence* 38, no. 1.

Daugherty, P.R., and Wilson, H.J. 2018b. *Human+ Machine: Reimagining Work in the Age of AI.* Harvard Business Press.

Dayong, Zhanga, Rongb, Zhao, and Jic, Qiang. 2019. "Green Innovation and Firm Performance: Evidence from Listed Companies in China." *Resources, Conservation & Recycling* 144: 48–55.

Du, K., Cheng, Y., and Yao, X. 2021. "Environmental Regulation, Green Technology Innovation, and Industrial Structure Upgrading: The Road to the Green Transformation of Chinese Cities." *Energy Economics* 98: 105247. https://doi.org/10.1016/j.eneco.2021.105247.

Domazlicky, B.R., and Weber, W.L. 2004. "Does Environmental Protection Lead to Slower Productivity Growth in the Chemical Industry"? *Environment Resource Economy* 28: 301–324.

Ederington, J., and Minier, J. 2003. "Is Environmental Policy a Secondary Trade Barrier? An Empirical Analysis." *Canadian Journal of Economics* 36: 137–154.

Elkington, J. 2018. "25 Years Ago I Coined the Phrase 'Triple Bottom Line.' Here's Why It's Time to Rethink It." *Harvard Business Review* 25, no. 2–5: 6p.

Fan, F., Zhang, X., and Lian, H. 2021. "Impact of Environmental Regulations on Green Innovation Efficiency of the Yangtze River Economic Belt: Intermediary Effect Test Based on Foreign Direct Investment." *Science Technology Management* 41: 191–196.

Fu, J., and Li, L. 2010. "A Case Study on the Environmental Regulation, the Factor Endowment and the International Competitiveness in Industries." *Management World* 10: 87–98.

Giunipero, L.C., Hooker, R.E., and Denslow, D. 2012. "Purchasing and Supply Management Sustainability: Drivers and Barriers." *Journal of Purchasing and Supply Management* 18, no. 4: 258–269.

Guo, J., Zhou, Y., Ali, S., Shahzad, U., and Cui, L. 2021. "Exploring the Role of Green Innovation and Investment in Energy for Environmental Quality: An Empirical Appraisal from Provincial Data of China." *Journal of Environmental Management* 292: 112779.

Gupta, H., and Barua, M.K. 2017. "Supplier Selection Among SMEs on the Basis of Their Green Innovation Ability Using BWM and Fuzzy TOPSIS." *Journal of Cleaner Production* 152: 242–258.

Huang, L., Wang, C., Chin, T., et al. 2022. "Technological Knowledge Coupling and Green Innovation in Manufacturing Firms: Moderating Roles of Mimetic Pressure and Environmental Identity." *International Journal of Production Economics* 248: 108482.

IRENA. 2019. *A New World: The Geopolitics of the Energy Transformation.* International Renewable Energy Agency.

Jaffe, A.B., and Palmer, K. 1997. "Environmental Regulation and Innovation: A Panel Data Study." *Review of Economics and Statistics* 79, no. 4: 610–619.

Jayant, A., and Azhar, M., 2014. "Analysis of the Barriers for Implementing Green Supply Chain Management (GSCM) Practices: An Interpretive Structural Modeling (ISM) Approach." *Procedia Engineering* 97: 2157–2166.

Jia, N., Huang, K.G., and Man Zhang, C. 2019. "Public Governance, Corporate Governance, and Firm Innovation: An Examination of State-Owned Enterprises." *Academy of Management Journal* 62, no. 1: 220–247.

Kaack, L.H., Donti, P.L., Strubell, E., et al. 2022. "Aligning Artificial Intelligence with Climate Change Mitigation." *Nature Climate Change* 12: 518–527. https://doi.org/10.1038/s41558-022-01377-7.

Kawai, N., Strange, R., and Zucchella, A. 2018. "Stakeholder Pressures, EMS Implementation, and Green Innovation in MNC Overseas Subsidiaries." *International Business Review* 27, no. 5: 933–946.

Lacy, P., and Rutqvist, J. 2015. *Waste to Wealth: The Circular Economy Advantage*. Palgrave Macmillan.

Li, C.J., Razzaq, A., Irfan, M., and Luqman, A. 2023. "Green Innovation, Environmental Governance and Green Investment in China: Exploring the Intrinsic Mechanisms under the Framework of COP26." *Technological Forecasting and Social Change* 194: 122708.

Li, W., and Bi, K. 2013. "Study on Impact of Environmental Regulation Intensity on Green Technology Innovation in Pollution Intensive Industries: An Empirical Test Based on Panel Data from 2003 to 2010." *R&D Management Journal* 6: 72–81.

Liu, Z., Wang, R., Gao, Y., et al. 2023. "Low-Cost Multi-Function Electrolyte Additive Enabling Highly Stable Interfacial Chemical Environment for Highly Reversible Aqueous Zinc Ion Batteries." *Advanced Functional Materials* 33, no. 49: 2308463.

Jin, P., Peng, C., and Song, M. 2019. "Macroeconomic Uncertainty, High-Level Innovation, and Urban Green Development Performance in China." *China Economic Review* 55: 1–18. https://doi.org/10.1016/j.chieco.2019.02.008.

Kneller, R., and Manderson, E. 2012. "Environmental Regulations and Innovation Activity in UK Manufacturing Industries." *Resource Energy Economics* 34: 211–235.

Lai, S., Wen, C., and Chen, Y. 2003. "The Exploration of the Relationship Between the Environmental Pressure and the Corporate Competitive Advantage. In *Proceedings of the CSMOT Academic Conference*, National Chiao Tung University, Hsin-Chu, Taiwan, 22 April.

Luthra, S., and Mangla, S.K. 2018. "When Strategies Matter: Adoption of Sustainable Supply Chain Management Practices in an Emerging Economy's Context." *Resource Conservative Recycle* 138: 194–206.

Nassani, A.A., Yousaf, Z., Radulescu, M., et al. 2023. "Green Innovation through Green and Blue Infrastructure Development: Investigation of Pollution Reduction and Green Technology in Emerging Economy." *Energies* 16, no. 4: 1944.

OECD. 2011. *Fostering Innovation for Green Growth.* OECD Green Growth Studies, OECD Publishing. http://dx.doi.org/10.1787/9789264119925-en

Peng, W., Cheng, F., and Lu, J. 2017. "Research on Threshold Effect of Environmental Regulation on Provincial Green Innovation Efficiency." *South China Journal of Economy* 9: 73–84.

Peng, X., and Li, B. 2016. "Research on Green Transformation of Chinese Industry Under Different Types of Environmental Regulations." *Journal of Financial Research* 7: 134–144.

Peuckert, J. 2014. "What Shapes the Impact of Environmental Regulation on Competitiveness? Evidence from Executive Opinion Surveys. *Environmental Innovation & Societal Transition* 10: 77–94.

Qi, G.Y., Shen, L.Y., Zeng, S.X., and Jorge, O.J. 2010. "The drivers for contractors' green innovation: An industry perspective." *Journal of Cleaner Production* 18: 1358–1365.

Raman, R., Pattnaik, D., Lathabai, H.H., et al. 2024. "Green and Sustainable AI Research: An Integrated Thematic and Topic Modeling Analysis." *Journal of Big Data* 11: 55.

Rui, Z., and Lu, Y. 2021. "Stakeholder Pressure, Corporate Environmental Ethics and Green Innovation." *Asian Journal of Technology Innovation* 29: 70–86.

Shahzad, M., Qu, Y., Zafar, A.U., Ding, X., and Rehman, S.U. 2020. "Translating Stakeholders' Pressure into Environmental Practices—The Mediating Role of Knowledge Management." *Journal of Cleaner Production* 275: 124–163.

Shao, J. 2019. "Sustainable Consumption in China: New Trends and Research Interests." *Business Strategy and the Environment* 28, no. 8: 1507–1517.

Shen, T., Li, D., Jin, Y., and Li, J. 2022. "Impact of Environmental Regulation on Efficiency of Green Innovation in China." *Atmosphere* 13: 767–782. https://doi.org/10.3390/atmos13050767.

Shu, C., Zhou, K.Z., Xiao, Y., et al. 2016. "How Green Management Influences Product Innovation in China: The Role of Institutional Benefits." *Journal of Business Ethics* 133 (3): 471–485. https://doi.org/10.1007/s10551-014-2401-7.

Silva, G.M., Gomes, P.J., and Sarkis, J. 2019. "The Role of Innovation in the Implementation of Green Supply Chain Management Practices." *Business Strategy and the Environment* 28, no. 5: 819–832.

Singh, A. K., Raza, S. A., Nakonieczny, J., and Shahzad, U. 2023. "Role of Financial Inclusion, Green Innovation, and Energy Efficiency for Environmental Performance? Evidence from Developed and Emerging Economies in the Lens of Sustainable Development." *Structural Change and Economic Dynamics* 64: 213–224.

Singh, S.K., Del Giudice, M., Chiappetta Jabbour, C.J., et al. 2022. "Stakeholder Pressure, Green Innovation, and Performance in Small and Medium-Sized Enterprises: The Role of Green Dynamic Capabilities." *Business Strategy and the Environment* 31, no. 1: 500–514.

Song, M., Tao, J., and Wang, S. 2015. "FDI, Technology Spillovers and Green Innovation in China: Analysis Based on Data Envelopment Analysis." *Annals of Operation Research* 228, no 1: 47–64.

Song, M., and Wang, S. 2018. "Market Competition, Green Technology Progress and Comparative Advantages in China." *Management Decision* 56, no. 1: 188–203.

Song, W., Yu, H., and Xu, H. 2021. "Effects of Green Human Resource Management and Managerial Environmental Concern on Green Innovation." *European Journal of Innovation Management* 24, no. 3: 951–967.

Takalo, S.K., and Tooranloo, H.S. 2021. "Green Innovation: A Systematic Literature Review." *Journal of Cleaner Production* 279: 122474.

Tang, M., Walsh, G., Lerner, D., Fitza, M.A., and Li, Q. 2018. "Green Innovation, Managerial Concern and Firm Performance: An Empirical Study." *Business Strategy and the Environment* 27: 39–51.

Tariq, A., Badir, Y.F., Tariq, W., and Bhutta, U.S. 2017. "Drivers and Consequences of Green Product and Process Innovation: A Systematic Review, Conceptual Framework, and Future Outlook." *Technology in Society* 51: 8–23.

Tseng, M.L., Wang, R., Chiu, A.S., et al. 2013. "Improving Performance of Green Innovation Practices Under Uncertainty." *Journal of Cleaner Production* 40: 71–82.

Ullah, S., Ahmad, N., Khan, F.U., et al. 2021. "Mapping Interactions Among Green Innovations Barriers in Manufacturing Industry Using Hybrid Methodology: Insights from a Developing Country." *International Journal of Environmental Research and Public Health* 18: 7885.

Uthayasankar Sivarajah, Kumar, Sachin, Kumar, Vinod, Chatterjee, Sheshadri, and Li, Jing. 2024." A Study on Big Data Analytics and Innovation: From Technological and Business Cycle Perspectives." *Technological Forecasting and Social Change* 202: 2024.

Wang, J., Xue, Y., Sun, X., and Yang, J. 2020. "Green Learning Orientation, Green Knowledge Acquisition and Ambidextrous Green Innovation." *Journal of Cleaner Production* 250: 119475.

Wang, Q., Sun, T., and Li, R. 2023. "Does Artificial Intelligence Promote Green Innovation? An Assessment Based on Direct, Indirect, Spillover, and Heterogeneity Effects." *Energy & Environment* 0(0).

Wang, Y., and Shen, N. 2016. "Environmental Regulation and Environmental Productivity: The Case of China." *Renewable Sustainable Energy Review* 62: 758–766.

Wang, S.Y., Li, B.B., and Zhang, S.J. 2018. "Research on Environmental Regulation, Spatial Spillover and Green Innovation Based on Spatial Measurement." *Areal Research and Development* 2: 138–144.

Willcocks, L.P., Lacity, M., and Craig, A. 2015. "Robotic Process Automation at Xchanging." https://eprints.lse.ac.uk/64519/1/OUWRPS_15_05_published.pdf.

Xia, D., Zhang, M., Yu, Q., and Tu, Y. 2019. "Developing a Framework to Identify Barriers of Green Technology Adoption for Enterprises." *Resources, Conservation and Recycling* 143: 99–110.

Xie, X., Huo, J., and Zou, H. 2019. "Green Process Innovation, Green Product Innovation, and Corporate Financial Performance: A Content Analysis Method." *Journal of Business Research* 101: 697–706.

Xue, Min, Boadu, Francis, and Xie, Yu. 2019. "The Penetration of Green Innovation on Firm Performance: Effects of Absorptive Capacity and Managerial Environmental Concern." *Sustainability* 11, no. 9: 2455. https://doi.org/10.3390/su11092455.

Yehia, Ibrahim Alzoubi, and Mishra, Alok. 2024. "Green Artificial Intelligence Initiatives: Potentials and Challenges." *Journal of Cleaner Production* 46, no. 8.

Yin, J., Gong, L., and Wang, S. 2018. "Large-Scale Assessment of Global Green Innovation Research Trends from 1981 to 2016: A Bibliometric Study." *Journal of Cleaner Production* 197: 827–841.

Yin, K., Cai, F., and Huang, C. 2023. "How Does Artificial Intelligence Development Affect Green Technology Innovation in China? Evidence from Dynamic Panel Data Analysis." *Environmental Science and Pollution Research* 30: 28066–28090. https://doi.org/10.1007/s11356-022-24088-0.

Zhang, J., Kang, L., Li, H., Pablo, B., and Zuo, J. 2020. "The Impact of Environmental Regulations on Urban GI Efficiency: The Case of Xi'an." *Sustainable Cities and Society* 102123.

Zhang, Z., and Zhao, G. 2012. "FDI, environmental regulation and technological progress." *Journal of Quantitative Economy* 29: 19–32.

Index

T

V

W